APOPTOSIS GENES

APOPTOSIS GENES

edited by

James W. Wilson
Catherine Booth
Christopher S. Potten

Patterson Institute for Cancer Research
Manchester, U.K.

KLUWER ACADEMIC PUBLISHERS
Boston / Dordrecht / London

Distributors for North, Central and South America:
Kluwer Academic Publishers
101 Philip Drive
Assinippi Park
Norwell, Massachusetts 02061 USA
Telephone (781) 871-6600
Fax (781) 871-6528
E-Mail <kluwer@wkap.com>

Distributors for all other countries:
Kluwer Academic Publishers Group
Distribution Centre
Post Office Box 322
3300 AH Dordrecht, THE NETHERLANDS
Telephone 31 78 6392 392
Fax 31 78 6546 474
E-Mail <services@wkap.nl>

Electronic Services <http://www.wkap.nl>

Library of Congress Cataloging-in-Publication Data
Apoptosis genes / edited by James W. Wilson, Catherine Booth,
Christopher S. Potten.
p. cm.
Includes bibliographical references and index.
ISBN 0-412-83860-5 (alk. paper)
1. Apoptosis. 2. Genes. 3. Genetic regulation. I. Wilson,
James W. II. Booth, Catherine. III. Potten, C. S., 1940- .
QH671. A657 1998
571.9'36--dc21 98-42725
CIP

Printed on acid-free paper.

Printed in the United States of America

CONTENTS

LIST OF CONTRIBUTORS

Catherine Booth
Epithelial Biology Laboratory, CRC Section of Cell and Tumour Biology, Paterson Institute for Cancer Research, Manchester M20 4BX.

Yue Eugene Chin
Department of Pathology, Yale University School of Medicine, 310 Cedar Street, New Haven, CT 06520-8023, USA.

Thomas Chittenden
Apoptosis Technology Inc, 148 Sidney Street, Cambridge, MA 02139-4239, USA.

Rollie J. Clem
Division of Biology, Kansas State University, Manhattan 66506,USA.

Xin-Yuan Fu
Department of Pathology, Yale University School of Medicine, 310 Cedar Street, New Haven, CT 06520-8023, USA.

Andrew P. Gilmore
School of Biological Sciences, University of Manchester, 3.239 Stopford Building, Oxford Road, M13 9PT, UK.

J. Marie Hardwick
Departments of Molecular Microbiology and Immunology, Neurology and Pharmacology and Molecular Sciences, John Hopkins School of Public Health and Medicine, Baltimore, MD, USA.

Usha Kasid
Departments of Radiation Medicine and Biochemistry and Molecular Biology, Lombardi Cancer Center, Georgetown University Medical Center, Washington D.C. 20007, USA.

Gary Ketner
Departments of Molecular Microbiology and Immunology, John Hopkins School of Public Health and Medicine, Baltimore, MD, USA.

Arnold J. Levine
Department of Molecular Biology, Princeton University, Princeton NJ 08544, USA.

Pedro Lowenstein
Molecular Medicine Unit, Room 1.302 Stopford Building, Department of Medicine, University of Manchester, Oxford Road, Manchester M13 9PT.

Masayuki Miura
Department of Neuroanatomy, Osaka University Medical School and Core Research for Evolutional Science and Technology (CREST), Japan.

Maureen Murphy
Department of Molecular Biology, Princeton University, Princeton NJ 08544, USA.

John R. Nambu
Biology Department and Molecular and Cell Biology Program, University of Massachusetts at Amherst, Amherst, MA 01003, USA.

Christopher S. Potten
Epithelial Biology Laboratory, CRC Section of Cell and Tumour Biology, Paterson Institute for Cancer Research, Manchester M20 4BX.

Charles H. Streuli
School of Biological Sciences, University of Manchester, 3.239 Stopford Building, Oxford Road, M13 9PT, UK.

Asako Sugimoto
Department of Biophysics and Biochemistry, Graduate School of Science, University of Tokyo and PRESTO, Japan Science and Technology Corporation, Japan.

Simeng Suy
Departments of Radiation Medicine and Biochemistry and Molecular Biology, Lombardi Cancer Center, Georgetown University Medical Center, Washington D.C. 20007, USA.

Alastair Watson
Department of Medicine, University of Manchester, Clinical Sciences Building, Hope Hospital, Eccles Old Road, Salford M6 8HD.
Current Address: Department of Medicine, Liverpool University, The Duncan Building, Daulby Street, Liverpool L69 3GA, UK.

James W. Wilson
Epithelial Biology Laboratory, CRC Section of Cell and Tumour Biology, Paterson Institute for Cancer Research, Manchester M20 4BX.

John P. Wing
Biology Department and Molecular and Cell Biology Program, University of Massachusetts at Amherst, Amherst, MA 01003, USA.

Chapter 1

Introduction

J.W. Wilson, C. Booth and C.S. Potten.
Epithelial Biology Laboratory, CRC Section of Cell and Tumour Biology, Paterson Institute for Cancer Research, Wilmslow Road, Manchester M20 4BX, UK.

Apoptosis is the controlled deletion of unwanted cells. It is a process of fundamental importance in the structural development of the embryo and tissue homeostasis in the adult animal. It is also important in the removal of damaged or infected cells. Damage to DNA, or infection with viruses which carry potentially oncogenic gene sequences, can result in the malignant transformation of cells. Auto-reactive immune cells are also removed through apoptosis. Dysregulation of apoptosis in an organism can, therefore, lead to abnormality, disease and death (see Jones and Gore 1997; Rickenburger and Korsmeyer 1997; Stefanis *et al.* 1997; Hetts 1998).

It is only in the last 25 years that we have started to investigate this process and the biochemical and genetic factors which control it. The landmark publication for the inception of apoptosis research is generally taken to be that by Kerr, Wyllie and Currie in 1972, who realised the general significance of this physiological process for human disease and coined the term apoptosis (Kerr *et al.* 1972). Prior to this, other researchers had made similar observations on cell death during the development of vertebrates (Glücksmann 1951) and insects (Lockshin and Williams 1965). Lockshin and Williams called the cell deletion they observed programmed cell death. However, the fundamental importance of this early work was not recognised at the time.

The key morphological features of apoptosis are nuclear condensation and fragmentation, reduced cytoplasmic volume, loss of cell-cell contacts and detachment from neighbouring cells (adherent cells in culture will detach from the substratum and float free in the medium), blebbing of the plasma membrane, fragmentation of the cell into several apoptotic bodies and phagocytosis of cells/apoptotic bodies by neighbouring cells or "professional" phagocytes. Depending on the cell/tissue type some or all of these features may be observed. One of the important features of apoptosis is the maintenance of plasma membrane integrity in order to prevent an inflammatory response. There are occasions, however, when phagocytic activity is unable to cope with overwhelming levels of apoptosis, and secondary necrosis occurs, probably as a result of release of lysosomal enzymes by phagocytes unable to engulf large aggregates of dead cells, and non-phagocytosed apoptotic bodies/cells losing plasma membrane integrity.

The genes and their protein products that control apoptosis soon became a focus of many areas of research. Major progress was made with the discovery of *p53* (see Murphy and Levine - this volume) and *bcl-2* (see Chittenden - this volume) and the crucial role that their protein products have in the apoptotic process. *p53* is now established as being the most commonly mutated gene in human cancers. The p53 protein functions to ensure that a given cell responds appropriately to DNA damage and other environmental stress agents such as hypoxia, deciding whether the fate of the cell is to arrest and repair or to undergo apoptosis. It, therefore, plays a protective role within body as a whole, by preventing the establishment of cells with DNA mutations that could initiate/promote malignancy. It has hence been dubbed the "guardian of the genome" (Lane 1992). Since the discovery of *bcl-2* (Tsujimoto *et al.* 1984; Tsujimoto and Croce 1986), the prototypic inhibitor of apoptosis, many new genes have been discoverd that code for Bcl-2-related proteins which may either inhibit or promote apoptosis.

The use of easily genetically manipulable organisms such as the nematode, *Caenorhabditas elegans* (see Sugimoto and Miura - this volume) and the fruitfly, *Drosophila Melanogaster* (see Wing and Nambu - this volume) have helped identify key genes that regulate the apoptotic pathway and also demonstrate that the cell death machinery is highly conserved between lower and higher organisms. Again, this demonstrates the fundamentally important role that apoptosis plays.

The search for the cellular executioner has led to the identification of a whole family of cellular proteases which play an essential role in the apoptotic process (see Sugimoto and Miura - this volume). These have been termed caspases, for cysteine proteases which cleave after asparate residues (Alnemri *et al.* 1996). Different sets of caspases may be activated in response to different apoptotic signals and the distribution of caspase isotypes may also be cell/tissue specific. Many substrates are now known to be cleaved including retinoblastoma protein (Jänicke *et al.* 1996), polyADP ribose polymerase (Kaufmann *et al.* 1991), focal adhesion kinase (Wen *et al.* 1997) and Bcl-2 (Cheng *et al.* 1997). Cleavage of DNA, which was one of the first molecules demonstrated to be degraded in apoptotic cells (Wyllie 1980), has now also been linked to caspase activation through the caspase-induced activation of an endonuclease (Erani *et al.* 1998 and Sakahira *et al.* 1998). Caspases have been shown to be activated via many different pathways including the cell death receptors Fas/CD95 and TNF-R1 (Fraser and Evan 1996) and also by cytochrome c (Li *et al.* 1997), which has been shown to be released from mitochondria early in the apoptotic process.

As mentioned previously, apoptosis plays an important role in the deletion of virus-infected cells. Viruses, however, have adapted to block this process and produce a range of proteins which are able to inhibit the action of key apoptotic proteins (See Hardwick *et al.*. - this volume). For instance, the adenovirus E1B 55kDa protein binds to, and inhibits the function of, p53 (Kao *et al.*. 1990) and the cowpox protein crmA acts a caspase inhibitor

(Ray *et al.*. 1992). These and other viral products have greatly aided in the dissection of apoptosis signalling pathways.

Research over the last decade is now enabling us to comprehend how many genes and their protein products interact to control apoptosis. This has led us to the current position where we may be able to directly modify the action of key proteins through gene therapy (Neilsen and Maneval 1998) and antisense oligonucleotides (Webb *et al.* 1997) approaches. This book hopes to present a current overview of key genes involved in the control of apoptosis and provide the reader with a comprehensive history of apoptosis research together with thoughts on the future prospects and clinical applications.

REFERENCES.

Alnemri,E.S., Livingston,D.J., Nicholson,D.W. *et al.* (1996) Human ICE/CED-3 protease nomenclature. *Cell* **87**: 171.

Cheng,E.H., Krisch,D,G,, Clem,R.J. *et al.* (1997) Conversion of Bcl-2 to a Bax-like death effector by caspases. *Science* **278**: 1966-1968.

Erani,M., Sakahira,H., Yokoyama,H. *et al.* (1998) A caspase-activated DNase that degrades DNA during apoptosis, and its inhibitor ICAD. *Nature* **391**: 43-50.

Fraser,A and Evan,G .(1996) A licence to kill. *Cell* **85**: 781-784.

Glücksmann,A. (1951) Cell deaths in vertebrate ontogeny. *Biol. Rev*. **26**: 59-86.

Hetts,S.W. (1998) To die or not to die: an overview of apoptosis and its role in disease. *JAMA* **279**: 300-307.

Jänicke,R.U., Walker,P.A., Lin,X.Y. and Porter,A.G. (1996) Specific cleavage of the retinoblastoma protein by an ICE-like protease in apoptosis. *EMBO J*. **15**: 6969-6978.

Jones,B.A. and Gores,G.J. (1997) Physiology and pathophysiology of apoptosis in epithelial cells of the liver, pancreas and intestine. *Am. J. Physiol*. **273**: G1174-1188.

Kao,C.C., Yew,P.R. and Berk,A.J. (1990) Domains required for *in vitro* association between the cellular p53 and the adenovirus 2 E1B 55K proteins. *Virology*. **179**: 806-814.

Kauffman,S.H., Desnoyers,S., Ottaviano,Y. *et al.* (1991) Specific proteolytic cleavage of poly(ADP-ribose) polymerase: an early marker of chemotherapy-induced apoptosis. *Cancer Res*. **53**: 3976-3985.

Kerr,J.F.R., Wyllie,A.H. and Currie,A.R. (1972) Apoptosis: a basic biological phenomenon with wide ranging implications in tissue kinetics. *Br. J Cancer*. **26**: 239-257.

Lane,D.P. (1992) p53, guardian of the genome. *Nature* **358**: 15-16.

Li,P., Nijhawan,D., Budihardjo,I. *et al.* (1997) Cytochrome c and dATP-dependent formation of apaf-1/caspase-9 complex initiates and apoptotic protease cascade. *Cell* **91**: 479-489.

Lockshin,R.A. and Williams,C.M. (1965) Programmed cell death: Cytology of degeneration in the intersegmental muscles of the silkmoth. *J. Insect Physiol*. **11**: 123-133.

Neilsen,L.L. and Maneval,D.C. (1998) p53 tumour suppressor gene therapy for cancer. *Cancer Gene Ther*. **5**: 52-63.

Ray,C.A., Black,R.A., Kronheim,S.R. *et al.* (1992) Viral inhibition of inflammation: cowpox virus encodes an inhibitor of the interleukin-1ß converting enzyme. *Cell* **69**: 597-604.

Rickenburger,J.L. and Korsmeyer,S.J. (1997) Errors in homeostasis and deregulated apoptosis. *Curr. Opin. Gen. Dev*. **7**: 589-596.

Sakahira,H., Erani,M. and Nagata,S. (1998) Cleavage of CAD inhibitor in CAD activation and DNA degradation during apoptosis. *Nature* **391**: 96-99.

Stefanis,L., Burke,R.E. and Greene,L.A. (1997) Apoptosis in neurodegenerative diseases. *Curr. Opin. Neurol*. **10**: 299-305.

Tsujimoto,Y., Finger,L.R., Yunis,J. *et al.* (1984) Cloning of the chromosome breakpoint of neoplastic B cells with the t(14;18) chromosome translocation. *Science* **226**: 1097-1099.

Tsujimoto,Y., Croce,C.M. (1986) Analysis of the structure, transcripts and protein products of *bcl-2,* the gene involved in human folicular lymphoma. *Proc. Natl. Acad. Sci. USA*. **83**: 5214-5218.

Webb,A., Cunningham,D., Cotter,F. *et al.* (1997) Bcl-2 antisense therapy in patients with non-Hodgkin lymphoma. *Lancet*. **349**: 1137-1141.

Wen,L.P., Fahrni,J.A., Troie,S. *et al.* (1997) Cleavage of focal adhesion kinase by caspases during apoptosis. *J. Biol. Chem*. **272**: 26056-26061.

Wyllie,A.H. (1980). Glucocorticoid-induced thymocyte apoptosis is associated with endogenous endonuclease activation. *Nature* **284**: 555-556.

Chapter 2.

The role of p53 in apoptosis

Maureen Murphy and Arnold J. Levine
Department of Molecular Biology, Princeton University, Princeton NJ 08544-1014, USA.

1.0 THE p53 TUMOUR SUPPRESSOR GENE: FUNCTIONAL DOMAINS AND ACTIVITIES

It has been said that no matter which direction cancer research turns, the p53 tumour suppressor protein comes into view. If there is truth in this statement, then it has much to do with the ability of this protein to induce programmed cell death, or apoptosis. First discovered in 1979 as a protein complexed with the large T antigen of SV40 virus (Lane and Crawford, 1979; Linzer and Levine, 1979), the complete cDNA sequence of *p53* was elucidated in 1984 (Matlashewski *et al.* 1984), and its activity as a tumour suppressor gene was first suspected in 1985 when deletions of the gene were detected in transformed cells infected with Friend leukemia virus (Mowat *et al.* 1985). Later it was found that expression of *p53* could inhibit the neoplastic transformation of primary cells (Finlay *et al.* 1989), and that this gene was subject to frequent mutations in human tumours (Baker *et al.* 1989; Hollstein *et al.* 1991). Germline mutations of the *p53* gene were found in individuals suffering from Li-Fraumeni disorder, which is an autosomal dominant disorder predisposing affected individuals to osteosarcomas as well as tumours of the brain, breast and adrenal cortex (Malkin *et al.* 1990). Tumours from affected individuals were found to invariably lose expression of the wild type *p53* allele, thereby identifying *p53* as a classical tumour suppressor gene. Analogously, mice genetically engineered to be nullizygous for *p53* (*p53* "knock-out" mice) died from tumours at 3-6 months of age, although these were chiefly lymphoid tumours (Donehower *et al.* 1992). Analyses of *p53* mutations in human tumour samples to date indicate that this tumour suppressor gene is mutated in over 50% of human tumours, derived from a wide variety of tissues. The *p53* gene continues to hold the distinction as the most frequently mutated gene in human cancer.

Although some cancers appear to rarely select for *p53* mutations, in many of these cases it appears that this protein has been inactivated through means other than mutation. For example, amplification of the *mdm2* oncogene, which encodes a protein that binds to and inactivates the transcriptional properties of p53 (Momand *et al.* 1992), occurs in a significant percentage of soft tissue sarcomas (Oliner *et al.* 1992), as well as in gliomas (Reifenberger *et al.* 1993) and some leukemias (Watanabe *et al.* 1996). Mdm2 protein is also over expressed, due to enhanced translation, in a variety of human tumour cell lines (Landers *et al.* 1994; 1997). Additionally, inactivation of p53 by cytoplasmic sequestration has been observed in some breast carcinomas, as well as in neuroblastomas (Moll *et al.* 1992; 1995). In testicular teratomas, high levels of transcriptionally inactive wild type p53 protein have been observed (Lutzger and Levine, 1996). It is not unreasonable to suspect that the vast majority of human tumours have abrogated the activity of this tumour suppressor protein in some way.

The p53 protein is a sequence-specific DNA binding protein capable of transactivating genes containing p53 binding sites in either upstream regulatory regions or introns (Kern *et al.* 1991; Zambetti *et al.* 1992; Kastan *et al*, 1992). Unlike the tumour suppressor genes RB (retinoblastoma susceptibility gene) and *APC* (adenomatous polyposis coli), which typically contain truncation or frameshift mutations in tumours, the majority of mutations in *p53* result in single amino acid substitutions in the DNA binding domain of the protein. Almost invariably the remaining wild type allele is lost by deletion, mutation, or homologous recombination to reduce this locus to homozygosity. These DNA binding domain point mutations typically alter both the conformation and half-life of p53, and eliminate the ability of p53 to function as a sequence-specific DNA binding protein and transcriptional activator (for review, see Zambetti and Levine 1993). Figure 1 illustrates the functional domains of *p53*, as well as the spectrum of mutations that occur in tumours. p53 contains a transactivation domain located at the amino terminus (approximately residues 1-43), a sequence-specific DNA binding domain from residues 110-290, and a proline-rich region separating them (residues 61-94). p53 binds to DNA as a tetramer, and the region responsible for oligomerisation is contained within residues 323-355, which are preceded by a flexible linker region extending from the DNA binding domain. The carboxyl-terminal residues 363-393 form a basic region that exhibits strong non-specific DNA binding (Bayle *et al.* 1995) and anti-helicase (Wu *et al.* 1995) activities. It is currently accepted that this region negatively regulates the activity of the DNA-binding domain, perhaps allosterically (Hupp and Lane 1994; Hupp *et al.* 1995). Approximately 80% of the missense mutations in *p53* cluster in the DNA binding domain of this protein, and several of these altered residues are known to be contact points with DNA (Cho *et al.* 1994). Therefore, there is a powerful selection in human cancer to eliminate the DNA binding activity of p53, and presumably also the ability of this protein to function as a transcription factor.

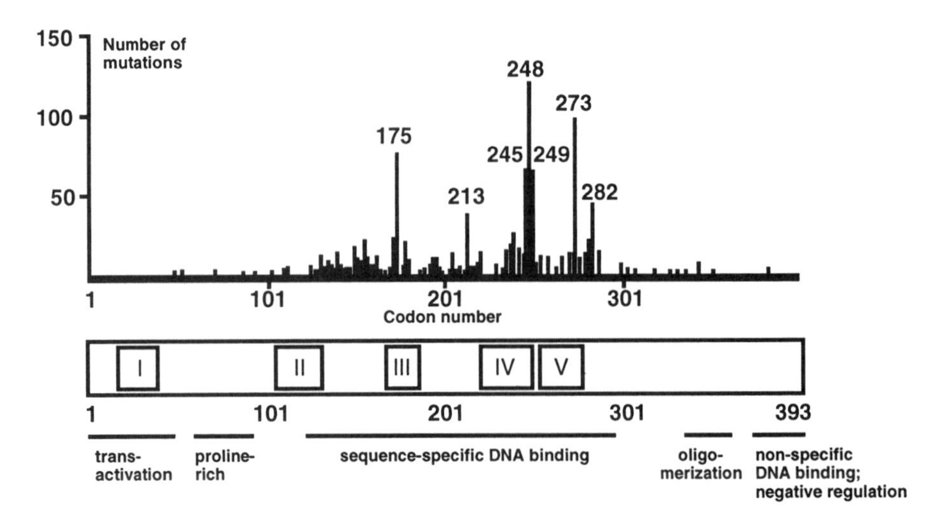

Figure 1. Schematic representation of human p53.
The spectrum of mutations in *p53* in human tumours is depicted, showing a clustering of mutations in the sequence-specific DNA binding domain of this protein in human tumour samples; an updated database of mutations in *p53* is available (Hollstein *et al.*, Nucleic Acids Research 22: 3551-3555, 1994). Below the mutational spectrum, the functional domains of *p53* are indicated, as well as the five evolutionarily conserved domains (i-v).

p53 was first implicated as a negative regulator of cell growth when it was found that expression of the wild type (wt) *p53* gene could inhibit the transformation of primary fibroblasts (Finlay *et al.* 1989). The normal physiological function of this protein was better understood when it was discovered that p53 protein levels were highly induced following irradiation of cells with ultra-violet or gamma radiation (Maltzman and Czyzyk 1984; Kastan *et al.* 1991). Concomitantly, it was found that induction of a temperature-sensitive mutant of *p53* at the permissive temperature (wild type p53 protein) led to a reversible growth arrest of cells in the G1 phase of the cell cycle (Michalovitz *et al.* 1990; Martinez *et al.* 1991). These results were followed by the discovery that growth arrest in G1 occurred in response to gamma irradiation only in tumour cells containing wt, but not mutant, *p53* (Kastan *et al.* 1992; Kuerbitz *et al.* 1992). The conclusion drawn from these studies was that *p53* functioned as a cell cycle checkpoint protein that responded to many different kinds of DNA damage. The existence of DNA damage was signalled to p53 via certain DNA damage detection protein(s), resulting in activation of this protein as a transcription factor and a concomitant growth arrest. A widely-held hypothesis stemming from these studies is that p53 induction following detection of DNA damage leads to a reversible growth arrest in G1 to allow cells time to repair damaged DNA. In this manner, replication of the damaged template and a high risk of mutation are prevented. It is now known that signals for p53 induction are not limited to agents that only inflict DNA damage. This protein is also activated by low oxygen concentrations (hypoxia), and limiting ribonucleotide pools (Linke *et al.* 1996), as well as other stimuli (reviewed in Gottlieb and Oren 1996; Ko and Prives 1996; Levine 1997).

p53 accomplishes many of its normal functions in the cell via its activity as a sequence-specific DNA binding protein that transactivates the expression of at least six or seven known genes containing two or more copies of the p53 consensus element 5' Pu Pu Pu C (A/T) (T/A) G Py Py Py 3'. The p53 transactivated genes identified to date include those coding for: Gadd45 (Kastan *et al.* 1992) which binds to the replication and repair protein PCNA and can itself inhibit cellular proliferation (Smith *et al.* 1994); mdm-2, which is a negative cellular regulator of p53 (Wu *et al.* 1993); p21/waf1/cip1/sdi1, which is a cyclin-dependent kinase inhibitor; cyclin G (Okamoto and Beach 1994); bax (Miyashita and Reed 1995); and IGF-BP3 (Buckbinder *et al.* 1995) - as well as other candidates (for review see Ko and Prives, 1996). Of particular note in this group is the gene $p21^{/waf1/cip1/sdi1}$, which was simultaneously identified as coding for a cyclin-dependent kinase inhibitor (Harper *et al.* 1993; Xiong *et al.* 1993), a p53-transactivated gene (El-Deiry *et al.* 1993), and a gene over expressed during cellular senescence (Noda *et al.* 1994). The contribution of p21 protein to the induction of G1 growth arrest by p53 was confirmed with the generation of mice genetically engineered to have deleted both copies of their *p21* gene. Fibroblasts from such *p21* "knockout" mice are partially deficient in their ability to undergo growth arrest in response to DNA damage (Deng *et al.* 1995), indicating that transactivation of this gene is partially, perhaps even largely, responsible for p53-mediated growth arrest. Given the role of p53 as a protein that arrests cell growth in response to DNA damage, it is perhaps not surprising that for some time it was accepted that the reason for this gene to be so frequently mutated in human cancer was due to its growth-arrest activity. In fact, it was not until 1991 that a role for p53 in the induction of apoptosis was identified (Yonish-Rouach *et al* 1991), and not until 1994 that it was fully appreciated that it is p53's activity as an inducer of programmed cell death that targets this gene for frequent mutation in human cancer (Symonds *et al* 1994).

2.0 p53 DIRECTLY INDUCES APOPTOSIS IN SOME CELL TYPES

Utilising the murine temperature-sensitive *p53* mutant protein described above (Ala135Val, where alanine is the wt amino acid at codon 135 and valine the mutant allele), Yonish-Rouach, Oren and colleagues generated stable transfectants of a murine myeloid leukemia cell line containing this *p53* mutant. Surprisingly, unlike mouse embryo fibroblasts containing the same mutation (which undergo growth arrest at the permissive temperature), culture at the permissive temperature caused this myeloid line to undergo programmed cell death, or apoptosis. Evidence for this included a loss of viability of cells at the permissive temperature, accompanied by chromatin condensation and DNA fragmentation (Yonish-Rouach *et al.* 1991; 1993). Interestingly, this group found that apoptosis induced by expression of wt p53 could be blocked by culturing cells in the presence of IL-6, a haematopoietic growth and survival factor. These observations suggested for the first time that p53 could sense the activation of signal transduction pathways, and in doing so modify the outcome of p53 induction (p53-dependent apoptosis or growth arrest). Shortly after this discovery a second group, also working with an inducible *p53* system, found that a human colorectal carcinoma cell line underwent extensive apoptosis following induction of wt p53 (Shaw *et al.* 1992). These studies indicated that certain cell types or genetic backgrounds might undergo growth arrest in response to p53 (for example, non-transformed fibroblasts), while other cell types (such as transformed epithelial or myeloid cells) were predisposed to undergo apoptosis. Delineating the parameters that influence the decision between growth arrest and apoptosis in transformed and normal cells has been a major focus of p53 research (reviewed in section VI).

A useful paradigm for p53-dependent apoptosis was uncovered by studies of the adenovirus *E1A* gene. E1A protein functions during adenoviral infection by commandeering control of the cell cycle of infected cells and inducing proliferation, chiefly by binding and inactivating the negative growth regulator RB (retinoblastoma susceptibility protein) and altering the p300 family of transcriptional co-activators and histone acetyl-transferases. Not surprisingly, the *E1A* gene is a potent transforming gene, and oncogenic transformation by adenovirus requires this gene product. Interestingly, while *E1A* can cooperate with other oncogenes to transform cells, introduction of this gene into primary cells containing temperature-sensitive *p53* leads to transformation at the restrictive temperature (mutant *p53*), but extensive cell death by apoptosis at the permissive temperature (wild type *p53*) (Debbas and White 1993). Therefore, *E1A* and wt *p53* cooperate to induce apoptosis (Debbas and White 1993; Lowe and Ruley 1993). The first cellular protein found to cooperate with p53 to induce apoptosis was E2F-1; like E1A, this RB-binding protein has a role in controlling entry into the synthesis (S) phase of the cell cycle and inducing proliferation. In immortalised murine fibroblasts containing temperature-sensitive *p53*, E2F-1 over expression induces apoptosis only at the permissive temperature (Wu and Levine, 1994). Perhaps not surprisingly based on these studies, other proteins with roles in the induction of S-phase have also been shown to cooperate with p53 to induce apoptosis, such as the c-myc oncoprotein (Wang *et al.* 1993; Hermeking and Eick 1994). The take-home lesson from these studies is that p53 protein is sensitive to perturbations of normal cell cycle controls, and this protein responds to unchecked cellular proliferation by inducing apoptosis. This almost certainly explains why transformed cells are more sensitive to DNA-damage induced apoptosis than non-transformed cells.

Apoptosis induced by E1A in a p53-dependent manner can be inhibited by co-expression of the adenovirus *E1B-55K* gene (Debbas and White 1993). The E1B-55K gene product has been shown to bind and inactivate p53 (Sarnow *et al.* 1982; Yew and Berk 1992). Indeed, several DNA viruses encode protein products involved in stimulating cell cycle transitions, e.g. the E7 protein of human papilloma virus and the adenoviral E1A protein. Such proteins invariably trigger p53, and induce this protein's apoptotic functions. In all cases, however, it has been discovered that viruses encode proteins that inactivate p53 in some manner, thus preventing apoptosis. Thus the E1B 55K protein of adenovirus and the large T antigen of SV40 virus each function to bind to p53 and inhibit p53-mediated apoptosis, and the E6 protein of human papilloma virus can bind to p53 and induce its degradation via the ubiquitin-proteasome pathway (Scheffner *et al.* 1993). Many other viruses encode proteins that are reported to either inactivate p53 or block apoptosis; these examples have been recently reviewed in Shen and Shenk (1995).

The significance of p53 to apoptosis increased considerably when two groups elucidated a physiological role for p53 in this process. These studies took advantage of a mouse that was genetically engineered to be null for *p53*. Offspring from such *p53* "knock-out" mice at first appeared to develop normally, but succumb at very early ages (3-6 months) to a variety of tumour types, chiefly lymphocytic lymphoma (Donehower *et al.* 1992). Utilizing cultures of resting thymocytes isolated from such *p53* "knockout" mice, two groups simultaneously found that *p53* null thymocytes were resistant to radiation-induced apoptosis, compared to thymocytes isolated from matched, wt *p53*-containing littermates (Clarke *et al.* 1993; Lowe *et al.* 1993). Interestingly, these studies indicated that, while *p53*-null thymocytes were resistant to radiation-induced apoptosis, they maintained normal sensitivity to glucocorticoid and T cell-receptor-mediated apoptosis. This differential sensitivity to radiation-induced apoptosis between *p53 +/+* and *p53 -/-* thymocytes was limited to non-proliferating cells; once stimulated to proliferate, thymocytes of both *p53* status are equally susceptible to radiation-induced apoptosis (Strasser *et al.* 1994). In a separate but equally compelling study, gamma irradiation of mice containing wt *p53* was found to result in extensive early apoptosis in the stem cell compartment of the small intestine; this apoptosis was not found in identically-irradiated intestinal cells from *p53*-null mice (Merritt *et al.* 1994). These studies indicated that p53 protein is a critical physiological mediator of DNA damage-induced apoptosis *in vivo*. Further, it was clear from these studies that there exist both p53-dependent and -independent pathways of apoptosis, in response to radiation as well as other stimuli.

3.0 p53-DEPENDENT APOPTOSIS CONTROLS TUMOUR INITIATION, PROGRESSION AND RESPONSE TO CHEMOTHERAPY

The finding that p53 was a physiological inducer of apoptosis led to elegant *in vivo* studies illustrating the importance of p53-mediated apoptosis as a critical controller of tumour progression. In particular, studies by Van Dyke and colleagues found that expression of the large T antigen (LTAg) of SV40 virus could mimic a *p53*-null scenario in transgenic mice, and induce tumour formation in either B and T cells (McCarthy *et al.* 1994), or the choroid plexus (Symonds *et al.* 1994), depending on the promoter construct driving expression of the LTAg transgene. The LTAg protein of SV40 binds to several cellular proteins with roles in cell cycle regulation in addition to p53; among them are the retinoblastoma protein (RB), the RB-related proteins p107 and p130, and the transcriptional co-activator protein p300 (see for review Howley 1995). Van Dyke and co-workers created transgenic mice containing wt LTAg, which succumbed to aggressively growing tumours of the choroid plexus epithelium after a mean of 6 weeks. These mice were compared to mice made transgenic for a mutant form of LTAg in which the p53-binding domain of LTAg is deleted. These latter mice (T121), which fail to inactivate the cellular *p53* protein, developed tumours that grew much more slowly. T121 mice had an average survival time of 26 weeks. Significantly, the growth rates of the tumours generated from these wt LTAg- and T121-transgenic mice were identical; however, tumours from the T121 mice contained areas of extensive apoptosis. Crossing T121 mice into a *p53 -/-* background effectively eliminated this apoptosis, indicating that it was almost exclusively p53-dependent (Symonds *et al.* 1994). Therefore, *p53* inactivation, which occurs in over 50% of tumours, was shown to directly contribute to tumour progression by eliminating the p53-dependent apoptosis that arises due to cell cycle perturbations accompanying transformation *in vivo*.

Inactivation of p53 protein can also be achieved by expression of mutant forms of p53, which can function as dominant-negative inhibitors by binding to wt p53 through the oligomerisation domain and inhibiting the sequence-specific DNA binding of this protein. In fact, expression of the p53-oligomerisation domain alone is sufficient to function in this dominant-negative fashion (Gottlieb *et al.* 1994), and inhibit apoptosis in the T121 mice described above (Bowman *et al.* 1996). Interestingly, crossing the T121 mice with mice made nullizygous for the cell cycle-regulated transcription factor E2F-1 effectively abolishes this *p53*-mediated apoptosis, indicating that p53-dependent apoptosis in the choroid plexus epithelium is almost exclusively E2F-1-dependent (T. Van Dyke, personal communication). These data, generated *in vivo*, validate the significance of results outlined previously obtained with cell lines (Wu and Levine 1994), which indicated that p53 over expression induces apoptosis in cooperation with the over expression of cell cycle proteins like E2F-1.

The suggestion that p53-dependent apoptosis is a rate-limiting step for tumour growth was further supported by studies indicating that regions of low oxygen concentration (hypoxia), which occur in central portions of unvascularised tumour tissue, are characterised by extensive apoptosis. Both p53 protein and activity are induced by hypoxia, and it has been shown that p53 induction occurs in hypoxic regions of tumours, concomitant with apoptosis (Graeber *et al.* 1994, 1996). Such apoptosis was not detected in tumours from genetically similar animals that lacked p53 . Significantly, it was demonstrated that adding a small percentage of *p53*-null transformed fibroblasts to a larger population of transformed fibroblasts containing wt *p53* led to a considerable overgrowth of *p53*-null cells in hypoxic regions of an implanted tumour (Graeber *et al.* 1996). Therefore, hypoxia-induced p53-dependent apoptosis occurs in non-vascularised regions of tumours, and mutations in *p53* in these regions of tumour are strongly selected for during tumour development. This strong selection for mutations in *p53* in hypoxic regions of tumours may synergise with a reported anti-angiogenic property of p53 - transactivation of the angiogenesis inhibitor thrombospondin (Dameron *et al.* 1994).

A role for p53-dependent apoptosis in normal mammalian development has been inferred from closer examination of the progeny of *p53*-null mice. While it was initially believed that such mice developed completely normally, closer examination has revealed a statistically significant increase in the number of birth defects in mice born null for *p53*. In particular, defects in normal closure of the neural tube, leading to abnormal proliferation of cells in the mid-brain (a condition known as exencephaly), has been noted in a number of female progeny that are *p53 -/-* (Sah *et al.* 1995). The reason for the predisposition in females is unclear. Another group has reported similar results, as well as a significant increase in *p53* null progeny of craniofacial malformations, ocular abnormalities and tooth defects (Armstrong *et al.* 1995). Furthermore, following irradiation of *p53 -/-* males, a significant increase in the percentage of exencephalic female embryos were found (23% to 60%), in addition to an increased incidence of other abnormalities, such as retinal dysplasia (Armstrong *et al.* 1995). A similar role for p53 in eliminating embryos with radiation-induced genetic defects has now been found by others (Norimura *et al.* 1996). In a separate study, a higher incidence of fetal resorptions of *p53 -/-* embryos following treatment of females with benzo(a)pyrene has been noted, indicating that the presence of p53 protects the developing embryo from DNA damaging chemicals and developmental oxidative stress (Nicol *et al.* 1995). Taken together these results support a role for p53 in monitoring the integrity of the genome during normal development, and preventing fetal abnormalities by the removal of cells with damaged DNA through either growth arrest and repair, or p53-dependent apoptosis.

Perhaps the most relevant role of p53-dependent apoptosis in terms of cancer development is in the cellular response to radiation and chemotherapeutic agents. In two studies, *p53* +/+ and -/- fibroblasts, transformed with either *E1A* or *E1A* and *c-Ha-ras* together, were treated with either irradiation or chemotherapy agents. The ability of these treatments to induce apoptosis in transformed fibroblasts that differed only in *p53* status was assessed. The absence of functional p53 was clearly correlated with a decreased apoptotic response to irradiation or chemotherapeutic agents; both cells in culture and transplanted tumours lacking p53 grew more aggressively, and failed to respond to radiation or chemotherapy (Lowe *et al.* 1993c; 1994). Additionally, tumours from *p53* +/+ cells that developed resistance to chemotherapy were found to have acquired mutations in *p53* during tumour progression (Lowe *et al.* 1994). The paradigm from these studies is that there is a strong selection for mutations in *p53* during tumour development: such mutations limit the apoptotic response to uncontrolled proliferation, limit the apoptotic response to hypoxia, increase the mutation frequency in the tumour (therefore allowing for mutations leading to more aggressive tumour development), and limit the apoptotic response to radiation and chemotherapy, (which are the most common forms of cancer treatment after surgery).

4.0 UPSTREAM MEDIATORS OF p53 ACTIVATION

4.1 DNA strand breaks are the signal for p53 induction

It is now commonly accepted that p53 protein responds to many, if not all, forms of DNA damage. This protein exists normally in cells in a latent or inactive state, with a half-life of 3-20 minutes, depending on cell type (See Zambetti and Levine 1993 for review). Treatment with any of a number of DNA damaging agents (outlined below) leads to an increase in p53 protein half-life and levels, as well as an activation of the sequence-specific DNA binding activity of the protein. It is currently believed that p53 stabilisation and activation for DNA binding are two separate but not necessarily independent events. In most cases, however, stabilisation and activation seem to occur concomitantly. Some of the agents that lead to increased p53 stability/activity are UV irradiation (Maltzman and Czyzyk 1984; Lu and Lane 1993), gamma irradiation (Kuerbitz *et al.* 1992), topoisomerase inhibitors such as etoposide and camptothecin, DNA alkylating agents such as MMS and mitomycin C, DNA intercalating agents such as actinomycin D and doxorubicin, and anti-metabolites such as methotrexate (see Fritsche *et al.* 1993 for review). Obviously, both the nature and kinetics of DNA damage induced by these agents are very different, but it seems clear that most, if not all, forms of DNA damage or damage repair are signalled to p53. There is some evidence that it is the generation of DNA strand breaks induced by these agents or by the repair pathway that is the stimulus for p53 activation. For example, UV-induced thymine dimers do not trigger p53 if their repair is inhibited, and topoisomerase inhibitors require replication-induced DNA stand breaks in order to induce p53 (Nelson and Kastan 1994). Along these lines, micro-injection of cells with restriction enzymes (Lu and Lane 1993; Nelson and Kastan 1994) or with as little as one molecule of linear DNA (Huang *et al.* 1996) appear to be sufficient to induce p53 protein activity. Kinetic studies indicate that a single DNA strand break generated by gamma irradiation may be sufficient to induce p53 stabilisation and activation (Huang *et al.* 1996).

The mechanism by which p53 protein detects or is triggered by DNA damage has been extensively studied. An emerging theme to this work is that p53 can respond to different types of DNA damage (for example, UV versus gamma irradiation) through discrete pathways (See Figure 2). There is also compelling evidence that p53 may detect DNA damage directly (the carboxyl terminus of p53 protein encodes several activities involved with damage detection and signalling). For example, this carboxyl-terminal domain of p53 (363-393) contains residues that bind to both single-stranded RNA and DNA molecules, as well as free DNA ends. This region also can catalyse the re-annealing of DNA and RNA (Bayle *et al.* 1995; Wu *et al.* 1995). This domain binds with high affinity to insertion-deletion loops (IDL mismatches), which are common intermediates in certain forms of DNA repair (such complexes of p53 and IDL loops are quite stable, with half-lives of up to 2 hours (Lee *et al.* 1995)). Significantly, the carboxyl terminus of p53 is also responsible for inhibiting the sequence specific DNA binding activity of this protein, by either sterically or allosterically interfering with the core DNA binding domain. This negative regulation of p53 activity can be relieved by deleting the p53 C-terminus, antibody binding in this region of p53 (Hupp *et al.* 1992), or by phosphorylation of residues in the last 26 amino acids by one of several kinases (see Gottlieb and Oren 1996 for review), as well as by binding to short strands of DNA (Jayaraman and Prives, 1995). These data suggest that direct binding of p53 to DNA-damage intermediates, or to molecules involved in DNA repair, may be an intermediate step in relieving the negative regulation of p53 by its C-terminus, and activating the transcriptional activity of this protein

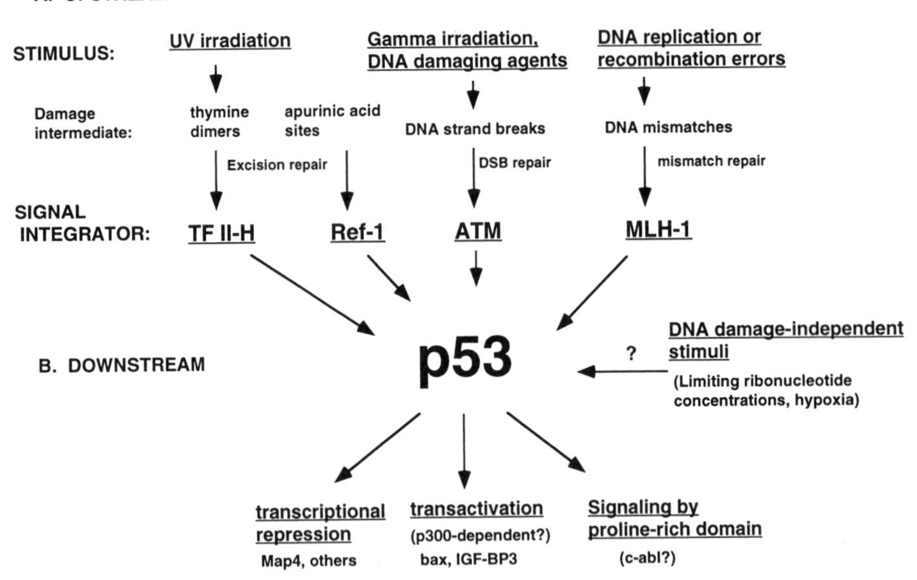

Figure 2. Diagram of p53 signalling pathways.
Upstream of p53 signalling are stimuli that induce DNA-damage and repair intermediates, such as ultraviolet (UV) irradiation, gamma irradiation, and DNA replication/recombination errors. In each case DNA repair intermediates from these pathways targets a signal integrator, such as TFII-H, Ref-1, ATM (ataxia-telangiectasia mutated), and MLH-1 (mutL homolog-1) to signal the presence of DNA damage to p53. These signal integrators appear to be specific for distinct types of DNA damage; for example, cells lacking the ATM gene product do not induce p53 protein in response to gamma irradiation, but p53 protein is induced normally in these cells following UV irradiation. (This figure is by no means inclusive of all forms of DNA damage that may occur in a cell, it is merely used to illustrate those types of damage for which the p53 response has been better characterised). The downstream functions of p53 that function in p53-dependent apoptosis are also depicted, as well as some genes with established or predicted roles in apoptosis.

4.2 The role of ATM (ataxia telangiectasia mutated) in the p53 response

Kastan and colleagues first showed that p53 protein was stabilised and activated for transcription in response to gamma irradiation, leading to a transient cell cycle arrest in the G1 phase of the cell cycle (Kastan *et al.* 1991; Kuerbitz *et al.* 1992). These studies were extended to include fibroblasts from individuals with the autosomal recessive disorder ataxia-telangiectasia (AT). Individuals suffering from AT show evidence of cerebellar degeneration (producing ataxia), sensitivity to gamma irradiation, and increased incidence of tumours of the lymphoid system. The gene responsible for this disorder (ATM) was identified by positional cloning and found to be homologous to a large family of kinases with roles in cell cycle control and DNA repair (see Morgan and Kastan 1997 for review). Interestingly, Kastan and colleagues showed that fibroblasts from individuals with AT had defects in the ability of p53 to respond to gamma irradiation. Specifically, p53 protein induction was delayed and the p53-response gene *gadd45* was induced by gamma irradiation very late or not at all (Kastan *et al.* 1992). These studies were supported by the generation of a knock-out mouse for the *ATM* gene. Like individuals with AT, mice null for the *ATM* gene have increased incidence of lymphoid cancer, as well as other defects similar to symptoms of AT patients. Fibroblasts from *ATM* knockout mice also showed a significant defect in the p53 response following gamma irradiation, although the p53-response to UV irradiation was normal (Xu and Baltimore 1996). These studies support the placement of the ATM kinase upstream of p53 in the signal transduction pathway for gamma irradiation, but not in the UV irradiation pathway (see Figure 2). How the ATM gene communicates the presence of DNA damage to p53 is currently unclear, although it is tempting to speculate that this putative protein kinase phosphorylates and activates p53.

4.3 TFII-H, Ref-1 and MLH-1

Several upstream and downstream mediators of the p53-dependent UV response have been identified. p53 is known to interact with polypeptide components of the nucleotide excision repair and transcription complex TFII-H. This complex consists of several proteins, among them the transcription-coupled repair factors XPD and XPB which have helicase activity, and the kinase components cdk 7, cyclin H and p36. p53 has been reported to associate with both XPB and XPD via its C-terminus, and to inhibit the helicase activities of these proteins (Wang *et al.* 1995). Additionally, cells from individuals with xeroderma pigmentosa, containing germline defects in the TFII-H components XPB or XPD, are resistant to p53-dependent apoptosis. In contrast, cells containing mutated versions of the *XPA* or *XPC* genes maintain susceptibility to p53-dependent apoptosis

(Wang *et al.* 1996). Other studies have found that highly purified preparations of TFII-H, as well as recombinant preparations of cyclin H/cdk 7/p36, can phosphorylate p53 at the C terminus, resulting in an enhancement of sequence-specific DNA binding and transactivation (Lu *et al.* 1997). These data support the hypothesis that p53 can detect some forms of DNA damage in association with the RNA polymerase-associated factor TFII-H, and that following damage detection by TFII-H, this enzyme phosphorylates and activates p53 (Figure 3).

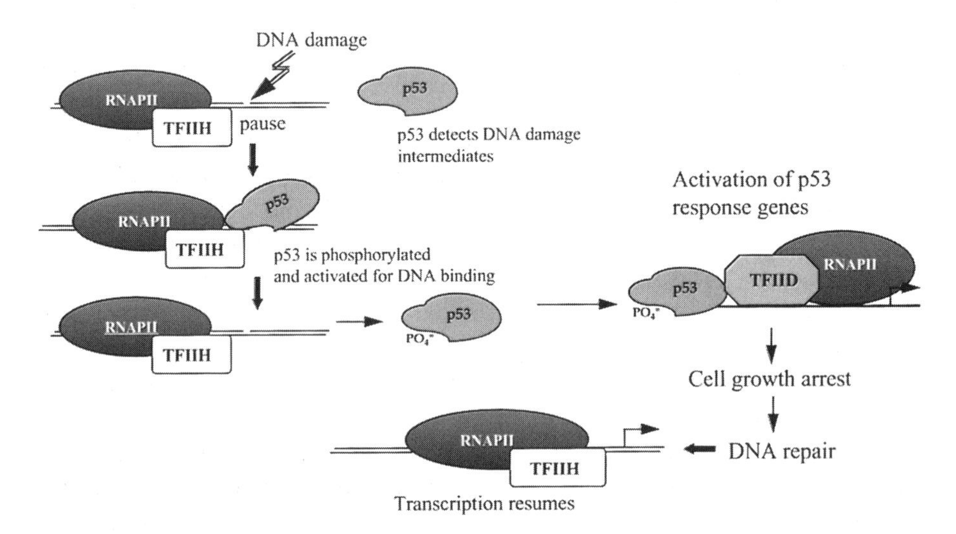

Figure 3. Model for p53 activation by TFII-H

UV-light induced DNA damage, and DNA repair induced-nicks, are detected by the RNA polymerase II-associated complex TFII-H. RNA polymerase II-directed transcription pauses following DNA damage detection, and p53, either associated with TFII-H or recruited into this complex, becomes phosphorylated by the cyclin H/cdk 7/ p36 (CAK, cyclin-activating kinase) component of TFII-H. This phosporylation activates p53 for sequence specific DNA-binding, and enhances the transactivation of p53-response genes, such as p21/waf1, thereby inducing p53-dependent growth arrest. Presumably, this G1 growth arrest allows for damaged DNA to be efficiently repaired.

Yet another protein that functions in DNA damage repair and activates the p53 protein is the cellular Ref-1 protein. Ref-1 is a dual-activity protein that functions both in controlling the redox state of proteins, as well as in DNA repair. This protein is an apurinic acid endonuclease that functions in excision repair. Significantly, Ref-1 has been shown to function as a non-covalent activator of sequence-specific DNA binding of p53, via both redox-dependent and independent mechanisms (Jayaraman *et al.* 1997). Therefore, while oxidised p53 is severely defective in DNA binding, reduction of p53 by Ref-1 enables the protein to bind DNA efficiently. Fully reduced p53 protein is also capable of being activated as a transcription factor by Ref-1, in a manner independent of its redox activity. This redox-independent activation of p53 by Ref-1 functions on the C-terminus of p53 (Jayaraman *et al.* 1997). Interestingly, a strong inducer of Ref-1 is hypoxia, indicating that Ref-1 may in part mediate the activation of p53 in regions of cellular hypoxia (Graeber *et al.* 1996).

Some protein components of the mismatch repair pathway may also function upstream of p53. Brown and colleagues studied cisplatin-resistance in ovarian carcinomas, and showed that acquisition of cisplatin-resistance corresponded to a loss in the expression of MLH1, a central component of the mismatch repair system (Anthoney *et al.* 1996). Consequently, these cancer cells had a higher rate of mismatch repair mutations. Significantly, while these cisplatin-resistant cell lines retained expression of wt *p53*, the p53-dependent response to DNA damage in these cells was found to be severely attenuated (Anthoney *et al.* 1996). One explanation for these observations is that the DNA damage repair protein MLH1, or an MLH1-associated protein (such as a kinase), may function upstream of p53 in its signalling pathway. While this remains an interesting speculation, several facts are clear: 1) different proteins are utilised in the cell to detect distinct types of DNA damage; 2) these detectors function in the assembly of the repair machinery; 3) these repair complexes signal to p53, the integrator of the DNA damage response, via redox changes, phosphorylation, or other post-translational modifications (see Figure 2). This results in activation of p53, and either growth arrest or apoptosis. How p53 accomplishes these two very different outcomes, and how the decision to undergo growth arrest or apoptosis is made, is the topic of the next two sections.

5.0 DOWNSTREAM MEDIATORS OF p53-DEPENDENT APOPTOSIS

5.1 p53-dependent transcriptional activation: bax and IGF-BP3

It now seems clear that p53 performs some, but certainly not all, of its functions in apoptosis through its ability to activate transcription of a subset of genes (reviewed in Ko and Prives, 1996). There is good evidence that p53 mutants with defective transactivation domains are compromised in their ability to induce apoptosis, indicating that transactivation is a significant component of this process. However, in some experimental systems transactivation-deficient mutants of p53 are effective inducers of apoptosis (see section V-B). Current evidence suggests that both transactivation-dependent and -independent pathways can contribute to p53-dependent apoptosis (see Figure 4)

Recently the transcriptional co-adaptor molecule p300, and the related molecule CBP (CREB-binding protein) have been implicated as proteins that bind to p53 and mediate transcriptional activation of p53-response genes (Avantaggiati *et al.* 1997; Gu *et al.* 1997; Lill *et al.* 1997). Expression of a dominant-negative form of p300 was found to compromise *p53*-dependent transactivation, and counteract both p53-mediated growth arrest and apoptosis (Avantaggiati *et al.* 1997). While a number of p53-response genes have been reliably identified, the significance of many of these to the process of p53-dependent apoptosis remains unclear. For example, the $p21^{/waf1/cip1/sdi1}$ gene encodes a cyclin-dependent kinase inhibitor whose induction by p53 is quite influential in p53-mediated growth arrest. However, this protein seems of little importance to the induction of apoptosis as both thymocytes and transformed fibroblasts from mice that are null for *p21* respond normally to gamma irradiation and p53 induction, undergoing apoptosis at rates indistinguishable from wt cells (Deng *et al.* 1995; Attardi *et al.* 1996). The contribution of other p53 response gene products, such as mdm2, gadd45, and cyclin G, to p53-dependent apoptosis are not obvious, and remain to be explored. Two genes regulated by p53 may play a clearer role in apoptosis: *bax* and *IGF-BP3.*

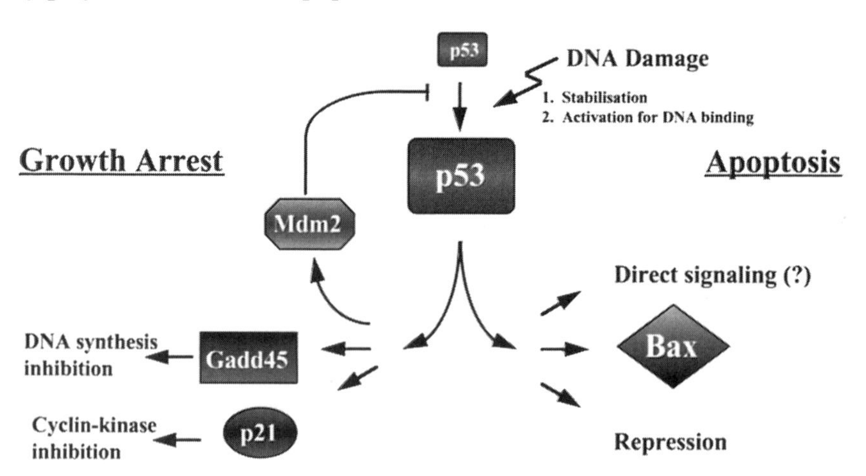

The *bax* gene is a homologue of the cellular *bcl2* gene, which codes for a membrane-associated intracellular protein that possesses anti-apoptotic activity. Bax and Bcl2 are capable of forming heterodimers *in vivo*, and a variety of experiments indicate that it is the ratio of these two proteins in a cell that determines cell death or survival: Bax homodimers can induce death, while Bcl2:Bax heterodimers promote survival. It has been speculated that these two proteins may influence mitochondrial membrane potential during apoptosis by forming pores in mitochondrial membranes. Alternatively, it has been found that Bcl2 can interfere with the binding of the mammalian ced-4 protein (Apaf-1) with certain members of the caspase family. The caspases comprise a group of evolutionarily-conserved cysteine proteases that are downstream effectors of apoptosis. The binding of ced-4 to caspase 8 activates the proteolytic activity of the latter, and Bcl2-binding is believed to inhibit this interaction (See Reed 1997 for review). While it is quite clear that Bax is a pro-apoptotic molecule, the contribution of Bax to p53-dependent apoptosis was initially controversial. While bax expression seems clearly inducible in cells containing temperature-sensitive mutants of *p53* (Miyashita and Reed 1995), several groups have failed to see induction of Bax mRNA or protein following p53 induction in response to ionizing radiation (Allday *et al.* 1995; Canman *et al.* 1995). Additionally, thymocytes from mice engineered to be null for Bax (Bax "knockout" mice) respond to p53-induced cell death following gamma irradiation in a manner indistinguishable from wt thymocytes (Knudson *et al.* 1995).

The best data arguing for the contribution of Bax to p53-mediated apoptosis has been generated from studies of the LTAg mutant T121 transgenic mice. As mentioned above, these mice get tumours of the choroid plexus, but such tumours have limited growth due to extensive p53-dependent apoptosis (Symonds *et al.* 1994). When T121 mice are crossed into a *bax* null background, the p53-mediated apoptosis in choroid plexus tumours from such mice (T121; bax -/-) was reduced approximately 50% (Yin *et al.* 1997). Analogous studies done in transformed cell lines derived from *bax -/-* fibroblasts support the conclusion that p53-mediated apoptosis is attenuated in the absence of Bax (McCurrach *et al.* 1997). While it might be argued that *bax* null cells may be resistant to many forms of apoptosis, both p53-dependent and independent forms, Yin and colleagues showed an eight-fold induction of Bax in choroid plexus cells undergoing p53-dependent apoptosis (Yin *et al.* 1997). Therefore, while the contribution of Bax to p53-mediated apoptosis seems clear, the transactivation of *bax* by p53 has not been seen in all cell types, so this contribution may be partially, or totally cell-type dependent. Why the induction of Bax by p53 may be tissue specific will be an important question to explore in the future.

p53 transactivates another gene with an expected role in apoptosis. Using a differential cloning approach, the gene encoding insulin-like growth factor binding protein 3 (IGF-BP3) was found to be transcriptionally activated by p53 (Buckbinder *et al.* 1995). Transactivation of this gene was found to be associated with increased secretion of a form of IGF-BP3 capable of binding to, and inhibiting, the insulin-like growth factor 1 (IGF-1) (Buckbinder *et al.* 1995). Interestingly, over expression of the IGF-1 receptor can inhibit p53-mediated apoptosis (Prisco *et al.* 1997), and over expression of IGF-BP3 has been shown to induce apoptosis in some cell types (Rajah *et al.* 1997). It seems reasonable to speculate that IGF-1 acts as a cell maintenance factor with anti-apoptotic activity. IGF-BP3 blocks its action and is made in response to p53. Thus, there is a clear potential role for transactivation of this gene in apoptosis. An interesting similarity in the transactivation of *IGF-BP3* and *bax* genes by p53 was recently established. Two groups uncovered mutant forms of p53 in human tumours that are defective in their ability to induce apoptosis, but when over expressed in cell lines can still induce growth arrest in G1 (Friedlander *et al.* 1996; Rowan *et al.* 1996). Interestingly, in both cases these p53 mutants were found to be capable of transactivating the *p21* promoter, but were defective in their ability to transactivate either the *IGF-BP3* promoter or the *bax* promoter *in vitro* (Friedlander *et al.* 1996; Ludwig *et al.* 1996). Both of these latter promoters contain p53 binding sites with many differences from the ideal consensus nucleotide sequence, and a lower affinity for p53 protein (Buckbinder *et al.* 1995; Miyashita and Reed 1995). These data raise the interesting possibility that genes with poor affinity binding sites for p53 protein may be influential in apoptosis. Thus, low concentrations of p53 cause growth arrest whereas higher levels maybe required to induce apoptosis. In support of this hypothesis is the observation that induction of low levels of p53 in Saos-2 cells results in growth arrest, but induction of higher levels of wt p53 leads to apoptosis in this same cell line (Chen *et al.* 1996a).

5.2 Transactivation-independent apoptosis mediated by p53

There is emerging evidence for a transactivation-independent role for p53 in apoptosis. In cells containing both an inducible *c-myc* gene, as well as temperature-sensitive *p53*, synthesis of high levels of both myc and wt p53 proteins leads to apoptosis. However, apoptosis in this system occurs in the presence of protein synthesis inhibitors, and without induction of p53-responsive genes, indicating the absence of a requirement for p53-mediated transactivation (Wagner *et al.* 1994). In a similar experimental system involving apoptosis, another group also found that p53-dependent apoptosis was not prevented by inhibitors of new RNA or protein synthesis (Caelles *et al.* 1994). Further, transient transfection studies in HeLa cells, followed by flow cytometric sorting of apoptotic cells, indicated that a truncated form of wt p53 (from amino acids 1-214) (incapable of functioning as a sequence-specific transcriptional transactivator), was still capable of inducing apoptosis in this cell line (Haupt *et al.* 1995b). This mutant was also capable of suppressing the transformation of primary cells, an activity of p53 generally presumed to rely on its apoptotic function. All the above data indicate the existence of a strong pro-apoptotic function of p53 that is independent of transactivation. The current hypothesis is that both transactivation-dependent and -independent pathways of p53-mediated apoptosis exist and that two pathways probably synergise toward apoptosis. Interestingly, the requirement for p53-dependent transactivation in apoptosis may be cell-type specific. For example, transactivation-deficient p53 is an effective inducer of apoptosis in some cell types (Haupt *et al.* 1995b) but not others (Sabbatini *et al.* 1995b; Yonish-Rouach *et al.* 1996). Additionally, the mdm2 oncoprotein, which binds to p53 and inhibits its transactivation functions (but presumably allows transactivation-independent functions), can inhibit p53-dependent apoptosis in some cell types but not others (Chen *et al.* 1996b; Haupt *et al.* 1996).

5.3 The proline-rich domain of p53 constitutes a novel signalling domain for apoptosis

Recently, apoptotic activity has been ascribed to a novel p53 domain. This domain, from amino acids 61-94, is proline-rich, and links the transactivation domain to the sequence-specific DNA binding domain in the p53 protein. In human p53 this proline-rich region contains five repeats of the sequence PXXP, where P represents proline and X any amino acid. This consensus motif has been shown to create a binding site for signal transduction proteins containing SH3 (src-homology domain 3) motifs (Yu *et al.* 1994) and may, therefore, constitute a unique domain of p53 capable of interacting with signalling molecules. Significantly, deletion of this domain compromised the ability of p53 to induce apoptosis, as measured in a growth suppression assay (Walker and Levine, 1996) although the protein remained competent for transactivation of endogenous p53-response genes. Furthermore, mutation of these proline residues rendered p53 incapable of cooperating with gas-1 (growth-arrest-specific-1) to suppress proliferation (Ruaro *et al.* 1997). It seems likely that the normal function of this domain is influential in transactivation-independent p53-mediated apoptosis. One possible mediator of this apoptotic function is the c-abl protein. c-abl is a nuclear kinase that contains an SH3 domain. This protein has been reported to interact with p53, and expression of c-abl can inhibit cell growth only in cells containing wt, but not mutant, p53 (Yuan *et al.* 1996b). Additionally, the kinase activity of c-abl is DNA damage-inducible, and expression of c-abl has been shown to induce apoptosis in some cell lines (Yuan *et al.*

1996a).

5.4 p53-mediated transcriptional repression

Transcriptional repression by p53 is an activity that only recently has been shown to be physiologically relevant and sequence-specific (Murphy *et al.* 1996; Murphy and Levine, unpublished results). The possibility that the transcriptional repression activity of p53 might be influential in apoptosis first came from studies on the adenoviral gene *E1B-19K*. This gene, which is essential for adenoviral replication and can cooperate with the adenovirus *E1A* gene to transform cells, is a viral homologue of *bcl-2*, a potent inhibitor of apoptosis. It was discovered that expression of either the *bcl-2* gene or the *E1B-19K* gene in human tumour cells could render such cells resistant to p53-dependent apoptosis (Chiou *et al.* 1994; Sabbatini *et al.* 1995a). Significantly, the only activity of p53 found to be inhibited in the presence of E1B-19K was its transcriptional repression activity; the ability of p53 to transactivate several different genes and induce G1 growth arrest was unperturbed in the presence of E1B-19K (Shen and Shenk 1994; Sabbatini *et al.* 1995a; Murphy *et al.* 1996). These results point to the possibility that repression of the expression of a subset of genes by p53 might be necessary for p53-dependent apoptosis. Indeed, the ability to inhibit p53-mediated transcriptional repression, and p53-dependent apoptosis, has also been ascribed to the protein products of the *bcl-2* gene (Shen and Shenk 1994), as well as the Wilms tumour suppressor gene *WT1* (Maheswaran *et al.* 1995). In the latter case, it is known that WT1 protein interacts with p53, and may inhibit repression and apoptosis by this interaction (Maheswaran *et al.* 1995). How Bcl2 and E1B-19K proteins inhibit p53-mediated transcriptional repression is currently unclear.

Several genes with decreased expression following *p53* induction have been identified (Amson *et al.* 1996; Werner *et al.* 1996; Prisco *et al.* 1997). The best characterised *p53*-repressed gene is that for the microtubule-associated protein, Map4. This ubiquitously expressed protein catalyses the polymerisation of microtubules from tubulin subunits, and stabilises them by direct binding. Transcriptional repression of *Map4* occurs during physiological induction of p53 leading to either growth arrest or apoptosis, but does not occur during p53-independent apoptosis. Significantly, transcriptional repression of the endogenous *Map4* gene by p53 is inhibited by E1B-19K and WT-1, both of which inhibit p53-mediated apoptosis but not transactivation or growth arrest (Murphy *et al.* 1996). Furthermore, over expression of Map4 in cells undergoing p53-mediated apoptosis can significantly delay the appearance of cells with fragmented DNA (Murphy *et al.* 1996), indicating that repression of this gene appears to contribute to the rapid progression of apoptosis. The identification of other p53-repressed genes is likely to be a fruitful step toward understanding the activity of this protein in apoptosis.

6.0 THE DECISION BETWEEN p53-MEDIATED GROWTH ARREST AND APOPTOSIS

6.1 The retinoblastoma susceptibility protein (RB) and E2F-1

Perhaps the first and physiologically most relevant protein found to influence the apoptotic activity of p53 is the product of the retinoblastoma susceptibility gene, *RB*. The RB protein is a critical component of the G1/S transition in mammalian cells. Inactivation of RB due to phosphorylation by G1 cyclin-dependent kinases is believed to be the rate limiting step for entry into the DNA synthesis (S) phase of the cell cycle. Phosphorylation inactivates the negative growth effect of RB, and frees the transcription factor E2F-1 (and other E2F-family members), presumably to activate the transcription of genes necessary for completion of S phase.

In transgenic mice expressing the human papillomavirus *E7* gene in the lens of the eye, the E7 protein binds and inactivates the RB protein, as well as Rb-family members p107 and p130. This results in the activation of E2F-transcription factors and extensive apoptosis in these cells (Howes *et al.* 1994; Pan and Griep 1994; 1995). Significantly, this apoptosis was found to be dependent on p53, as crossing these mice into a *p53* -/- background (Howes *et al.* 1994), or co-expressing the HPV E6 protein (which inactivates p53) resulted in lens cell hyperplasia or tumorigenesis instead of apoptosis (Pan and Griep 1994). p53-dependent apoptosis is also evident in the developing mouse lens in the *Rb* "knockout" mouse (Morgenbesser *et al.* 1994). In cell culture systems, *Rb* status has been shown to directly control the decision between p53-mediated growth arrest and apoptosis. Treatment of *p53* +/+, *Rb* +/+ mouse embryo fibroblasts with DNA damaging agents leads to cell cycle arrest whereas the same treatment of *p53* +/+, *Rb* -/- fibroblasts leads to extensive apoptosis (Almasan *et al.* 1995). The combined data indicate that the presence of inactive RB, through phosphorylation, deletion, or binding of viral inhibitors, can tip the scales toward apoptosis instead of growth arrest. The corollary to this is that active RB (bound to E2F) might inhibit p53-mediated apoptosis, and lead to growth arrest instead. In fact, it has been shown that transfection of HeLa cells with excess pRb can partially protect these cells from p53-mediated apoptosis (Haupt *et al.* 1995). It is perhaps not surprising that many of the viral and cellular proteins that cooperate with p53 to induce apoptosis (such as E7 and *E1A*) all affect RB protein function, and result in activation of E2F. It now appears that active or excessive E2F-1, but not E2F-2, -3, -4 or -5, is the signal that mediates p53's choice to induce apoptosis and not cell cycle arrest (DeGregori *et al.* 1997). This dual activity of E2F-1, inducing cellular proliferation but also cooperating with p53 in apoptosis, may explain why *E2F-1* can alternately function as an oncogene when over expressed in the absence of p53 (Johnson *et al.* 1994) or a tumour suppressor when deleted by homologous recombination in a *p53* wt mouse (Yamasaki *et al.* 1996).

6.2 Cytokines, Bcl-2, E1B-19K and WT-1

In addition to Rb status, a variety of signal transduction pathways can also impact on the cellular decision between p53-mediated growth arrest and apoptosis. For example, p53-dependent apoptosis is inhibited in the presence of cellular growth factors such as interleukin-3 (IL-3), IL-6, erythropoietin, steel factor, IGF-I and -II and others (Yonish-Rouach *et al.* 1991; Abrahamson *et al.* 1995; Canman *et al.* 1995; Lin and Benchimol 1995). However, as in the case for Bcl-2, E1B-19K and WT-1, this inhibition of apoptosis does not include inhibition of p53-transactivation. Hence the result of p53 induction in the presence of cytokines is a p53-dependent growth arrest (Abrahamson *et al.* 1995; Canman *et al.* 1995; Lin and Benchimol 1995; Maheswaran *et al.* 1995; Wang *et al.* 1995). It seems clear that in the presence of activated p53, growth factors function as survival factors, protecting cells from death until the stimulus for p53 activation is removed. It will be interesting to determine the extent to which incubation in these "survival" factors affects the phosphorylation state of RB, and the transcriptional repression function of p53. All the above indicate that p53 is able to "read" signal transduction pathways, and RB/E2F-1 status, and in doing so influence the “growth arrest or apoptosis” decision. How p53 monitors RB and E2F-1 status and signal transduction pathways is currently unclear. One intriguing possibility is that this function is mediated by the proline-rich region of p53, containing the SH3-binding elements. The possibility that this domain mediates p53 interaction with an SH3-containing protein important in signal transduction and influenced by growth factor pathways is presently an area of extensive investigation.

6.3 Tissue type and developmental stage

The earliest studies on p53-dependent apoptosis indicated the existence of certain cell types that growth arrest in response to p53, and others in which p53 induction led directly to apoptosis. These studies raised the possibility that either cell type or the accumulated subset of genetic mutations might influence the decision between growth arrest and apoptosis. It is now clear that both of these factors have significant impact on this decision. In particular, cell type has a significant contribution to activation and outcome of p53 induction, and this cell type-dependent regulation of the p53 response is developmentally regulated. To date, three laboratories have generated mice that are transgenic for the b-galactosidase (*lacZ*) gene driven by a p53-responsive promoter, and have examined the response of p53 to gamma irradiation and the DNA damaging agent doxorubicin. These studies examined p53 response using immunostaining for p53 to detect protein stabilisation and induction, as well as X-gal staining to measure p53 activity, and TUNEL analysis (a terminal transferase-based method to detect fragmented DNA ends) to measure apoptosis induced in each tissue. The take home message from such studies has been surprising: irradiation of mice early in development (the first 8-9 days of foetal life) led to an induction of p53 protein in almost all tissues, with concomitant X-gal staining and some measure of apoptosis (MacCallum *et al.* 1996). However, this response becomes more and more restricted to particular tissue types through development, birth and adulthood, where there is significant heterogeneity in both the transcriptional and the apoptotic response. There appear to be three classes of p53 response to gamma irradiation in the adult mouse (summarised from Gottlieb *et al.* 1997; Komarova *et al.* 1997; MacCallum *et al.* 1997):

1) Tissues in which p53 protein is induced, *lacZ* expression is found, and apoptosis occurs. These tissues include cells from the spleen, thymus, haematopoietic bone marrow, intestine and ependyma. Interestingly, in this subset, the spleen shows X-gal staining (ie. p53 transactivation) only in the white pulp, but apoptosis that is p53-dependent in both the red and white pulp. This supports a role for transactivation-independent apoptosis by p53 in the spleen.

2) Tissues in which p53 protein and activity is induced, but there is no evidence for apoptosis. Such tissues include the salivary gland, myocardium, adrenal gland, kidney, osteocytes of bone, and choroid plexus. Here, p53 transactivation (X-gal staining) was found in all of the aforementioned tissues except myocardium, where p53 protein was induced but no evidence for *lacZ* expression was found. In the remaining tissues it is presumed that p53 induction resulted in growth arrest rather than apoptosis.

3) Tissues in which no p53 protein was induced, there was no evidence for X-gal staining, and no evidence for apoptosis. These tissues include liver, muscle and brain.

The conclusion from such studies is that the p53 response *in vivo* is more complicated than previously appreciated. It is both developmentally and cell-type regulated, and there is evidence for both transactivation-dependent and -independent apoptosis. One unexpected result from such studies indicated that the induction of p53 by DNA damage was far less efficient in *p53* +/- mice, compared to wt mice (Gottlieb *et al.* 1997). This may indicate why Li Fraumeni individuals (heterozygous for mutant *p53*) have such a high cancer penetrance, and why some human tumours maintain a wt *p53* allele along with a single mutant copy.

7.0 RECENT DEVELOPMENTS IN p53-MEDIATED APOPTOSIS

7.1 Cancer therapy

Progress has been made toward using p53-mediated apoptosis as an avenue for therapy of human tumours. In particular, efforts have focused on using retroviral vectors as vehicles to deliver the *p53* gene to tumours . Other efforts have focused on attempts to re-activate mutant p53, or to liberate wt p53 bound to mdm2. Gene therapy using *p53* has involved the use of retroviral vectors expressing wt *p53* which are injected directly into tumours (Fujiwara *et al.* 1993; Roth *et al.* 1996). In one study, tumour regression was noted in three of nine patients in which a retrovirus encoding wt *p53* was injected into the tumour, and tumour growth was found to be stabilised in three other cases (Roth *et al.* 1996). An unrelated gene therapy protocol uses the absence of wild type *p53* in tumours as a signal for cell killing. This is accomplished using an adenovirus mutant that is defective in the *E1B-55K* gene. The product of *E1B-55K* is essential for adenoviral replication in part due to the fact that this protein binds to and inhibits p53, thereby preventing apoptosis and allowing viral replication. Therefore, only tumour cells null for *p53*, or in which p53 protein has been otherwise inactivated, are permissive for viral infection; cells with wt p53 are not productively infected with adenovirus mutants lacking E1B-55K. One study demonstrated that primary normal human cell cultures with wild type *p53* tolerated doses of adenovirus mutant 100 times greater than the dose that lysed several human tumour cell lines of different histological type containing inactivated *p53*. Significantly, this virus was very efficient at reducing the growth of implanted tumours containing *p53* mutations (Heise *et al.* 1997).

The *mdm2* oncogene is transactivated by p53, but is also a negative regulator of p53, therefore constituting part of a negative autoregulatory loop for p53 (Momand *et al.* 1992; Wu *et al.* 1993). The importance of this interaction during development is underscored by the finding that *mdm* null or "knockout" embryos do not survive the period of gestation. However, this lethal defect is rescued in p53 null embryos, suggesting that mdm2 is required to limit p53 activity during implantation or embryogenesis (Montes de Oca Luna *et al.* 1995). It has been shown that the mdm2 protein inactivates p53 function by binding and concealing the transactivation domain (Oliner *et al.* 1993). Furthermore, this mdm2:p53 complex can actively repress p53-responsive genes (Thut *et al.* 1997). The binding of mdm2 may also target p53 for ubiquitin-mediated degradation, thereby reducing expression of this protein (Haupt *et al.* 1997; Kubbutat *et al.* 1997). Recent efforts have focused on the disruption of mdm2:p53 complexes in tumours containing over expressed mdm2, with subsequent activation of the p53-dependent apoptotic pathway. Such therapy would not only be useful in tumours that contain over-expressed mdm2 due to the process of gene amplification, such as soft tissue sarcomas and gliomas (Oliner *et al.* 1992; Reifenberger *et al.* 1993), but also to an emerging class of human tumours that contain increased levels of mdm2 protein in the absence of gene amplification (Landers *et al.* 1997). Utilizing phage display protocols, Lane and colleagues have identified several peptides, as well as an mdm2 monoclonal antibody, that can strongly inhibit the interaction between mdm2 and p53, and can disrupt existing complexes in a cell (Bottger *et al.* 1997). It is anticipated that disruption of these complexes will either directly induce apoptosis in a tumour cell, or make the cell more susceptible to radiation and chemotherapeutic agents. The development of small molecular weight molecules that enter cells and disrupt p53:mdm2 complexes via competition for the p53 binding site on mdm2 protein is another example of rational drug discovery for cancer therapy.

Recently, efforts have been made to activate the DNA binding and transcription functions of certain mutant p53 molecules that frequently occur in human tumours. For example, the common human mutant containing histidine at codon 273 (His273) encodes a protein that is not transcriptionally active, but that maintains a wild type conformation *in vivo*. Treatment of human tumour cells containing this mutant with a peptide identical to residues 369-383 of p53 serves to activate this mutant for DNA binding (Selivanova *et al.* 1997). This peptide is believed to interact with the DNA binding domain of p53, and relieve the allosteric negative regulation of the C-terminus of p53 (Hupp *et al.* 1995). Interestingly, treatment of tumour cells containing either wt or His273 p53 protein with this small peptide results in extensive apoptosis in these cells, indicating that such measures may enhance cell killing by p53, or be used to target and remove tumours containing His273 mutations (Selivanova *et al.* 1997). Undoubtedly future studies in this field will continue to ask how p53-mediated cell death can be used to kill a tumour.

7.2 Future goals and conclusions

In 1991 it was demonstrated that p53 could induce apoptosis in tumour cells (Yonish-Rouach *et al.* 1991). Since that time several critical facts have become clear: (1) Mutation of the *p53* gene in 50-60% of all cancers inactivates p53-mediated apoptosis of tumour cells. (2) In model systems p53-mediated apoptosis is clearly responsible for slowing tumour development; (3) p53 mediates apoptosis in the presence of activated oncogenes (*E1A, c-myc, E7*) and a free and active E2F-1 transcription factor appears to be critical to p53's choice between growth arrest and apoptosis. (4) p53-mediated apoptosis results from both p53-transcriptional and non-transcriptional activities. Cell or tissue-type specificities are also important variables, with transcriptional activation of both the *bax* and *IGF-BP3* gene playing some role in apoptosis. (5) Radiation and chemotherapy rely heavily upon p53-mediated apoptosis to kill cells, and preferentially kill cells with activated oncogenes. (6) DNA damage detectors and repair proteins signal to p53, the integrator of this response, with the result being either growth arrest or apoptosis. (7) p53 plays a central role in DNA damage surveillance in the fetus, and it employs apoptosis to eliminate defective embryos. (8) The p53-mediated apoptotic pathway remains a central target for future attempts to destroy cancer cells even when *p53* mutations have occurred in such cells.

While much has been learned, much remains to be studied: (1) Are there additional p53-activated genes that function in apoptosis? (2) How does p53 communicate with the members of the caspase family, whose proteolytic activation is believed to be the penultimate step in apoptosis? (3) What is the contribution of p53-dependent transcriptional repression to apoptosis? (4) How does the proline-rich domain of p53 function in apoptosis? Does it communicate with signal transduction pathways that activate (e.g. fas or TNF-receptor-mediated signals) or block (e.g. IL-3, IL-6, etc.) apoptosis? These and other such questions will need to be answered if we are able to rationally design agents which kill cancer cells selectively using apoptosis.

Acknowledgments

The authors would like to thank members of the Levine and Shenk laboratories, particularly Dan Notterman and Paul Bailey, for critical input and advice. References in this chapter were not chosen to be exhaustive, and MM and AJL hope that the authors of works not cited will understand that space limitations prevented the citation of many pertinent papers. MM is a fellow of the Jane Coffin Childs Memorial Fund for Medical Research.

8.0 REFERENCES

Abrahamson,J.L., Lee,J.M. and Bernstein,A. (1995) Regulation of p53-mediated apoptosis and cell cycle arrest by Steel factor. *Mol.Cell Biol.* **15**: 6953-6960.

Allday,M.J., Inman,G.L., Crawford,D.H. and Farrell,P.J. (1995) DNA damage in human B cells can induce apoptosis, proceeding from G1/S when p53 is transactivation competent and G2/M when it is transactivation defective. *EMBO J.* **14**: 4994-5005.

Almasan,A., Yin,Y, Kelly,R.E., *et al.* (1995) Deficiency of retinoblastoma protein leads to inappropriate S-phase entry, activation of E2F-responsive genes, and apoptosis. *Proc. Natl. Acad. Sci. USA* **92**: 5436-5440.

Amson,R.B., Nemani,M., Roperch,J.P., *et al.* (1996) Isolation of 10 differentially expressed cDNAs in p53-induced apoptosis: activation of the vertebrate homologue of the *drosophila* seven-in-absentia gene. *Proc. Natl. Acad. Sci. USA* **93**: 3953-3957.

Anthoney,D.A., McIlwrath,A.J., Gallagher,W.M., *et al.* (1996) Microsatellite instability, apoptosis and loss of p53 function in drug-resistant tumour cells. *Cancer Res.* **56**: 1374-1381.

Armstrong,J.F., Kaufman,M.H., Harrison,D.J. and Clarke,A.R. (1995) High-frequency of developmental abnormalities in *p53*-deficient mice. *Curr. Biol.* **5**: 931-936.

Attardi,L.D., Lowe,S.W., Brugarolas,J. and Jacks,T. (1996) Transcriptional activation by p53, but not induction of the *p21* gene, is essential for oncogene-mediated apoptosis. *EMBO J.* **15**: 3702-3712.

Avantaggiati,M.L., Ogryzko,V., Gardner,K., *et al.* (1997) Recruitment of p300/CBP in p53-dependent signal repair pathways. *Cell* **89**: 1175-1184.

Baker,S.J., Fearon,E.R., Nigro,J.M., *et al.* (1989) Chromosome 17 deletions and *p53* gene mutations in colorectal carcinomas. *Science* **244**: 217-221.

Bayle,J.H., Elenbaas,B. and Levine,A.J. (1995) The carboxyl-terminal domain of the p53 protein regulates sequence-specific DNA binding through its non-specific nucleic acid-binding activity. *Proc. Natl. Acad. Sci. USA* **92**: 5729-5733.

Bottger.A., Bottger,V., Garcia-Echeverria,C., *et al.* (1997) Molecular characterisation of the hdm2-p53 interaction. *J. Mol. Biol.***269**: 744-756.

Bowman ,T., Symonds,H., Gu,L., *et al.* (1996) Tissue-specific inactivation of p53 tumour suppression in the mouse. *Gen. Dev.* **10**: 826-835.

Buckbinder,L., Talbott,R., Velasco-Miguel,S., *et al.* (1995) Induction of the growth inhibitor IGF-binding protein 3 by p53. *Nature* **377**: 646-649.

Caelles,C., Helmberg,A., and Karin,M. (1994) p53-dependent apoptosis in the absence of transcriptional activation of p53-target gene. *Nature* **370**: 220-223.

Canman,C.E., Gilmer,T.M., Coutts,S.B. and Kastan,M.B. (1995) Growth factor modulation of p53-mediated growth arrest versus apoptosis. *Gen. Dev.* **9**: 600-611.

Chen,J., Wu,X., Lin,J. and Levine,A.J. (1996b) mdm-2 inhibits the G1 arrest and apoptosis functions of the p53 tumour suppressor protein. *Mol. Cell Biol.* **16**: 2445-2452.

Chen,X., Ko,L.J., Jayaraman,L. and Prives,C. (1996a) p53 levels, functional domains, and DNA damage determine the extent of the apoptotic response of tumour cells. *Gen. Dev.* **10**: 2438-2451.

Chiou,S.K., Rao,L. and White,E. (1994) Bcl-2 blocks p53-dependent apoptosis. *Mol. Cell Biol.* **14**: 2556-2563.

Cho,Y., Gorina,S., Jeffrey,P.D. and Pavletich,N.P. (1994) Crystal structure of a p53 tumour suppressor-DNA complex: understanding tumorigenic mutations. *Science* **265**: 346-355.

Clarke,A.R., Purdie,C.A., Harrison,D.J., *et al.* (1993) Thymocyte apoptosis induced by p53-dependent and independent pathways. *Nature* **362**: 849-852.

Dameron,K.M., Volpert,O.V., Tainsky,M.A. and Bouck,N. (1994) Control of angiogenesis in fibroblasts by p53 regulation of thrombospondin-1. *Science* **265**: 1582-1584.

Debbas,M. and White,E. (1993) Wild type p53 mediates apoptosis by E1A, which is inhibited by E1B. *Gen. Dev.* **7**: 546-554.

DeGregori,J., Leone,G., Miron,A., *et al.* (1997) Distinct roles for E2F proteins in cell growth control and apoptosis. *Proc. Natl. Acad. Sci. USA* **94**: 7245-7250.

Deng,C., Zhang,P., Harper,J.W., *et al.* (1995) Mice lacking p21/waf1/cip1 undergo normal development, but are defective in G1 checkpoint control. *Cell* **82**: 675-684.

Donehower,L.A., Harvey,M., Slagle,B.L., *et al.* (1992) Mice deficient for *p53* are developmentally normal but susceptible to spontaneous tumours. *Nature* **356**: 215-221.

El-Deiry,W.S., Tokino,T., Velculescu,V.E., *et al.* (1993) WAF1, a potential mediator of p53 tumour suppression. *Cell* **75**: 817-825.

Finlay,C.A., Hinds,P.W. and Levine,A.J. (1989) The *p53* proto-oncogene can act as a suppressor of transformation. *Cell* **57**: 1083-1093.

Friedlander,P., Haupt,Y., Prives,C. and Oren,M. (1996) A mutant p53 that discriminates between p53-responsive genes cannot induce apoptosis. *Mol. Cell Biol.* **16**: 4961-4971.

Fritsche,M., Haessler,C. and Brandner,G. (1993) Induction of nuclear accumulation of the tumour suppressor protein p53 by DNA damaging agents. *Oncogene* **8**: 307-318.

Fujiwara,T., Grimm,E.A., Mukhopadhyay,T., *et al.* (1993) A retroviral wild type *p53* expression vector penetrates human lung cancer spheroids and inhibits growth by inducing apoptosis. *Cancer Res.* **53**: 4129-4133.

Gottlieb,E., Haffner,R., von Ruden,T., *et al.* (1994) Down-regulation of wild-type p53 activity interferes

with apoptosis of IL-3-dependent hematopoietic cells following IL-3 withdrawal. *EMBO J.* **13**: 1368-1374.
Gottlieb,E., Haffner,R., King,A., *et al.* (1997) Transgenic mouse model for studying the transcriptional activity of the p53 protein: age- and tissue-dependent changes in radiation-induced activation during embryogenesis. *EMBO J.* **16**: 1381-1390.
Gottlieb,T.M. and Oren,M. (1996) p53 in growth control and neoplasia. *Biochem. Biophys. Acta* **1287**: 77-102.
Graeber,T.G., Peterson,J.F., Tsai,M., *et al.* (1994) Hypoxia induces accumulation of p53 protein, but activation of a G1 phase checkpoint by low-oxygen conditions is independent of p53 status. *Mol. Cell Biol.* **14**: 6264-6277.
Graeber,T.G., Osmanian,C., Jacks,T., *et al.* (1996) Hypoxia-mediated selection of cells with diminished apoptotic potential in solid tumours. *Nature* **379**: 88-91.
Gu,W., Shi,X-L. and Roeder,R.G. (1997) Synergistic activation of transcription by CBP and p53. *Nature* **387**: 819-823.
Harper,W., Adami,G.R., Wei,N., *et al.* (1993) The p21 cdk-interacting protein Cip1 is a potent inhibitor of G1 cyclin-dependent kinases. *Cell* **75**: 805-816.
Haupt,Y., Rowan,S. and Oren,M. (1995) p53-mediated apoptosis in HeLa cells can be overcome by excess pRB. *Oncogene* **10**: 1563-1571.
Haupt,Y., Rowan,S., Shaulian,E., *et al.* (1995) Induction of apoptosis in HeLa cells by transactivation-deficient p53. *Gen. Dev.* **9**: 2170-2183.
Haupt,Y., Barak,Y. and Oren,M. (1996) Cell type-specific inhibition of p53-mediated apoptosis by mdm2. *EMBO J.* **15**: 1596-1606.
Haupt,Y., Maya,R., Kazaz,A. and Oren,M. (1997) Mdm2 promotes the rapid degradation of p53. *Nature* **387**: 296-299.
Heise,C., Sampson-Johannes,A., Williams,A., *et al.* (1997) ONYX-015, and E1B gene-attenuated adenovirus, causes tumour-specific cytolysis and antitumoural efficacy that can be augmented by standard chemotherapeutic agents. *Nature Med.* **3**: 639-645.
Hermeking,H. and Eick,D. (1994) Mediation of c-myc-induced apoptosis by p53. *Science* **265**: 2091-2093.
Hollstein,M., Sidransky,D., Vogelstein,B. and Harris,C.C. (1991) *p53* mutations in human cancers. *Science* **253**: 49-53.
Hollstein,M., Rice,K., Greenblatt,M.S., *et al.* (1994) Database of p53 gene somatic mutations in human tumors and cell lines. *Nuc. Acids Res.* **22**: 3551-3555.
Howes,K.A., Ransom,N., Papermaster,D.S., *et al.* (1994) Apoptosis or retinoblastoma: alternative fates of photoreceptors expressing the HPV-16 E7 gene in the presence or absence of p53. *Gen. Dev.* **8**: 1300-1310.
Howley,P.M. (1995) Viral Carcinogenesis. In: The Molecular Basis of Human Cancer (eds. Mendelsohn, P. Howley, M. Israel and L. Liotta), WB Saunders Co, Philadelphia.
Huang,L.C., Clarkin,K.C. and Wahl,G.M. (1996) Sensitivity and selectivity of the DNA damage sensor responsible for activating p53-dependent G1 arrest. *Proc. Natl. Acad. Sci. USA* **93**: 4827-4832.
Hupp,T.R., Meek,D.W., Midgely,C.A. and Lane,D.P. (1992) Regulation of the specific DNA binding function of p53. *Cell* **71**: 875-886.
Hupp,T.R., and Lane,D.P. (1994) Allosteric activation of latent p53 tetramers. *Curr. Biol.* **4**: 865-875.
Hupp,T.R., Sparks,A., and Lane,D..P. (1995) Small peptides activate the latent sequence specific binding function of p53. *Cell* **83**: 237-245.
Jayaraman,L. and Prives,C. (1995) Activation of p53 sequence-specific DNA binding by short single strands of DNA requires the p53 C-terminus. *Cell* **81**: 1021-1029.
Jayaraman,L., Murthy,K.G.K., Zhu,C, *et al.* (1997) Identification of redox/repair protein Ref-1 as a potent activator of p53. *Gen. Dev.* **11**: 558-570.
Johnson,D.G., Cress,W.D., Jakoi L. and Nevins,J.R. (1994) Oncogenic capacity of the *E2F1* gene. *Proc. Natl. Acad. Sci. USA* **91**: 1 2823-12827.
Kastan,M.B., Onyekwere,O., Sidransky,D., *et al.* (1991) Participation of p53 protein in the cellular response to DNA damage. *Cancer Res.* **51**: 6304-6311.
Kastan,M.B., Zhan,Q., El-Deiry,W.S., *et al.* (1992) A mammalian cell cycle checkpoint pathway utilizing p53 and GADD45 is defective in ataxia telangiectasia. *Cell* **71**: 587-597.
Kern,S.E., Kinzler,K.W., Bruskin,A., *et al.* (1991) Identification of p53 as a sequence-specific DNA binding protein. *Science* **252**: 1708-1711.
Knudson,C.M., Tung,S.K.T., Tourtellotte,W.G., *et al.* (1995) *Bax*-deficient mice with lymphoid hyperplasia and male germ cell death. *Science* **270**: 96-99.
Ko,L.J. and Prives,C. (1996) p53: puzzle and paradigm. *Gen. Dev.* **10**: 1054-1072.
Komarova,E.A., Chernov,M.V., Franks,R., *et al.* (1997) Transgenic mice with p53-responsive *lacZ*: p53 activity varies dramatically during normal development and determines radiation and drug sensitivity *in vivo*. *EMBO J.* **16**: 1391-1400.
Kubbutat,M.H.G., Jones,S.N. and Vousden,K.H. (1997) Regulation of p53 stability by mdm2. *Nature* **387**: 299-303.
Kuerbitz,S.J., Plunkett,B.S., Walsh,W.V. and Kastan,M.B. (1992) Wild-type p53 is a cell cycle checkpoint determinant following irradiation. *Proc. Natl. Acad. Sci. USA* **89**: 7491-7495.
Landers,J.E., Haines,D.S., Strauss,J.F., and George,D.L. (1994) Enhanced translation: a novel mechanism of mdm2 overexpression identified in human tumor cells. *Oncogene* **9**: 2745-2750.

Landers,J.E., Cassel,S.L. and George,D.L. (1997) Translational enhancement of *mdm2* oncogene expression in human tumor cells containing a stabilised wild-type p53 protein. *Cancer Res.* **57**: 3562-3568.
Lane,D.P. and Crawford,L.V. (1979) T antigen is bound to a host protein in SV40-transformed cells. *Nature* **278**: 261-263.
Lee,S., Elenbaas,B., Levine,A. and Griffith,J. (1995) p53 and its 14 kDa C-terminal domain recognise primary DNA damage in the form of insertion/deletion mismatches. *Cell* **81**: 1013-1020.
Levine,A.J. (1997) p53, the cellular gatekeeper for growth and division. *Cell* **88**: 323-331.
Lill,N.L., Grossman,S.R., Ginsberg,D., *et al.* (1997) Binding and modulation of p53 by p300/CBP coactivators. *Nature* **387**: 823-827.
Lin,Y. and Benchimol,S. (1995) Cytokines inhibit p53-mediated apoptosis but not p53-mediated G1 arrest. *Mol. Cell Biol.* **15**: 6045-6054.
Linke,S.P., Clarkin,K.C., Di Leonardo,A, *et al.* (1996) A reversible, p53-dependent G0/G1 cell cycle arrest induced by ribonucleotide depletion in the absence of detectable DNA damage. *Genes Dev.* **10**: 934-947.
Linzer,D.I. and Levine,A.J. (1979) Characterisation of a 54 K dalton cellular SV40 tumor antigen present in SV40-transformed cells and uninfected embryonal carcinoma cells. *Cell* **17**: 43-52.
Lowe,S.W., Schmitt,E.M., Smith,S.W., *et al.* (1993a) p53 is required for radiation-induced apoptosis in mouse thymocytes. *Nature* **362**: 847-849.
Lowe,S.W. and Ruley,H.E. (1993b) Stabilisation of the p53 tumor suppressor is induced by adenovirus 5 E1A and accompanies apoptosis. *Gen. Dev.* **7**: 535-545.
Lowe,S.W., Ruley,H.E., Jacks,T., and Housman,D.E. (1993c) p53-dependent apoptosis modulates the cytotoxicity of anticancer agents. *Cell* **74**: 957-967.
Lowe,S.W., Bodis,S., McClatchey,A., *et al.* (1994) p53 status and the efficacy of cancer therapy *in vivo*. *Science* **266**: 807-810.
Lu,H., Fisher,R.P., Bailey,P. and Levine,A.J. (1997) The cdk7/cyclin H/p36 complex of TFIIH phosphorylates p53 and enhances its sequence-specific DNA binding activity *in vitro*. *Mol. Cell Biol.*, In Press.
Lu,X. and Lane,D.P. (1993) Differential induction of transcriptionally active p53 following UV or ionizing radiation: defects in chromosome instability syndromes? *Cell* **75**: 765-778.
Ludwig,R.L., Bates,S., and Vousden,K.H. (1996) Differential activation of target cellular promoters by p53 mutants with impaired apoptotic function. *Mol. Cell Biol.* **16**: 4952-4960.
Lutzker,S.G. and Levine,A.J. (1996) A functionally inactive p53 protein in teratocarcinoma cells is activated by either DNA damage or cellular differentiation. *Nature Med.* **2**: 804-810.
MacCallum,D.E., Hupp,T.R., Midgely,C.A., *et al.* (1996) The p53 response to ionising radiation in adult and developing murine tissues. *Oncogene* **13**: 2575-2587.
Maheswaran,S., Englert,C., Bennett,P., *et al.* (1995) The WT1 gene product stabilises p53 and inhibits p53-mediated apoptosis. *Gen. Dev* **9**: 2143-2156.
Malkin,D., Li,F.P., Strong,L.C., *et al.* (1990) Germ line *p53* mutations in a familial syndrome of breast cancer, sarcomas, and other neoplasms. *Science* **250**: 1233-1238.
Maltzman,W. and Czyzyk,L. (1984) UV irradiation stimulates levels of p53 cellular tumor antigen in nontransformed mouse cells. *Mol Cell Biol* **4**: 1689-1694.
Martinez,J., Georgoff,I., Martinez,J. and Levine,A.J. (1991) Cellular localisation and cell cycle regulation by a temperature-sensitive p53 protein. *Gen. Dev.* **5**: 151-159.
Matlashewski,G., Lamb,P., Pim,D., *et al.* (1984) Isolation and characterisation of a human *p53* cDNA clone: expression of the human *p53* gene. *EMBO J.* **13**: 3257-3262.
McCarthy,S.A., Symonds,H.S. and Van Dyke,T. (1994) Regulation of apoptosis in transgenic mice by simian virus 40 T-antigen-mediated inactivation of p53. *Proc. Natl. Acad. Sci.. USA* **91**: 3979-3983.
McCurrach,M.E., Connor,T.M., Knudson,C.M., *et al.* (1997) *bax*-deficiency promotes drug resistance and oncogenic transformation by attenuating p53-dependent apoptosis. *Proc. Natl. Acad. Sci. USA* **94**: 2345-2349.
Merritt,A.J., Potten,C.S., Kemp,C.J., *et al.* (1994) The role of p53 in spontaneous and radiation-induced apoptosis in the gastrointestinal tract of normal and *p53*-deficient mice. *Cancer Res.* **54**: 614-617.
Michalovitz,D., Halevy,O. and Oren,M. (1990) Conditional inhibition of transformation and of cell proliferation by a temperature-sensitive mutant of p53. *Cell* **62**: 671-680.
Miyashita,T. and Reed,J.C. (1995) Tumor suppressor p53 is a direct transcriptional activator of the human *bax* gene. *Cell* **80**: 293-299.
Moll UM, Riouo G and Levine AJ. (1992) Two distinct mechanisms alter p53 in breast cancer: mutation and nuclear exclusion. *Proc. Natl. Acad. Sci. USA* **89**: 7262-7266.
Moll,U., LaQuaglia,M., Benard,J. and Riou,G. (1995) Wild-type p53 undergoes cytoplasmic sequestration in undifferentiated neuroblastomas but not in differentiated tumors. *Proc. Natl. Acad. Sci. USA* **92**: 4407-4411.
Momand,J., Zambetti Olson,D.C, George,D. and Levine,A.J. (1992) The mdm2 oncogene product forms a complex with the p53 protein and inhibits p53-mediated transactivation. *Cell* **69**: 1237-1245.
Montes de Oca Luna,R., Wagner,O. and Lozano,G. (1995) Rescue of early embryonic lethality in *mdm2*-deficient mice by deletion of *p53*. *Nature* **378**: 203-206.
Morgan,S.E. and Kastan,M.B. (1997) p53 and ATM: Cell cycle, cell death and Cancer. In: Advances in Cancer Research pp1-25, Academic Press, New York.
Morganbesser,S.D., Williams,B.O., Jacks,T., and DePinho,R.A. (1994) p53-dependent apoptosis

produced by Rb-deficiency in the developing mouse lens. *Nature* **371**: 72-74.
Mowat,M., Cheng,A., Kimura,N., *et al.* (1985) Rearrangements of the cellular p53 gene in erythroleukemic cells transformed by Friend virus. *Nature* **314**: 633-636.
Murphy,M., Hinman,A. and Levine,A.J. (1996) Wild type p53 negatively regulates the expression of a microtubule-associated protein. *Gen. Dev.* **10**: 2971-2980.
Nelson,W.G. and Kastan,M.B. (1994) DNA strand breaks: the DNA template alterations that trigger p53-dependent DNA damage response pathways. *Mol. Cell Biol.* **14**: 1815-1823.
Nicol,C.J., Harrison,M.L., Laposa,R.R., *et al.* (1995) A teratologic suppressor role for p53 in benzo[a]pyrene-treated transgenic *p53*-deficient mice. *Nature Gen.* **10**: 181-187.
Noda,A., Ning,Y., Venable,S.F., *et al.* (1994) Cloning of senescent cell-derived inhibitors of DNA synthesis using an expression screen. *Exp. Cell Res.* **211**: 90-98.
Norimura,T., Nomoto,S., Katsuki,M., *et al.* (1996) p53-dependent apoptosis suppresses radiation-induced teratogenesis. *Nature. Med.* **2**: 577-580.
Okamoto,K and Beach,D. (1994) Cyclin G is a transcriptional target of the p53 tumor suppressor protein. *EMBO J.* **13**: 4816-4822.
Oliner,J.D., Kinzler,K.W., Meltzer,P.S., *et al.* (1992) Amplification of a gene encoding a p53-binding protein in human sarcomas. *Nature* **358**: 80-83.
Oliner,J.D., Pietenpol,J.A., Thiagalingam,S., *et al.* (1993) Oncoprotein mdm2 conceals the activation domain of tumor suppressor p53. *Nature* **362**: 857-860.
Pan,H. and Griep,A.E. (1994) Altered cell cycle regulation in the lens of HPV-16 E6 or E7 transgenic mice: implications for tumor suppressor gene function in development. *Gen. Dev.* **8**: 1285-1299.
Pan,H. and Griep,A.E. (1995) Temporally distinct patterns of p53-dependent and p53-independent apoptosis during mouse lens development. *Gen. Dev.* **9**: 2157-2169.
Prisco,M., Hongo,A., Rizzo,M.G., *et al.* (1997) The Insulin-like growth factor I receptor as a physiologically relevant target of p53 in apoptosis caused by interleukin-3 withdrawal. *Mol. Cell Biol.* **17**: 1084-1092.
Rajah,R., Valentinis,B. and Cohen,P. (1997) Insulin-like growth factor (IGF-binding protein 3 induces apoptòsis and mediates the effects of transforming growth factorl on programmed cell death through a p53 and IGF-independent mechanism. *J. Biol. Chem.* **272**: 12181-12188.
Reed,J.C. (1997) Double identity for proteins of the Bcl-2 family. *Nature* **387**: 773-776.
Reifenberger,G., Liu,L., Ichimura,K., *et al.* (1993) Amplification and overexpression of the *mdm2* gene in a subset of malignant gliomas without *p53* mutations. *Cancer Res.* **53**: 2736-2739.
Roth,J.A., Nguyen,D., Lawrence,D.D., *et al.* (1996) Retrovirus-mediated wild type *p53* gene transfer to tumors of patients with lung cancer. *Nature Med.* **2**: 985-991.
Rowan,S., Ludwig,R.L., Haupt,Y., *et al.* (1996) Specific loss of apoptotic but not cell cycle arrest function in a human tumor derived p53 mutant. *EMBO J.* **15**: 827-838.
Ruaro,E.M., Collavin,L., Del Sal,G., *et al.* (1997) A proline-rich motif in p53 is required for transactivation-independent growth arrest as induced by Gas1. *Proc. Natl. Acad. Sci. USA* **94**: 4675-4680.
Sabbatini,P., Chiou,S-K., Rao,L. and White,E. (1995a) Modulation of p53-mediated transcriptional repression and apoptosis by adenovirus E1B 19K protein. *Mol. Cell Biol.* **15**: 1060-1070.
Sabbatini,P., Lin,J., Levine,A.J. and White,E. (1995b) Essential role for p53-mediated transcription in E1A-induced apoptosis. *Gen. Dev.* **9**: 2184-2192.
Sah,V.P., Attardi,L.D., Mulligan,G.J., *et al.* (1995) A sub-set of p53 deficient embryos exhibit exencephaly. *Nature Gen.* **10**: 175-180.
Sarnow,P., Ho,Y.S., Williams,J. and Levine,A.J. (1982) Adenovirus E1B-58kD tumor antigen and SV40 large tumor antigen are physically associated with the same 54 kd cellular protein. *Cell* **28**: 387-394.
Scheffner,M., Huibregtse,J.M., Vierstra,R.D. and Howley,P.M. (1993) The HPV-16 E6 and E6-AP complex functions as a ubiquitin-protein ligase in the ubiquitination of p53. *Cell* **75**: 495-505.
Selivanova,G., Iotsova,V., Okan,I., Fritsche,M., *et al.* (1997) Restoration of the growth suppressive function of mutant p53 by a synthetic peptide derived from the p53 C-terminal domain. *Nature Med.* **3**: 632-638.
Shaw,P., Bovey,R., Tardy,S., *et al.* (1992). Induction of apoptosis by wild type p53 in a human colon tumor-derived cell line. *Proc. Natl. Acad. Sci. USA* **89**: 4495-4499.
Shen,Y. and Shenk,T.E. (1994) Relief of p53-mediated transcriptional repression by the adenovirus E1B-19K protein or the cellular bcl2 protein. *Proc. Natl. Acad. Sci. USA* **91**: 8940-8944.
Shen,Y. and Shenk,T.E. (1995) Viruses and apoptosis. *Curr. Opin. Genet. Dev.* **5**: 105-111.
Smith,M.L., Chen,I.T., Zhan,Q., *et al.* (1994) Interaction of the p53-regulated protein Gadd45 with proliferating cell nuclear antigen. *Science* **266**: 1376-1380.
Strasser,A., Harris,A.W., Jacks,T. and Cory,S. (1994) DNA damage can induce apoptosis in proliferating lymphoid cells via p53-independent mechanisms inhibitable by bcl-2. *Cell* **79**: 329-339.
Symonds,H., Krall,L., Remington,L., *et al.* (1994) p53-dependent apoptosis suppresses tumor growth and progression *in vivo*. *Cell* **78**: 703-711.
Thut,C., Goodrich,J.A. and Tjian,R. (1997) Repression of p53-mediated transcription by mdm2: a dual mechanism. *Gen. Dev.* **11**: 1974-1986.
Wagner,A.J., Kokontis,J.M. and Hay,N. (1994) Myc-mediated apoptosis requires wild type p53 in a manner independent of cell cycle arrest and the ability of p53 to induce p21waf1/cip1. *Gen. Dev.* **8**: 2817-2830.
Walker,K.K. and Levine,A.J. (1996) Identification of a novel p53 functional domain that is necessary for

efficient growth suppression. *Proc. Natl. Acad. Sci. USA* **93**: 15335-15340.
Wang,X.W., Yeh,H., Schaeffer,L, *et al.* (1995) p53 modulation of TFIIH-associated nucleotide excision repair activity. *Nature Gen.* **10**: 188-195.
Wang,X.W., Vermeulen,W., Coursen,J.D., *et al.* (1996) The XPB and XPD DNA helicases are components of the p53-mediated apoptosis pathway. *Gen. Dev.* **10**: 1219-1232.
Wang,Y., Ramqvist,T., Szekely,L, *et al.* (1993) Reconstitution of wild type *p53* expression triggers apoptosis in a *p53* negative *v-myc* retrovirus-induced T cell lymphoma line. *Cell Growth Diff.* **4**: 467-473.
Wang,Y., Okan,I., Szekely,L., *et al.* (1995) bcl-2 inhibits wild type p53-triggered apoptosis but not G1 cell cycle arrest and transactivation of WAF1 and bax. *Cell Growth Diff.* **6**: 1071-1075.
Watanabe,T., Ichikawa,A., Saito,H. and Hotta,T. (1996) Overexpression of the *mdm2* oncogene in leukemia and lymphoma. *Leuk. Lymphoma* **21**: 391-397.
Werner,H., Karnieli,E., Rauscher,F.J. and LeRoith,D. (1996) Wild type and mutant p53 differentially regulate transcription of the insulin-like growth factor I receptor gene. *Proc. Natl. Acad. Sci. USA* **93**: 8318-8323.
Wu,L., Bayle,J.H., Elenbaas,B, *et al.* (1995) Alternatively spliced forms in the carboxy-terminal domain of the p53 protein regulate its ability to promote annealing of complementary single strands of nucleic acids. *Mol. Cell Biol.* **15**: 497-504.
Wu,X., Bayle,J.H., Olson,D.C. and Levine,A.J. (1993) The p53-mdm2 autoregulatory feedback loop. *Gen. Dev.* **7**: 1126-1132.
Wu,X. and Levine,A.J. (1994) p53 and E2F-1 cooperate to mediate apoptosis. *Proc. Natl. Acad. Sci. USA* **91**: 3602-3606.
Xiong,Y, Hannon,G.J., Zhang,H, *et al.* (1993) p21 is a universal inhibitor of cyclin kinases. *Nature* **366**: 701-704.
Xu,Y., and Baltimore,D. (1996) Dual roles of ATM in the cellular response to radiation and in cell growth control. *Gen. Dev.* **10**: 2401-2410.
Yamasaki,L., Jacks,T., Bronson,R., *et al.* (1996) Tumor induction and tissue atrophy in mice lacking E2F-1. *Cell* **85**: 537-548.
Yew,P.R. and Berk,A.J. (1992) Inhibition of p53 transactivation required for transformation by adenovirus early 1B protein. *Nature* **357**: 82-85.
Yin,C., Knudson,C.M., Korsmeyer,S.J. and Van Dyke,T. (1997) Bax suppresses tumorigenesis and stimulates apoptosis *in vivo*. *Nature* **385**: 637-640.
Yonish-Rouach,E., Resnitzky,D., Lotem,J., *et al.* (1991) Wild type p53 induces apoptosis of myeloid leukemic cells that is inhibited by IL-6. *Nature* **352**: 345-347.
Yonish-Rouach,E., Grunwald,D., Wilder,S., *et al.* (1993) p53-mediated cell death: relationship to cell cycle control. *Mol. Cell Biol.* **13**: 1415-1423.
Yonish-Rouach,E., Deguin,V., Zaitchouk,T., *et al.* (1996) Transcriptional activation plays a role in the induction of apoptosis by transiently transfected wild type *p53*. *Oncogene* **11**: 2197-2205.
Yu,H., Chen,J.C., Feng,S., *et al.* (1994) Structural basis for the binding of proline-rich peptides to SH3 domains. *Cell* **76**: 933-945.
Yuan,Z.M., Huang,Y., Ishiko,T., *et al.* (1996a) Regulation of DNA damage-induced apoptosis by the c-abl tyrosine kinase. *Proc. Natl. Acad. Sci. USA* **94**: 1437-1440.
Yuan,Z.M., Huang,Y, Whang,Y, *et al.* (1996b) Role for c-abl tyrosine kinase in growth arrest response to DNA damage. *Nature* **382**: 272-274.
Zambetti,G.P., Bargonetti,J., Walker,K., *et al.* (1992) Wild-type p53 mediates positive regulation of gene expression through a specific DNA sequence element. *Gen. Dev.* **6**: 1143-1152.
Zambetti,G.P. and Levine,A.J. (1993) A comparison of the biological activities of wild-type and mutant p53. *FASEB J.* **7**: 855-865.

Chapter 3

Mammalian *bcl-2* family genes

Thomas Chittenden
Apoptosis Technology Inc, 148 Sidney Street, Cambridge MA 02139-4239, USA.

1.0 INTRODUCTION

Research on the molecular action of oncogenes has often yielded important insights into key signal transduction pathways that regulate the growth of cells. The *bcl-2* (for B-cell lymphoma-2) gene was originally identified as the oncogene activated by the characteristic t(14;18) translocation in non-Hodgkin's follicular lymphomas (Cleary *et al.*, 1986; Tsujimoto and Croce, 1986). By contrast to many well-studied oncogenes which stimulate cell division, *bcl-2* was found to promote malignancy by inhibiting apoptosis, the cell-intrinsic suicide program (Vaux *et al.*, 1988; Hockenbery *et al.*, 1990). The discovery of *bcl-2* and its novel biological activity generated intense interest because it provided a foothold into the poorly understood and previously intractable pathway in mammalian cells that controls physiologic cell death. Furthermore, the role of activated *bcl-2* in follicular lymphoma provided direct evidence that suppression of apoptosis contributes to tumourigenesis and revealed a new class of aberrant signals in cancer cells that contributes to their unrestrained proliferation.

The significance of the control of apoptosis in tumourigenesis is underscored by the surprising discovery that many growth-stimulatory oncogenes, such as *myc*, are themselves potent inducers of apoptosis when expressed in a deregulated fashion (Evan *et al.*, 1992). The pro-apoptotic and proliferative activities of oncogenes appear to be tightly linked, suggesting the existence of a built-in safeguard against the evolution of malignant cells. In an otherwise normal cell, mutational activation of an oncogene such as *myc* may trigger apoptosis and thereby eliminate the potential outgrowth of a cancerous cell. This altruistic cellular response can be blocked, however, by cooperating genetic lesions that suppress apoptosis, such as the translocation and activation of *bcl-2*, and allow tumourigenesis to progress. In support of this scenario, *bcl-2* synergises with myc both in *in vitro* cell transformation assays and in transgenic mouse models of lymphomagenesis (Strasser *et al.*, 1990; McDonnell and Korsmeyer, 1991; Fanidi *et al.*, 1992).

The essential contribution of anti-apoptotic signals such as activated Bcl-2 to tumourigenesis provides the rationale for a promising new approach to eliminate cancer cells. Instead of seeking to block cell division pathways, it may be possible to interfere with the compensatory anti-apoptotic signals that tumour cells depend upon, thereby re-enabling cancerous cells to self-destruct. The feasibility of such an approach with respect to Bcl-2 requires a mechanistic understanding of how it and related proteins regulate the apoptotic pathway in mammalian cells. Here, the structure and function of the mammalian Bcl-2 family are reviewed, including recent progress in understanding how these proteins function to suppress apoptosis and how their activity is regulated and deregulated in tumour cells. The particular focus will thus be on molecular aspects of the Bcl-2 protein family and their potential involvement in malignancy. The important role that these proteins normally play in physiological processes, such as development and maintenance of the immune and nervous systems, have been the subject of separate recent reviews (Cory, 1995; Hawkins and Vaux, 1997; Merry and Korsmeyer, 1997).

2.0 STRUCTURE AND FUNCTION OF MAMMALIAN BCL-2 HOMOLOGUES

Bcl-2 exhibits a remarkable capacity to inhibit apoptosis induced by a vast array of diverse agents and treatments, including deprivation of essential cytokines, heat shock, exposure to DNA damaging agents, calcium ionophores, and staurosporine (reviewed by Reed, 1994). In addition to inhibiting apoptosis induced by over-expression of oncogenes, Bcl-2 suppresses cell death induced by many other apoptotic signals that are relevant to the evolution and/or chemoresistance of tumour cells. Elevated expression of Bcl-2 provides protection against ionising radiation and many commonly used chemotherapeutic drugs, conferring a multidrug resistant phenotype to cells (Reed *et al.*, 1996). Bcl-2 expressing cells exhibit enhanced protection against apoptosis induced by hypoxia (Jacobson and Raff, 1995; Shimizu *et al.*, 1995) and loss of substrate attachment (Frisch and Francis, 1994), two apoptotic stimuli which may be particularly important to the survival of metastatic tumour cells. The ability of Bcl-2 to suppress apoptosis elicited by a multitude of seemingly unrelated stimuli is believed to reflect its action at a key decision point in a central cell death pathway, downstream of where many apoptotic signals converge. This interpretation is supported by genetic studies of programmed cell death in C. elegans, which demonstrated that Bcl-2 is a structural and functional homologue of the Ced-9 protein (Hengartner and Horvitz, 1994). Ced-9 functions as a negative regulator of the C. elegans death pathway, through which all programmed cell deaths in the nematode occur (see Sugimoto and Miura - this volume).

In mammalian cells, however, there are several additional layers of complexity to the regulation of the cell death pathway by Ced-9/Bcl-2 homologues. First, Bcl-2 represents just one member of an emerging family of related proteins, numbering at least a dozen, that also modulate apoptosis. Second, certain Bcl-2 related proteins are potent suppressors of apoptosis, whereas other Bcl-2 homologues function in a diametrically opposed fashion by promoting apoptosis and antagonising the protective function of Bcl-2. Finally, pro-apoptotic and anti-apoptotic Bcl-2 homologues heterodimerise with each other, modulating their respective activities. The susceptibility of a cell to undergo apoptosis is governed, at least in part, by these molecular interactions and the relative abundance of Bcl-2 homologues of opposing function. Therefore, in order to understand how cell death is regulated in mammalian cells, the relative contributions of these various players and their interactions must be considered.

The mammalian Bcl-2 related proteins can be divided, at a simplistic level, into two functional classes: (1) the anti-apoptotic homologues, which include Bcl-2, Bcl-xL, Bcl-w, Mcl-1 and A1, and (2) the pro-apoptotic family members, consisting of Bax, Bak, Bok, Bik, Bid, Hrk, Bim and Bad. Much of the homology among the Bcl-2 family members is principally confined to relatively short non-contiguous domains, termed Bcl-2 homology (BH) domains (illustrated schematically in Figure 1). The most highly conserved of these domains, BH1 and BH2, are located in the C-terminal portion of these proteins; two other motifs, BH3 and BH4, are more loosely conserved. The sequence homologies within BH1, BH2 and BH4 have been independently noted and referred to by different names, including either conserved domains S2, S3 and S1, or Bcl-2 domains B, C and A, respectively (Sato *et al.*, 1994; Cory, 1995; Yang and Korsmeyer, 1996). The presence of one or more of these conserved domains has served to define membership in the Bcl-2 family, with a supporting criteria being the ability of the protein to modulate apoptosis in functional assays.

2.1 Anti-apoptotic Bcl-2 homologues

2.1.1 Bcl-2

The prototypic family member, Bcl-2, is a 25kD protein that is localised principally to the cytosolic face of outer mitochondrial membranes, although it is also found at the endoplasmic reticulum and outer nuclear envelope (Chen-Levy *et al.*, 1989; Hockenbery *et al.*, 1990; Krajewski *et al.*, 1993). A stretch of 19 hydrophobic residues at the C-terminal end of the protein functions to localise Bcl-2 to intracellular membranes (Nguyen *et al.*, 1993). Deletion of this C-terminal membrane anchor results in a diffuse cytoplasmic pattern of localisation and diminishes, but typically does not eliminate, the anti-apoptotic activity of Bcl-2 (Hockenbery *et al.*, 1993; Tanaka *et al.*, 1993; Borner *et al.*, 1994). The full anti-apoptotic potency of mutants lacking the C-terminal tail can be restored by the substitution of heterologous membrane anchor sequences, that direct localisation to the outer mitochondrial membrane (Nguyen *et al.*, 1994; Zhu *et al.*, 1996). Taken as a whole, these studies indicate that the proper localisation of Bcl-2 to intracellular membranes facilitates its anti-apoptotic function, a theme that has played out with the discovery of other Bcl-2 homologues.

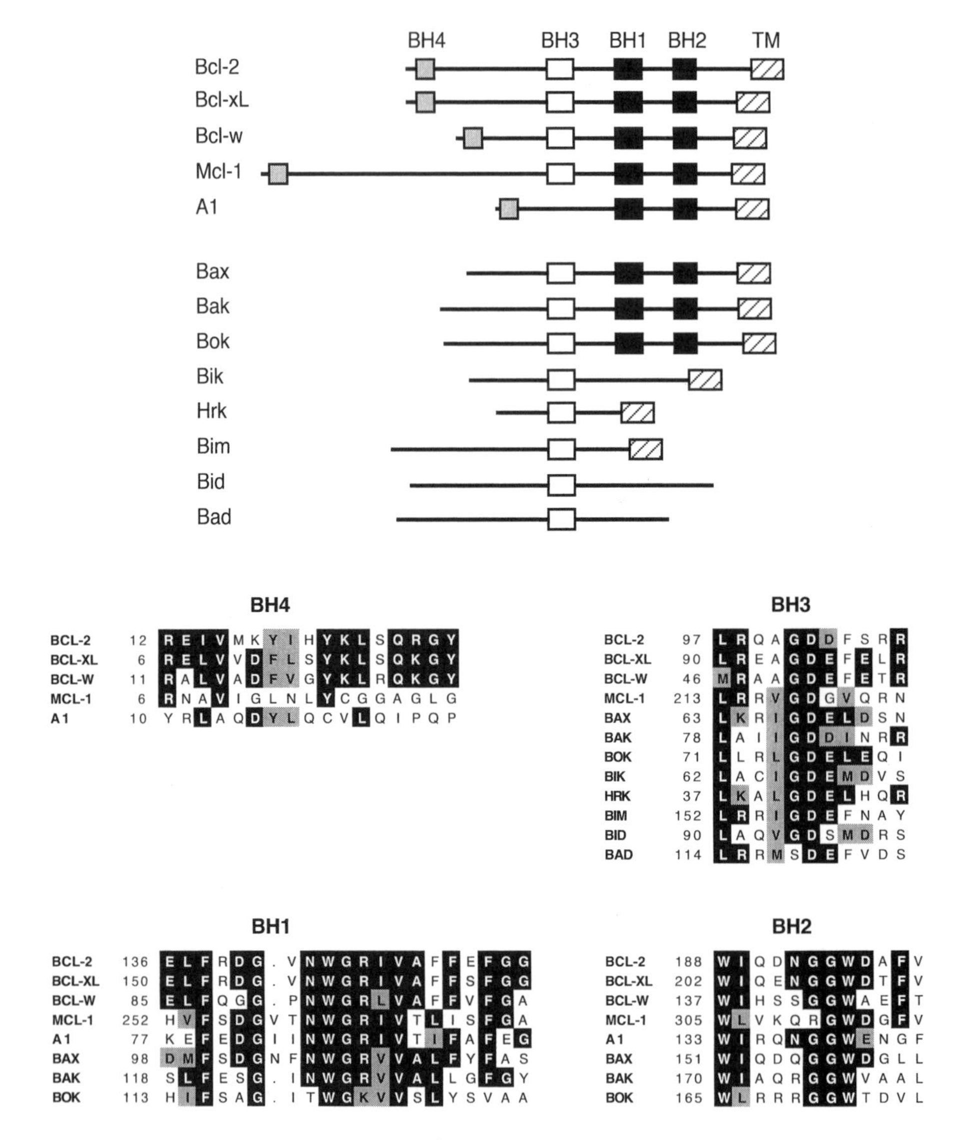

Figure 1. Structure of mammalian Bcl-2 family proteins.

Much of the sequence homology shared by Bcl-2-related proteins is confined to conserved Bcl-2 homology (BH) domains. The arrangement of these domains are shown schematically in the linear structures of both the anti-apoptotic and pro-apoptotic Bcl-2 family members. A trans-membrane anchor sequence (TM) is located at the C-terminus of many Bcl-2 homologues. Amino acid sequence alignments within the four BH domains are shown at the bottom; sequences correspond to the human homologue for each family member, where available.

The 25kD form of Bcl-2, termed Bcl-2α, is encoded by a 6 kb mRNA and represents the predominant *bcl-2* gene product (Cleary *et al.*, 1986; Tsujimoto and Croce, 1986). An additional Bcl-2 isoform, Bcl-2ß, is generated by an alternatively spliced transcript arising from the failure to splice the second intron, resulting in an altered reading frame at the C-terminus that effectively replaces the hydrophobic tail present in Bcl-2α with an unrelated sequence (Tsujimoto and Croce, 1986). Bcl-2ß exhibits cytoplasmic intracellular localisation, similar to artificially generated deletion mutants of the C-terminus, and diminished anti-apoptotic activity (Tanaka *et al.*, 1993). The physiological importance of Bcl-2ß is uncertain since this isoform is rarely detected in cells.

2.1.2 Bcl-x

Bcl-x is the family member most closely related to Bcl-2 and is homologous at both the nucleotide and amino acid level (Boise *et al.*, 1993). In fact, the similar genomic organisation of introns and exons in *bcl-2* and *bcl-x* argues that these genes are close relatives arising from a gene duplication event (Grillot *et al.*, 1997). The predominant product of the *bcl-x* gene, Bcl-xL (for Bcl-x long) is a 233 amino acid protein that exhibits 44% overall amino acid identity to Bcl-2, with the highest degree of homology localised to the BH1 and BH2 domains in the C-terminal portion of the molecules, and the BH3 and BH4 elements in the amino terminal half of the protein. Bcl-xL also contains a COOH hydrophobic tail element and is similarly localised to mitochondrial membranes (Gonzalez-Garcia *et al.*, 1994). Bcl-xL is functionally similar to Bcl-2 in that it is a robust suppressor of apoptosis, blocking death induced by diverse agents and in many experimental systems (e.g. see Chao *et al.*, 1995). An isoform produced by alternative splicing, Bcl-xß, is analogous to Bcl-2ß in that the failure to splice intron two results in a novel C-terminal sequence lacking the hydrophobic tail. A second protein produced by alternatively splicing, Bcl-xS (for Bcl-x short) results in a 63 amino acid in-frame deletion that removes both the BH1 and BH2 conserved domains (Boise *et al.*, 1993). This deletion of BH1 and BH2 has drastic functional consequences since, by contrast to Bcl-xL, expression of Bcl-xS accelerates apoptosis and antagonises the protective effect of Bcl-xL (or Bcl-2). The pro-apoptotic activity of Bcl-xS may reflect a dominant interfering activity, possibly through formation of inactive complexes with Bcl-x, Bcl-2 or their downstream effectors. However, the biological significance of Bcl-xS is unclear, since it is not detected in most tissues and cell lines, including those that express abundant levels of Bcl-xL (Gonzalez-Garcia *et al.*, 1994; Krajewski *et al.*, 1994). Bcl-xS differs in this respect from other "professional" pro-apoptotic Bcl-2 family members that have been described (see below).

2.1.3 Bcl-w

The *bcl-w* cDNA was isolated by a degenerate PCR approach that was based on the assumption that other Bcl-2 homologues would retain conserved BH1 and BH2 domains (Gibson *et al.*, 1996). Structurally, the 193 amino acid Bcl-w protein most closely resembles Bcl-2 and Bcl-xL and, in addition to BH1 and BH2, contains homologous BH3 and BH4 domains. The amino terminal BH3 and BH4 domains of Bcl-w, however, are more closely spaced, accounting for the smaller size of the protein. Bcl-w contains a C-terminal hydrophobic tail and exhibits an intracellular localisation pattern similar to Bcl-2. Expression of Bcl-w suppresses apoptosis induced by multiple cytotoxic treatments, including IL3 withdrawal, dexamethasone, and irradiation (Gibson *et al.*, 1996). Xenopus leavis cDNA clones designated *XR*I and *XR*II encode proteins closely related to Bcl-w and Bcl-xL, respectively, indicating these proteins are well conserved across different species (Cruz-Reyes and Tata, 1995).

2.1.4 Mcl-1

Two additional Bcl-2-related genes, *mcl-1* and *A1*, were identified by their induction of expression in response to specific extrinsic stimuli. *Mcl-1* was originally cloned by a differential screening approach in a search for mRNAs induced in a differentiating haematapoeitic cell line (Kozopas *et al.*, 1993). Treatment of the human ML-1 myeloid leukaemia cell line with the phorbol ester, TPA, initiates differentiation of the cells along the monocyte/macrophage lineage. The expression of the *mcl-1* mRNA is rapidly elevated (within 1-3 hours) after TPA treatment and precedes the appearance of differentiation markers. The 350 amino acid Mcl-1 protein shows significant homology to Bcl-2, principally in the regions surrounding BH1 and BH2 in the C-terminal portion of the protein, and Mcl-1 contains a hydrophobic carboxyl tail element that mediates an intracellular distribution similar to Bcl-2 (Yang,T *et al.*, 1995). The amino terminal half of Mcl-1 bears little homology to Bcl-2 and contains a highly charged region flanked by domains rich in glycine/alanine, as well as two PEST sequences characteristic of certain proteins that are degraded rapidly. Enforced expression of Mcl-1 suppresses apoptosis induced by growth factor withdrawal, DNA damage, and cytotoxic agents such as staurosporine and calcium ionophores (Reynolds *et al.*, 1996; Zhou *et al.*, 1997).

2.1.5 A1

The mouse *A1* gene was also first identified by differential screening for early response genes in haematopoeitic cells, in this case by its rapid induction following treatment of bone marrow derived macrophages with GM-CSF (Lin *et al.*, 1993). The homology of A1 to Bcl-2 is mostly restricted to a segment encompassing BH1 and BH2. Although a potential membrane anchoring span at its carboxyl terminus is interrupted by a glutamic acid residue, A1 continues to exhibit an intracellular membrane localisation pattern similar to Bcl-2 (Carrió *et al.*, 1996a). Expression of A1 suppresses apoptosis induced by growth factor withdrawal, tumour necrosis factor-α, and a variety of cytotoxic drugs, indicating it functions as a potent anti-apoptotic Bcl-2 homologue (Karsan *et al.*, 1996a; Lin *et al.*, 1996). A human gene encoding a closely related protein was cloned independently by several groups and referred to by different names: *Bfl-1* (for Bcl-2 related gene expressed in foetal liver), *GRS* (Glasgow rearranged sequence) and human *A1* (Choi *et al.*, 1995; Karsan *et al.*, 1996b; Kenny *et al.*, 1997). Given the close similarity to murine A1 structure and expression patterns, this gene is likely to represent the human homologue of mouse *A1*.

2.1.6 Bcl-2-like viral proteins

In addition to these cellular Bcl-2 family members, several unrelated DNA viruses that infect mammalian cells encode genes homologous to Bcl-2 and represent one of the many mechanisms that viruses have evolved to suppress apoptosis during infection (see Chapter 10). The adenovirus protein, E1B19kD, lacks overt sequence homology to Bcl-2 yet is functionally equivalent (Chiou *et al.*, 1994; Huang *et al.*, 1997a), acting as a powerful suppressor of apoptosis and interacting with many of the same cellular proteins as Bcl-2 (Boyd *et al.*, 1994). Viral proteins that exhibit clear homology to Bcl-2 include the Epstein-Barr virus BHRF-1 protein, herpesvirus saimiri ORF16, Kaposi sarcoma-associated virus (human herpesvirus 8) KSbcl-2, and African swine fever virus LMW5-HL, and have been shown to function as potent anti-apoptotic proteins (Henderson *et al.*, 1993; Afonso *et al.*, 1996; Nava *et al.*, 1997; Sarid *et al.*, 1997). As is the case with the cellular anti-apoptotic Bcl-2 homologues, the sequence homology among these viral proteins is mostly confined to BH1 and BH2.

2.1.7 Nr-13

Additional anti-apoptotic Bcl-2 homologues are likely to function in mammalian cells. A chicken *bcl-2*-related gene, designated *Nr-13*, was identified as an mRNA induced by the v-*src* oncogene and functions as a cell death suppressor when expressed in both avian and mammalian cells (Gillet *et al.*, 1995; Mangeney *et al.*, 1996). Although Nr-13 contains well-conserved BH1 and BH2 domains, sequences outside these domains do not resemble any particular mammalian family member described to date. In view of the conservation of other family members such as Bcl-2 and Bcl-x in chicken, it seems probable that an as yet unidentified mammalian counterpart to Nr-13 exists.

Why do mammalian cells encode such a diversity of Bcl-2 related genes? Analysis of expression profiles and gene knockout studies suggest that multiple Bcl-2 homologues have evolved to play specific roles in particular tissues, developmental stages, or in response to different cytokines or cytotoxic insults. Although many of the Bcl-2 family members appear to be widely expressed in diverse tissues and cell lines, clear differences have emerged in their distribution in primary tissues when surveyed at either the mRNA or protein level (Krajewski *et al.*, 1995b; Ohr *et al.*, 1995; Carrió *et al.*, 1996; Krajewski *et al.*, 1996). Even within a cell lineage, the two most closely related Bcl-2 homologues, Bcl-2 and Bcl-xL, appear to play distinct roles. Bcl-2 and Bcl-xL exhibit reciprocal expression patterns during thymocyte development and Bcl-xL, but not Bcl-2, is induced upon activation of peripheral T-cells (Boise *et al.*, 1995; Hawkins and Vaux, 1997). Bcl-2 expression is detected early in neurogenesis and declines after birth, whereas Bcl-xL levels increase after birth and are abundant in the adult brain (Merry and Korsmeyer, 1997). Moreover, the distinct phenotypes of knockout mice deficient for either Bcl-2 or Bcl-xL illustrates that their functions are not wholly redundant and are essential to the regulation of cell death in different cell lineages. *Bcl-2*-null mice develop normally and are viable but, by four weeks of age, succumb to a massive loss of both T and B lymphocytes and polycystic kidney disease (Veis *et al.*, 1993). Deficiency of Bcl-xL results in lethality at around embryonic day 13 characterised by massive apoptosis of immature neurons and haematapoeitic cells (Motoyama *et al.*, 1995). Thus, these closely related genes play essential, non-redundant roles in the maintenance and/or development of the immune and nervous systems.

2.2 Pro-apoptotic family members

2.2.1 Bax

The pro-apoptotic family member, Bax (Bcl-2-associated X protein), was identified biochemically as a 21kD protein that co-immunoprecipitates with Bcl-2 in cell lysates (Oltvai *et al.*, 1993). Structurally, Bax contains many of the hallmarks of Bcl-2 family members, including conserved BH1, BH2, and BH3 sequences and a hydrophobic C-terminal tail element. The 21kD Bax protein, designated Baxα, is encoded by a widely expressed 1.0 kb mRNA; alternative splicing produces at least two additional, less abundant variants, lacking the hydrophobic tail (Baxß and Baxγ). Despite its clear homology to Bcl-2, expression of Baxα promotes apoptosis and counteracts the anti-apoptotic function of Bcl-2. Bax heterodimerises with Bcl-2 and other cellular anti-apoptotic Bcl-2 homologues including Bcl-xL, Mcl-1 and A1, and can also form homodimers (Oltvai *et al.*, 1993; Sato *et al.*, 1994; Sedlak *et al.*, 1995). The physical interaction between Bcl-2 family members of opposing function suggested a molecular "rheostat" model for control of cell death, where the set point for apoptosis is determined in part by the interaction between Bax and Bcl-2 and their relative levels (Oltvai and Korsmeyer, 1994). When Bax is in excess, Bax homodimers predominate and susceptibility to apoptosis is enhanced. With an over-abundance of Bcl-2, however, most or all of the Bax in the cell exists in a complex with Bcl-2 and susceptibility to apoptosis is suppressed. This rheostat can be artificially altered in experimental systems, where enforced expression of Bax leads to an increase in the Bax/Bcl-2 ratio and enhanced susceptibility to apoptotic stimuli such as growth factor withdrawal, chemotherapeutic drugs and irradiation (Strobel *et al.*, 1996; Wagener *et al.*, 1996), essentially the same cell death signals that are suppressed by Bcl-2.

2.2.2 Bak

The discovery of additional pro-apoptotic Bcl-2 homologues indicated that the set-point for apoptosis in a given cell is not dictated solely by Bcl-2 and Bax but by a complex interplay of multiple pro- and anti-apoptotic Bcl-2 homologues. Bak (for Bcl-2 antagonist/killer) was identified by virtue of its interaction with adenovirus E1B19kD in the yeast two hybrid system and by degenerate PCR cloning approaches (Chittenden *et al.*, 1995b; Farrow *et al.*, 1995; Kiefer *et al.*, 1995). Bak contains well-conserved BH1, BH2 and BH3 domains, and a characteristic hydrophobic C-terminal tail. Both Bak and Bax show an intracellular distribution similar to their anti-apoptotic relatives, localising prominently to mitochondrial membranes. High level expression of Bak accelerates apoptosis following cytokine withdrawal, and induces apoptosis in serum starved fibroblasts and transiently transfected cell lines. Bak heterodimerises with multiple cellular anti-apoptotic proteins including Bcl-xL, Bcl-2, Mcl-1 and A1. Surprisingly, structure/function analysis indicated that much of the Bak molecule, including BH1 and BH2, is dispensable for Bak's ability to induce apoptosis and bind to Bcl-xL (Chittenden *et al.*, 1995a). Only mutations in the BH3 domain of Bak impairs its proapoptotic function and interaction with Bcl-xL. Short truncated forms of Bak encompassing the BH3 domain retain Bcl-xL binding and pro-apoptotic function, indicating that BH3 is both necessary and sufficient for these activities. Similar mutational studies demonstrated that the BH3 domain of Bax also mediates its pro-apoptotic function and interaction with Bcl-2 and E1B19kD (Han *et al.*, 1996a; Hunter, 1996; Zha *et al.*, 1996a).

2.2.3 Bik

The unique importance of the BH3 domain to the function of pro-apoptotic Bcl-2 family members was further established by the characterisation of Bik (for Bcl-2 interacting killer) (Boyd *et al.*, 1995). Bik was cloned by its interaction with Bcl-2 in the two hybrid system, and independently (designated Nbk) by its binding to the E1B19kD protein (Han *et al.*, 1996b). Bik also binds specifically to other cellular and viral anti-apoptotic Bcl-2 homologues, including Bcl-xL and BHRF-1, and induces cell death in transient transfection assays. The deduced Bik amino acid sequence harbours a hydrophobic C-terminal tail but lacks canonical BH1 and BH2 elements that typically define membership in the Bcl-2 family. However, Bik contains a 12 amino acid element that shows substantial homology to the BH3 domains of Bak and Bax, and other Bcl-2 relatives. Mutation of the Bik BH3 domain abrogates its ability to induce apoptosis and bind cell death suppressors such as Bcl-2, and a truncated form of Bik encompassing BH3 retains these activities (Chittenden *et al.*, 1995a; Elangovan and Chinnadurai, 1997). Apart from the conserved BH3 domain, Bik does not show significant homology to other Bcl-2 family members. Thus, Bik represents the first example of a "BH3-only" sub-class of Bcl-2 family members that bind anti-apoptotic Bcl-2 homologues and promote apoptosis through a conserved BH3 domain.

2.2.4 Hrk, Bim, Bid

Three additional BH3-only cell death agonists, Hrk, Bim and Bid, have been identified by their physical interaction with Bcl-2. Like Bik, the structures of these proteins share a conserved BH3 domain, but otherwise lack homology to other Bcl-2 family members. Hrk (Harakiri) is a small 91 amino acid protein that binds to Bcl-2 and Bcl-xL and induces cell death in transient transfection assays (Inohara *et al.*, 1997). Both of these activities depend on the integrity of its BH3 domain. Similarly, Bim (Bcl-2 interacting mediator of cell death) and Bid (BH3 interacting domain death agonist) bind to multiple anti-apoptotic Bcl-2 homologues, including Bcl-2 and Bcl-xL, and promote cell death (Wang *et al.*, 1996c; O'Connor *et al.*, 1998). Again, the heterodimerisation and cell killing activities of these proteins are mediated by their respective BH3 domains. Bid distinguishes itself from the other BH3-only proteins, Bik, Hrk and Bim, in that it does not appear to contain a hydrophobic C-terminal membrane anchor. Furthermore, Bid also heterodimerises with Bax, providing the first example of an interaction between different pro-apoptotic proteins (Wang,K *et al.*, 1996). Mutational analysis of the Bid BH3 domain indicates that the cell death agonist function of Bid correlates with its ability to interact with Bax, rather than Bcl-2, suggesting that Bid may promote apoptosis by modulating the function of Bax.

2.2.5 Bad

Bad (Bcl-2 associated death promoter) is a pro-apoptotic family member that heterodimerises with Bcl-2 and Bcl-xL, and counteracts their anti-apoptotic function in cells (Yang,E. *et al.*, 1995). The Bad protein does not contain a hydrophobic membrane anchor sequence typically found at the C-terminus of Bcl-2 family members. Two appropriately-spaced sequence elements that show weak homology to BH1 and BH2 were noted in the C-terminus of the protein, suggesting that Bad is a distantly related Bcl-2 homologue. Subsequent studies localised the sequences in Bad required for binding to Bcl-xL/Bcl-2 to a short peptide domain which overlaps with the putative BH1 element in Bad, but bears significant homology to the BH3 domains defined in other pro-apoptotic family members such as Bak, Bax, and Bid (Kelekar *et al.*, 1997; Ottilie *et al.*, 1997b; Zha *et al.*, 1997). Truncated derivatives, or synthetic peptides encompassing this putative BH3 element of Bad are sufficient for binding to Bcl-2 or Bcl-xL. Furthermore, an alanine substitution mutant of Bad at leucine residue 151, corresponding to an invariant leucine found in the BH3 domains of pro-apoptotic family members, abrogates binding to Bcl-xL and Bcl-2 and largely eliminates the ability of Bad to promote apoptosis. These findings indicate that the cell death and protein binding functions of Bad are mediated principally by a conserved BH3 domain and suggest that Bad comprises an additional "BH3 only" cell death agonist.

2.2.6 Bok

A recently described pro-apoptotic Bcl-2 homologue, Bok (Bcl-2-related ovarian killer), was isolated by its interaction with the Mcl-1 cell death suppressor in a yeast two hybrid screen (Hsu *et al.*, 1997). Expression of Bok appears to be restricted primarily to reproductive tissues including ovary, uterus and testis. Furthermore, Bok is less promiscuous than other pro-apoptotic homologues in its interactions with cell death suppressors, binding preferentially to Mcl-1 and Bfl-1 but not to Bcl-2, Bcl-xL or Bcl-w in the two hybrid system. By contrast, Bax, Bak and Bik each bind strongly to all of these cell death suppressors in equivalent assays. Transient transfection of Bok induces cell death in CHO cells, which is attenuated by co-transfection of Mcl-1. Bok contains a conserved BH3 domain but is not a BH3 only pro-apoptotic Bcl-2 homologue since BH1 and BH2 domains are also present along with a span of hydrophobic amino acids at its C-terminus. By conservation of these structural features, Bok more closely resembles the pro-apoptotic homologues Bax and Bak and it is likely that the BH3 domain of Bok, as in Bax and Bak, will prove to mediate Bok's interaction with Mcl-1 and its pro-apoptotic function.

Several general principles can be distilled from the characterisation of this complex array of Bcl-2 homologues. It is clear that the conservation, and hence functional importance, of BH domains differs significantly between the pro-apoptotic and anti-apoptotic family members. Structurally, the pro-apoptotic homologues fall into two sub-categories, defined either by the conservation of BH1, BH2 and BH3 domains (Bax, Bak Bok), or BH3 only (Bik, Hrk, Bid, and Bad), but in each case appear to bind cell death suppressors in a mechanistically similar fashion (via their BH3 domains). For all known cellular and viral anti-apoptotic Bcl-2 homologues, conservation of BH1 and BH2 domains is paramount, and mutational studies have established that BH1 and BH2 are indeed critical to the anti-apoptotic function of these proteins. Additionally, the more loosely-conserved BH4 domain is found exclusively at the amino-terminus of anti-apoptotic, but not pro-apoptotic family members, and may contribute to the suppression of apoptosis. Deletion of BH4 in fact eliminates the protective function of Bcl-2 and converts it into a transdominant inhibitor that accelerates apoptosis, much like Bak or Bax (Hunter *et al.*, 1996). The Bcl-2 homologues that suppress cell death exhibit similar intracellular membrane localisation, and where tested, can inhibit cell death triggered by many of the same stimuli (although their relative potency may vary). In view of the similarities in structure and activity, it is likely that anti-apoptotic Bcl-2 homologues are variants of a common theme, and operate in a mechanistically similar fashion.

While it is convenient from a conceptual standpoint to group Bcl-2 homologues into either pro- or anti-apoptotic functional classes, in actuality their functions may not be so precisely delineated. Alternative splicing of cell death suppressor genes may generate pro-apoptotic products (e.g. Bcl-xS). Bak and Bax, which generally act to promote apoptosis, may at least in some settings, function to inhibit cell death. Expression of Bak suppresses apoptosis following serum withdrawal or treatment with a calcium ionophore in a lymphoblastoid cell line (Kiefer *et al.*, 1995). Over-expression of Bax can transiently inhibit the death of microinjected sensory neuronal cells following deprivation of neurotrophic factors (Middleton *et al.*, 1996) and there is evidence for enhanced levels of apoptosis in germ cells of Bax-deficient mice (Knudson *et al.*, 1995). The biological effect of Bak and Bax expression may therefore vary depending on the apoptotic stimulus or cell type, for reasons that are not yet understood.

2.3 Molecular interactions among Bcl-2 homologues

Anti-apoptotic and pro-apoptotic Bcl-2 homologues heterodimerise in an inherently non-symmetrical fashion and require different structural elements for interaction. For the pro-apoptotic homologues, the BH3 domain appears to be sufficient to mediate binding, whereas the requirements for cell death suppressors are complex and involve at least several distinct domains. The heterodimerisation function of Bcl-2 is adversely affected by mutations spanning much of the molecule. Deletions or amino acid substitutions at conserved residues within the BH1 and BH2 domains of Bcl-2 eliminate binding to Bax. Additional sequences within the amino and carboxyl terminal portions of Bcl-2 also appear to be required (Yin *et al.*, 1994; Hanada *et al.*, 1995). Similarly, point mutations in the BH1 and BH2 domains of Bcl-xL disrupt its interaction with Bax (Sedlak *et al.*, 1995; Cheng *et al.*, 1996).

The mutational data suggest a lock and key type interaction wherein multiple elements, including BH1 and BH2, within the anti-apoptotic protein combine to act as a receptor for a BH3 domain presented by the pro-apoptotic family member. This interpretation has been validated by the solved crystal and solution structure of Bcl-xL (Muchmore *et al.*, 1996). The core of Bcl-xL is formed by two alpha helices of largely hydrophobic amino acids which are arranged in anti-parallel fashion and encompassed by five amphipathic helices. Sequences within the BH1, BH2 and BH3 domains, including highly conserved core residues, are juxtaposed and contribute to the formation of a solvent-exposed hydrophobic groove in Bcl-xL. The solution structure of Bcl-xL in complex with a 16 amino acid peptide corresponding to the Bak BH3 domain indicates that this cleft acts as a docking site for the BH3 domain of pro-apoptotic Bcl-2 homologues, and possibly other proteins (Sattler *et al.*, 1997). The Bak BH3 peptide, in its bound form, is an alpha helix that lies across the elongated cleft formed by BH1, BH2 and BH3 residues in Bcl-xL. The complex is stabilised by hydrophobic interactions with apolar residues that are highly conserved in BH3 domains, as well as by electrostatic interactions between charged side chains. Synthetic peptides derived from either the Bak or Bax BH3 domain are sufficient to disrupt heterodimerisation of the full length pro-apoptotic proteins with Bcl-xL or Bcl-2, confirming the essential role of the BH3 binding cleft. In light of the contribution of conserved BH domains in this interaction, the data argue that similar BH3-binding clefts will exist in other anti-apoptotic Bcl-2 homologues and provides a general model for cell-death suppressor/promoter interactions.

BH3 sequences also appear to be important for homodimerisation of certain Bcl-2 family members, although the significance of homodimerisation remains uncertain since the point mutations in BH1 and BH2 of Bcl-2 that eliminate anti-apoptotic function do not affect the ability to homodimerise (Yin *et al.*, 1994). *In vitro*, Bcl-2 and Bcl-xL homodimerisation is blocked by BH3 peptides derived from either Bak or Bax, arguing that these homotypic protein interactions also depend on a BH3-binding cleft (Diaz *et al.*, 1997). The homodimerisation of the pro-apoptotic homologues Bax and Bak also requires an intact BH3 domain, and based on their structural homology to Bcl-xL, the conserved BH1 and BH2 sequences retained in Bax and Bak may contribute to a similar BH3 binding cleft. The BH3 domain of Bid mediates its interaction with Bax, supporting the possibility of a BH3-binding cleft in Bax (Wang,K *et al.*, 1996). BH3-only death promoters such as Bik and Hrk lack the BH1 and BH2 elements that contribute to the binding cleft and do not appear to homodimerise. Thus, many of the hetero- and homotypic protein interactions among Bcl-2 homologues may prove to be mediated by a BH3 ligand presented by one partner, and a receptor domain formed by multiple elements, including BH1 and BH2, in the second molecule.

Pro-apoptotic Bcl-2 homologues may bind preferentially to specific cell death suppressors, suggesting a hierarchy for interactions among Bcl-2 related proteins. The pro-apoptotic protein Bok avidly binds to Mcl-1 but does not interact with Bcl-2, Bcl-xL or Bcl-w (Hsu *et al.*, 1997). Although Bak binds to multiple anti-apoptotic Bcl-2 homologues, it also exhibits some preferences, interacting strongly with Bcl-xL but relatively poorly with Bcl-2 (Chittenden *et al.*, 1995a; Farrow *et al.*, 1995). This preference for binding Bcl-xL over Bcl-2 is confirmed by binding studies of Bak BH3 peptides to Bcl-xL and Bcl-2 *in vitro* (Diaz *et al.*, 1997). Bax and Bik BH3 domains appear to bind Bcl-2 and Bcl-xL equally well. The Bad BH3 domain exhibits strong binding to a Bcl-xL mutants that have lost binding to the BH3 domains of both Bak and Bax (Ottilie *et al.*, 1997b). These data demonstrate that BH3 domains from different Bcl-2 homologues are functionally distinct, and may contribute a level of specificity to the interactions between Bcl-2 family members.

2.4 Functional relevance of heterodimerisation of Bcl-2 family members

As first postulated by the rheostat model, the function of Bcl-2 family members and susceptibility to apoptosis is determined in part by the heterodimerisation among cell death promoters and death suppressors (Oltvai and Korsmeyer, 1994). A central question that remains unanswered, is which one of the two components, cell death promoters or suppressors, is the active effector species and which one is the regulator? One possibility is that death suppressors such as Bcl-2 actively inhibit cell death by a mechanism independent of their interaction with pro-apoptotic Bcl-2 homologues. If so, the cell death promoters may accelerate apoptosis by simply binding cell death suppressors and interfering with their function. Alternatively, cell death promoters such as Bax may actively trigger apoptosis through an independent mechanism and Bcl-2 (and its equivalents) may suppress apoptosis by binding and sequestering the pro-apoptotic effectors. Experimental attempts to resolve these possibilities have provided conflicting data probably because the two cases are not necessarily mutually exclusive. That is, both anti- and pro-apoptotic Bcl-2 homologues could have active cell death functions that are independent of their interaction with each other (illustrated in Figure 2).

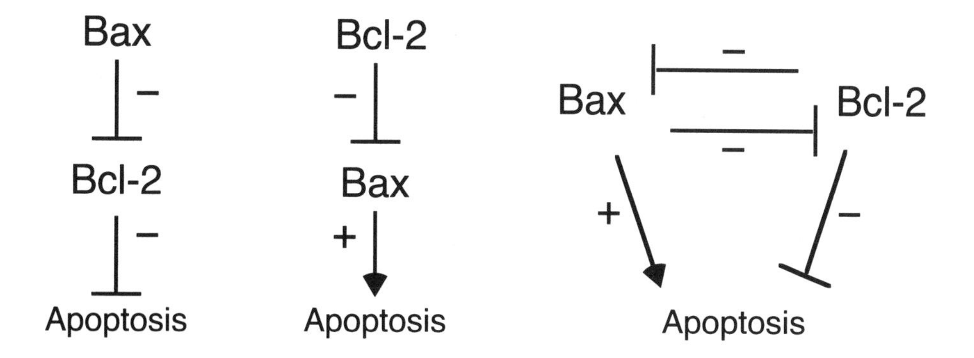

Figure 2. Regulatory interactions between pro-apoptotic and anti-apoptotic Bcl-2 homologues.

Several possibilities have been considered for the role of heterodimerisation between Bcl-2 family members with opposing functions. Left: Pro-apoptotic proteins such as Bax may function as negative regulators of Bcl-2 and related anti-apoptotic proteins, which actively suppress the cell death programme. Apoptosis is triggered by Bax through the neutralisation of Bcl-2 function. Centre: Bax may trigger apoptosis directly, but be subject to negative regulation by interaction with Bcl-2. Right: Bax and Bcl-2 may each contribute a cell death effector function that is negatively regulated through heterodimerisation.

A number of point mutations in Bcl-2 (e.g. G145A within BH1) disrupt interaction with Bax and abrogate its anti-apoptotic function (Yin *et al.*, 1994). The Bcl-2 G145A mutant, however, retains the ability to bind to a truncated form of Bax, that has lost BH1 and BH2 but retains BH3 (Ottilie *et al.*, 1997a). The Bcl-2 mutant is still able to suppress apoptosis induced by this truncated Bax (but not wild type Bax, to which it fails to bind). The genetic linkage of anti-apoptotic function with the ability to bind Bax supports the possibility that Bcl-2 works by binding and inactivating Bax.

However, results from other studies have undermined this argument. Bcl-xL mutants which harbour certain amino acid substitutions within BH1 or BH2 are unable to bind either Bax or Bak, yet retain most of their anti-apoptotic function (Cheng *et al.*, 1996). The viral Bcl-2 homologue KsBcl-2 likewise inhibits apoptosis without detectably binding to either Bax, or Bak (Cheng *et al.*, 1997). These data suggest that Bcl-xL and KsBcl-2 actively suppress apoptosis independently of their binding to cell death promoters. Some uncertainty remains, however, since KsBcl-2 and these Bcl-xL mutants may still bind to pro-apoptotic homologues that were either not tested or have yet to be identified. Indeed, the same BH1 and BH2 mutants of Bcl-xL that fail to bind Bax and Bak are unaffected in their ability to bind Bad, indicating that the BH3 binding cleft in these Bcl-xL mutants is still functional (Ottilie *et al.*, 1997b).

In some cases, analysis of death promoters has supported the possibility that they function principally as regulatory molecules, simply binding and antagonising anti-apoptotic Bcl-2 homologues. Mutation of the BH3 domains of Bak, Bax, Bik, Hrk or Bad, eliminates binding to cell death suppressors such as Bcl-2, and abolishes their ability to accelerate apoptosis. The ability of Bad to promote apoptosis and antagonise the protective function of Bcl-2 or Bcl-xL appears to depend entirely on its binding to these proteins via its BH3 domain. Certain mutants of Bcl-xL (e.g. Val 126 to Gly) are no longer able to bind to Bad, but retain full anti-apoptotic function (Kelekar *et al.*, 1997). If Bad triggers cell death by a mechanism independent of heterodimerisation, co-expression of these Bcl-xL mutants should not affect the ability of Bad to promote apoptosis. When tested, however, Bad was unable to accelerate apoptosis in cells over-expressing a mutant Bcl-xL deficient in binding Bad (Kelekar *et al.*, 1997), strongly supporting the view that Bad and possibly other BH3 containing pro-apoptotic proteins, promote apoptosis by binding and neutralising anti-apoptotic Bcl-2 homologues.

Other studies, however, suggest that Bax and Bak can deliver pro-apoptotic signals unrelated to their heterodimerisation with cell death antagonists. In at least some experimental settings, BH3 mutants of Bax and Bak are still able to accelerate apoptosis induced by chemotherapeutic drugs, despite their inability to bind Bcl-2 or Bcl-xL (Simonian *et al.*, 1996,1997). An active killing function of Bak and Bax is also suggested from the cytotoxicity these proteins exhibit when expressed in yeast (Sato *et al.*, 1994; Hanada *et al.*, 1995; Ink *et al.*, 1997; Jürgensmeier *et al.*, 1997). It is inferred that cell death induced by Bak and Bax in yeast can not result from simply binding and inactivating Bcl-2 homologues, since yeast presumably lack endogenous Bcl-2 related proteins. S. pombe cells killed by Bak and Bax exhibit extensive vacuolisation, chromatin condensation and dissolution of nuclear membranes, the latter two being morphological features shared with apoptotic mammalian cells (Ink *et al.*, 1997; Jürgensmeier *et al.*, 1997). Toxicity of Bak and Bax in yeast can be suppressed by co-expression of death suppressors such as Bcl-2 and Bcl-xL (Sato *et al.*, 1994) and the ability of Bak and Bax to induce cell death in yeast depends on an intact BH3 domain, analogous to their proapoptotic function in mammalian cells (Zha *et al.*, 1996b; Ink *et al.*, 1997). If yeast cells truly lack endogenous Bcl-2 homologues, these findings suggest that Bak and Bax have an intrinsic cytotoxic function distinct from their interaction with anti-apoptotic family members. Experiments in yeast provide similar evidence for Bcl-2 actively suppressing cell death independently of its binding to cell death promoters. Toxicity of Bax in S. cerevisiae can be suppressed by Bcl-2 and Bcl-xL mutants that fail to bind Bax detectably (Tao *et al.*, 1997). Additionally, S. cerevisiae mutants deficient in superoxide dismutase can be rescued by expression of Bcl-2 (Kane *et al.*, 1993; Longo *et al.*, 1997), presumably through a mechanism independent of endogenous yeast Bcl-2 homologues.

Genetic studies in mice have established that, under certain conditions, both Bcl-2 and Bax can modulate apoptosis independently of each other. Thymocytes from transgenic mice that over-express Bcl-2 under the control of the Lck promoter are resistant to apoptosis induced by irradiation and show prolonged survival when placed in culture (Sentman *et al.*, 1991). This protective effect is still observed in a Bax-deficient genetic background, demonstrating that anti-apoptotic activity is independent of dimerisation with Bax (Knudson and Korsmeyer, 1997). Similar results were reported for transgenic mice expressing a BH1 mutant of Bcl-2 which fails to bind Bax or Bak, yet protects immature thymocytes from apoptosis (St. Clare *et al.*, 1997). Interestingly, expression of this Bcl-2 BH1 mutant does not suppress apoptosis in peripheral T-cells indicating the importance of heterodimerisation with Bax (or other death agonists) to the function of Bcl-2 function in this specific setting. Conversely, the marked loss of thymocytes characteristic of *bcl-2*-knockout mice is largely reversed in a *bax*-null background (Knudson and Korsmeyer, 1997). The presence of a single *bax* allele is sufficient to restore the thymic hypoplasia phenotype in bcl-2 deficient mice, demonstrating that Bax can promote apoptosis in the absence of Bcl-2 function.

Taken as a whole, there are compelling data to support a conclusion that death agonists and antagonists can either promote or suppress cell death independently of physical interaction with each other. Both classes of Bcl-2 homologues may actively modulate apoptosis, at least under certain conditions. This conclusion is entirely compatible with heterodimerisation of Bcl-2 homologues also playing a role in the regulation of cell death by these proteins. The relative importance of the effector or regulatory component (heterodimerisation-independent or -dependent) to the function of a particular Bcl-2 homologue may depend on the cellular context, such as the relative expression levels of other Bcl-2 family members. Certainly, this generalisation regarding hetero-dimerisation dependent or independent functions may not hold for all Bcl-2 homologues. It is quite possible that certain BH3-only death promoters, such as Bad, act exclusively as titrators of anti-apoptotic Bcl-2 homologues, whereas others (Bax and Bak) possess additional functions that actively promote apoptosis.

3.0 MECHANISM OF ACTION OF BCL-2

The interactions among Bcl-2 family members appears to contribute principally to the regulation of their activity. Still open is the "effector" question of how, at a molecular level, Bcl-2 and its relatives actively modulate apoptosis. This mechanistic question has been the subject of intensive research since the discovery of Bcl-2, yet clear answers have remained elusive. In order to gain insights into its mechanism, the biological and biochemical effects of Bcl-2 expression have been examined in a host of experimental systems, and are wide-ranging. Bcl-2 blocks lipid peroxidation and the generation of reactive oxygen species and impacts cellular redox potentials, suggesting that it operates in an anti-oxidant pathway (Hockenbery *et al.*, 1993; Kane *et al.*, 1993). Bcl-2 also alters intracellular ion fluxes that occur during apoptosis, including changes in the partitioning of Ca^{2+} in the endoplasmic reticulum, nucleus, and mitochondria (Baffy *et al.*, 1993; Lam *et al.*, 1994; Marin *et al.*, 1996). Other reported effects of Bcl-2 on mitochondria (where Bcl-2 resides) include the inhibition of a mitochondrial permeability transition, an early or possibly pre-apoptotic event that leads to a loss of membrane potential (Marchetti *et al.*, 1996; Petit *et al.*, 1996; Kroemer *et al.*, 1997). Activation of caspases (e.g. Cpp32), proteases that comprise part of the cell death machinery, and endonucleases are also inhibited by Bcl-2 expression (Chinnaiyan *et al.*, 1996).

Many of these studies have been plagued by difficulties in establishing whether the biological effect in question is directly related to the mechanism by which Bcl-2 suppresses apoptosis, or is instead an indirect consequence of Bcl-2 preventing cell death. Correlation of a biological effect with Bcl-2 expression does not establish cause or effect since, if cell death is prevented by Bcl-2, a broad spectrum of downstream biochemical and biological events associated with apoptosis will be prevented. More direct evidence for the molecular mechanism of Bcl-2's anti-apoptotic function has recently emerged from several lines of research. The implication from these findings is that Bcl-2 is a multifunctional protein that modulates apoptosis through distinct mechanisms, including an intrinsic biochemical activity related to its ability to form membrane channels, and by interacting with, and modulating the function of, other components of the cell death pathway. Indeed, both of these functions may eventually prove to contribute the suppression of apoptosis by Bcl-2.

3.1 Membrane pore-forming activity

The primary sequences of Bcl-2 and its relatives have not provided clues as to their mechanism, since they do not resemble any proteins of known function. However, the crystal and solution structure of Bcl-xL revealed that the arrangement of its seven α-helices are markedly similar to the three dimensional structure of diphtheria toxin and other bacterial colicins that act as membrane pore-forming molecules (Muchmore *et al.*, 1996). By analogy to the bacterial toxins, the core of Bcl-xL consisting of two central α helices (α5 and α6) of largely hydrophobic residues, may insert across lipid bilayers and contribute to the formation of a membrane pore. This suggests that Bcl-xL (and other Bcl-2 family members) may form membrane pores that act as channels for ions, or possibly other molecules, and that this activity contributes to the regulation of apoptosis by these molecules.

Inspired by the structural clues, several laboratories subsequently demonstrated that Bcl-xL, Bcl-2, and the pro-apoptotic homologue Bax, can all form membrane pores *in vitro* (Antonsson *et al.*, 1997; Minn *et al.*, 1997; Schendel *et al.*, 1997; Schlesinger *et al.*, 1997). Recombinant forms of both Bcl-xL and Bcl-2 form ion channels at low pH in planar lipid bilayers or synthetic lipid vesicles, which are cation selective at physiological pH. The acidic pH dependence of Bcl-2/Bcl-xL channel formation and the requirement for acidic lipids resemble the properties of pore-forming domains of bacterial toxins, and is consistent with the proposed structure/function homology. Bax appears to form pores under a broader pH range, operating at both acidic and neutral pH. Additionally, membrane channels formed by Bax have biophysical characteristics distinct from Bcl-2 channels, displaying a mild Cl- selectivity in contrast to the mild K+ selectivity exhibited by Bcl-2. At physiological pH, the channel-forming activity of Bax is inhibited by the presence of Bcl-2 (Antonsson *et al.*, 1997). The details regarding the diameter of the pores formed by Bcl-2 homologues and their molecular composition are currently uncertain.How might channel formation contribute to the regulation of apoptosis by Bcl-2 homologues? It is likely that the membrane pore function of Bcl-2/Bcl-xL acts to suppress key initiating events at the mitochondria, possibly the mitochondrial permeability transition, that would otherwise lead to the cytosolic release of apoptosis-triggering factors. The characteristic mitochondrial permeability transition associated with apoptosis involves the opening of a large inner membrane pore and the dispersal of mitochondrial proteins and Ca^{2+} stores into the cytosol (Petit *et al.*, 1996). Bcl-2/Bcl-xL channels may inhibit the opening of these holes, perhaps by maintaining the integrity of ion gradients and mitochondrial membrane potential, and preventing the catastrophic release of apoptotic factors that activate the cell death machinery in the cytoplasm. Such factors have now been identified biochemically and include cytochrome c and an apoptogenic protease (AIF), both of which strongly activate cytoplasmic caspases (Liu *et al.*, 1996; Susin *et al.*, 1996). In keeping with this possibility, Bcl-2 prevents the release of cytochrome c from mitochondria and caspase activation both in cell-free apoptosis systems and in intact cells (Kluck *et al.*, 1997; Yang *et al.*, 1997), although this activity appears to be independent of mitochondrial permeability transition. In contrast, Bax promotes mitochondrial permeability transition and cytochrome c release, which may account for its cytotoxicity when expressed in yeast and its active death effector function that is independent of heterodimerisation with death suppressors (Xiang *et al.*, 1996; Manon *et al.*, 1997). Based on their structures, Bak and Bok may also form pores whereas other BH3-only death promoters presumably would not since they appear to lack the central membrane spanning helices.

The membrane pore-forming function is the first intrinsic biochemical activity identified for Bcl-2 homologues. However, direct evidence that these proteins form channels in cells is still lacking and the key question of whether pore-forming activity is essential for the regulation of apoptosis by Bcl-2 homologues remains unresolved. Deletion of the two central hydrophobic helices abolishes both the formation of ion channels *in vitro* and anti-apoptotic function in transfected cells (Schendel *et al.*, 1997). The pore-forming activity of other mutants have not yet been examined so there is no additional genetic data available to support the functional relevance of pore-formation. Recombinant Bax demonstrates a lytic activity when added to intact neurons or red blood cells, consistent with a possible membrane pore-forming activity (Antonsson *et al.*, 1997). Additional indirect support for the membrane channel hypothesis is the observation that many apoptotic stimuli cause marked swelling of mitochondria leading to rupture of the outer membrane, release of cytochrome c and subsequent depolarisation of the inner membrane (Heiden *et al.*, 1997). Expression of Bcl-xL in cells inhibits these mitochondrial pathologies that are invariably associated with apoptosis. Moreover, Bcl-xL inhibits the swelling of mitochondria induced by inhibitors of oxidative phosphorylation, arguing that Bcl-xL does not modulate apoptosis at a point upstream of the mitochondrial events (Heiden *et al.*, 1997). These data, combined with the evidence that Bcl-xL localises to mitochondria in cells and forms membrane pores *in vitro*, build a strong circumstantial case for Bcl-xL ion channels suppressing apoptosis by preventing osmotic swelling and regulating membrane potential of mitochondria.

3.2 Interaction with components of the cell death pathway

Mechanistic clues as to how Bcl-2 might inhibit apoptosis have also come from the analysis of programmed cell death in C. elegans where the nematode equivalent of Bcl-2, Ced-9, operates at an upstream point to inhibit a cell death pathway defined by the cell death effectors Ced-4 and Ced-3 (see Chapter 2). Regulation of the central death pathway by Ced-9 appears to be through its direct physical interaction with Ced-4, since Ced-9 binds to Ced-4 in the yeast two hybrid system, *in vitro*, and in transfected mammalian cells (Chinnaiyan *et al.*, 1997b; James *et al.*, 1997; Spector *et al.*, 1997; Wu *et al.*, 1997b). Ced-4, in turn, promotes apoptosis by binding Ced-3 and stimulating its proteolytic activity (Chinnaiyan *et al.*, 1997a; Seshagiri and Miller, 1997; Wu *et al.*, 1997a). The interaction of Ced-9 does not appear to prevent the association of Ced-4 with Ced-3, but inhibits the ability of Ced-4 to activate Ced-3. Inhibition of caspase activation by Ced-9 may be related to its ability to redirect the localisation of Ced-4 or Ced-3/Ced-4 complexes from the cytoplasm to intracellular membranes, or by inducing a conformational change that interferes with the ability of Ced-4 to activate caspases.

Since the components of the cell death pathway appear to be conserved between nematode and mammalian cells, the clear implication of these data is that Bcl-2 and its homologues suppress apoptosis, at least in part, by binding to a mammalian Ced-4 equivalent(s) and inhibiting its ability to activate caspases, the mammalian Ced-3 equivalents (Figure 3). This is consistent with the finding that Bcl-2/Bcl-xL appear to regulate apoptosis at a point upstream of caspases in mammalian cells (Chinnaiyan *et al.*, 1996; Shimizu *et al.*, 1996). More direct support for this scenario has come from the discovery that Bcl-xL can associate with the nematode Ced-4 protein in transfected cells, which in turn interacts with the mammalian caspases Flice and Ice (Chinnaiyan *et al.*, 1997b). Transient high-level expression of Ced-4 is cytotoxic to certain mammalian cell lines and this activity can be suppressed by co-expression of Bcl-xL. In the absence of Ced-4, complexes between Bcl-xL and caspases can still be detected in transfected cells, implying the existence of an endogenous mammalian Ced-4 activity that links the mammalian Ced-9 (Bcl-xL) and Ced-3 (caspase) homologues (Chinnaiyan *et al.*, 1997b). Interestingly, these complexes are not detected in the presence of co-transfected pro-apoptotic Bcl-2 homologues Bax, Bak and Bik suggesting that heterodimerisation prevents Bcl-xL interaction with Ced-4, and that Ced-4 may also interact with the pocket in Bcl-xL that

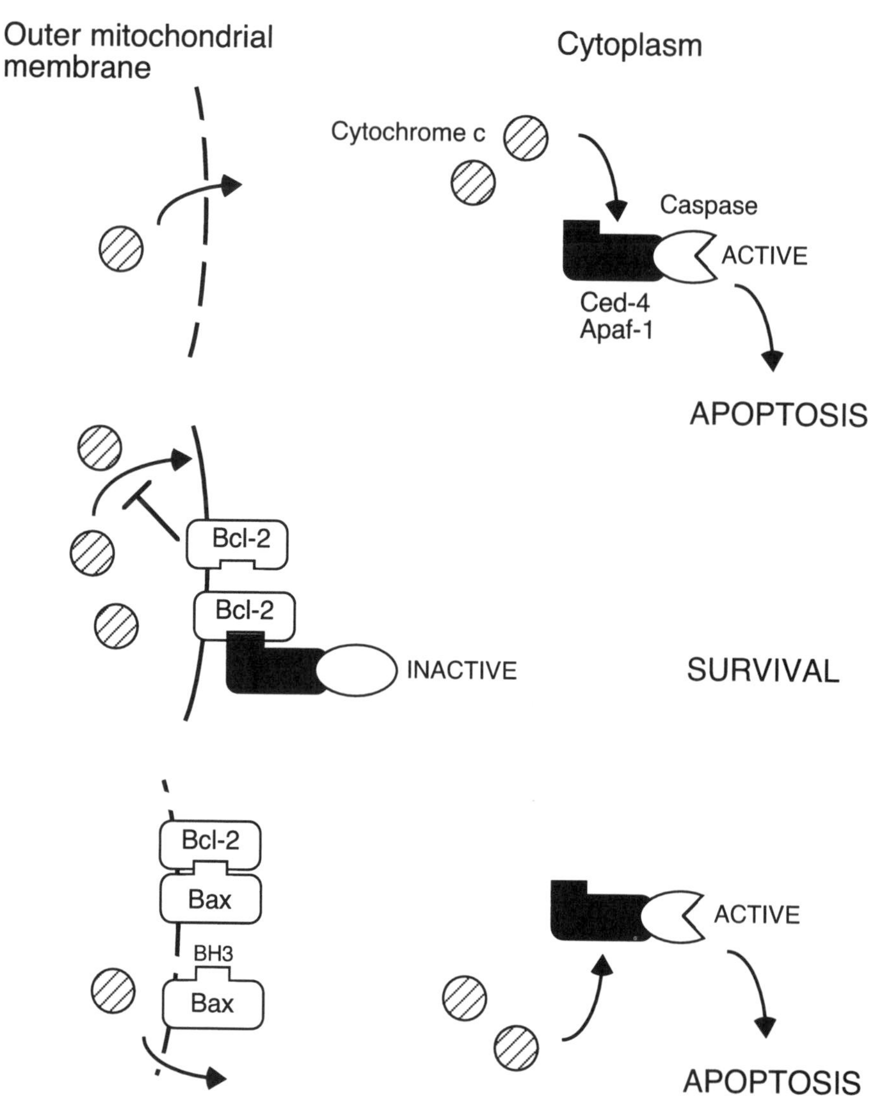

Figure 3. Model for the regulation of apoptosis by Bcl-2 family members.

Top: In an apoptotic cell, Apaf-1 or other mammalian Ced-4 homologues, bind specific caspases and activate their proteolytic function, initiating an apoptotic protease cascade in the cytoplasm. The pro-apoptotic function of Apaf-1 is activated by its interaction with cytochrome c which is released from mitochondria in apoptotic cells, due to ruptures in the outer mitochondrial membrane.

Centre: Bcl-2 may suppress apoptosis by interacting with mammalian Ced-4 homologues, sequestering them at mitochondrial membranes and inhibiting their ability to activate caspases. The membrane channel function of Bcl-2 may also contribute to the suppression of apoptosis by promoting mitochondrial homeostasis and preventing the release of apoptosis-inducing factors such as cytochrome c.

Bottom: Bax induces apoptosis by binding Bcl-2 and preventing the interaction with Ced-4 homologues. Additionally, Bax may promote the loss of mitochondrial integrity and release of cytochrome c through its pore-forming function.

serves as a receptor for BH3 domains. The framework of interactions predicted by the model that Bcl-2/Bcl-xL suppress apoptosis by interacting with Ced-4 can, therefore, be recapitulated in transfected mammalian cells, at least with the nematode Ced-4 and mammalian Ced-3/Ced-9 equivalents (caspases and Bcl-2).

Validation of this model awaits the direct demonstration that Bcl-2 homologues can interact with and modulate the pro-apoptotic function of the mammalian Ced-4 homologue(s). Despite the relative abundance of mammalian Ced-3 (caspase) and Ced-9 (Bcl-2) family members, a mammalian Ced-4 homologue has until recently eluded discovery. Apaf-1 (for apoptosis protease activating factor 1) represents the first candidate mammalian Ced-4 homologue and was identified using a biochemical approach to identify factors in Hela cell extracts required to trigger apoptosis in a cell-free assay (Zou *et al.*, 1997). Apaf-1 contains a modular structure consisting of several distinct domains including an amino terminal segment homologous to certain caspase prodomains, followed by a larger domain with significant homology to Ced-4. In addition to Apaf-1, two other proteins required for apoptosis in this *in vitro* model have been identified and include cytochrome c and caspase-9 (Liu *et al.*, 1996; Li *et al.*, 1997). The available data suggest that cytochrome c binds to Apaf-1 in a dATP dependent fashion and facilitates the interaction of Apaf-1 with caspase 9, via their respective amino terminal prodomain elements. Binding of Apaf-1 to caspase-9 causes activation of the protease, which subsequently cleaves and activates caspase 3 and initiates a protease cascade (Li *et al.*, 1997). By analogy to the interaction of Ced-9 with Ced-4, Bcl-2/Bcl-xL may bind directly to Apaf-1 and prevent its ability to activate caspase-9. Since the pro-apoptotic activity of Apaf-1 depends on its interaction with cytochrome c, Bcl-2/Bcl-xL may inhibit Apaf-1 indirectly, by maintaining mitochondrial integrity and preventing the release of cytochrome c, or, as suggested by one study, by binding to cytochrome c directly (Kharbanda *et al.*, 1997). A broader question remains as to whether Apaf-1 represents the sole Ced-4 homologue that functions in mammalian cells, particularly in view of the complexity of the Bcl-2 and caspase protein families. A recently described protein, p28 Bap31, lacks sequence homology to Ced-4 but exhibits some Ced-4-like properties since it functionally interacts with both Bcl-2 and caspase family members (Ng *et al.*, 1997). p28 Bap31 is a novel integral membrane protein localised to the endoplasmic reticulum that directly binds to Bcl-2/Bcl-xL and also associates with caspase-8 (Flice) *in vivo*. The p28 protein is cleaved by caspase-8 to a 20 kD isoform which exhibits pro-apoptotic activity in transfected cells.

3.3 Interactions with other cellular proteins

A physical connection with mammalian Ced-4 homologues may ultimately prove to be the interaction most relevant to the anti-apoptotic mechanism of Bcl-2 based on the analogy to C. elegans. Other cellular proteins, apart from Ced-4 and Bcl-2 family members, have been found to specifically interact with Bcl-2. At least some of these binding proteins are likely to act as regulators of Bcl-2, facilitating or inhibiting its function, as opposed to comprising key effectors of its anti-apoptotic mechanism. Interaction of Bcl-2 with the protein product of the *spinal muscular atrophy* gene (*SMN*) is an example of a positive or synergistic interaction (Iwahashi *et al.*, 1997). Function of the *SMN* gene is invariably lost in spinal muscular atrophy, a motor neuron degenerative disease with a pathology consistent with aberrant apoptosis. The SMN protein alone exhibits weak anti-apoptotic activity, which is greatly augmented by co-expression and interaction with Bcl-2. This interaction may be important for the survival of motor neurons that express low levels of Bcl-2 (Iwahashi *et al.*, 1997). Several proteins that bind to Bcl-2 and adenovirus E1B19kD, designated Nip1, Nip2 and Nip3, appear to be common targets for these distantly related cellular and viral anti-apoptotic proteins (Boyd *et al.*, 1994). Nip3 is a dimeric protein found in mitochondrial membranes and promotes apoptosis and antagonises the protective function of Bcl-2 in transfection studies (Chen *et al.*, 1997). Nip1 and Nip2 may also be membrane localised, although their biological functions remain as yet unknown.

Bcl-2 interacts with the Raf-1 kinase, through the BH4 domain, and re-directs its intracellular localisation to mitochondrial membranes (Wang *et al.*, 1996a). The delivery of Raf-1 to mitochondria by interaction with Bcl-2 may facilitate the phosphorylation of proteins involved in regulating apoptosis, possibly modulating their function. A synergistic effect of Raf-1 and Bcl-2 in suppressing apoptosis has been observed, and dominant interfering mutants of Raf-1 can diminish the anti-apoptotic function of Bcl-2 (Wang *et al.*, 1996a). However, the activity of Raf-1 is not essential for the survival function of Bcl-2 in at least some settings (Olivier *et al.*, 1997). A Bcl-2 binding protein termed Bag-1 enhances the anti-apoptotic function of Bcl-2 (Takayama *et al.*, 1995) and also binds to Raf-1, activating its kinase activity (Wang *et al.*, 1996b). The co-operative activity of Bag-1 may reflect its function as a regulator of Hsc70, a molecular chaperone involved in protein folding and maturation pathways (Hohfeld and Jentsch, 1997; Takayama *et al.*, 1997). Interactions between Bcl-2 and the Ras family proteins, R-Ras and H-Ras, have also been detected again implicating a connection to cellular signal transduction pathways (Fernanadez-Sarabia and Bischoff, 1993; Chen and Faller, 1996). Where tested, Bcl-2 interacting proteins do not bind pro-apoptotic homologues such as Bax, however, it is not yet clear whether these interactions are important to the cell death effector function of Bcl-2.

High level expression of Bcl-2 exerts a growth inhibitory effect that has been observed in multiple cell types. Bcl-2 expression delays entry into the cell cycle following addition of cytokine in IL-3 dependent cells and accelerates withdrawal into G0 in leukaemic cells induced to differentiate (Marvel *et al.*, 1994; Vairo *et al.*, 1996; Huang *et al.*, 1997b). Typically, Bcl-2 expression alters the kinetics of entry or exit from a G0 state without altering the cell cycle profiles of growing, asynchronous cultures. Enforced Bcl-2 expression likewise delays entry of activated T-cells into S-phase, whereas its absence accelerates exit from G0 (Linette *et al.*, 1996). In keeping with its ability to antagonise Bcl-2, Bax expression has the opposing effect in T cells and promotes rapid entry into S phase (Brady *et al.*, 1996). Others have reported that Bcl-2 has a more global impact on cell growth, for example, slowing proliferation by prolonging the G1 phase of the cell cycle (Borner, 1996; Mazel *et al.*, 1996). A Bcl-2 point mutant that has lost anti-apoptotic function also does not inhibit cell division (Borner, 1996). However, the anti-proliferative function of Bcl-2 appears not to be essential for its anti-apoptotic activity, at least under some conditions (Vairo *et al.*, 1996; Huang *et al.*, 1997b). These functions of Bcl-2 are genetically separable; mutation of a conserved tyrosine in the BH4 domain of Bcl-2 has no effect on the ability of Bcl-2 to suppress apoptosis in IL-3 dependent cells, yet abolishes the capacity of Bcl-2 to delay entry into S-phase following cytokine addition (Huang *et al.*, 1997b). The specific effect of this mutation implicates the involvement of the BH4 in cell cycle function and may reflect its ability to mediate the interaction of Bcl-2 with Raf-1 or other signal transduction proteins. In the case of T-cells, the growth inhibitory effect of Bcl-2 correlates with the inhibition of NFAT activity, a transcriptional activator needed for optimal expression of IL-2 (Linette *et al.*, 1996). This particular effect of Bcl-2 appears to be mediated by its binding to calcineurin, a protein phosphatase necessary for nuclear translocation of NFAT. Bcl-2 binds calcineurin through its BH4 domain and suppresses NFAT activation by re-localising calcineurin to intracellular membranes (Shibasaki *et al.*, 1997). Additional cell cycle effects of Bcl-2 might be mediated by its interaction with 53BP2, a protein originally discovered through its interaction with the p53 tumour suppressor protein. While the interaction has not been confirmed *in vivo*, 53BP2 co-localises with Bcl-2 in transiently transfected cells, and expression of 53BP2 results in increased numbers of cells in the G2/M phase (Naumovski and Cleary, 1996). Additionally, Bcl-2 expression has also been reported to alter the cell cycle arrest function of p53 and its subcellular localisation (Ryan *et al.*, 1994; Beham *et al.*, 1997).

4.0 REGULATION OF BCL-2 RELATED PROTEINS

The key role that Bcl-2 family members play in governing life or death decisions necessitates that their activities be precisely controlled in cells. Failure to properly regulate the activity of Bcl-2 could potentially have disastrous consequences for either the cell, causing it to needlessly self-destruct, or for the organism, causing inappropriate cell survival and malignancy. As discussed above, one important level of regulation is imparted by a complex network of molecular interactions between Bcl-2 related proteins. At least one other control mechanism, phosphorylation, operates to fine-tune the function of Bcl-2 and its relatives at a post-translational level. In addition, the expression of *bcl-2*-related genes are themselves under tight regulation and this transcriptional control is likely the principle factor responsible for the varied expression profiles of Bcl-2 homologues in different tissues and in response to different stimuli.

4.1 Control of gene expression

The transcriptional regulation of only a few *bcl-2* gene family members have been examined thus far, and of these, *bcl-2* has been the most intensively studied and best characterised. Transcription of *bcl-2* mRNA is initiated at two promoters, P1 and P2, located approximately 1400 bp or 80 bp upstream of the translational start site, respectively (Seto *et al.*, 1988). P1 functions as the major transcriptional promoter and is characterised by multiple binding sites for the Sp1 transcription factor, GC rich sequences, an absence of a canonical TATA element, and multiple transcription initiation sites. By contrast, P2 contains both a TATA element, a CCAAT box, and an octomer motif also found in B-cell immunoglobulin promoters - yet produces only a small fraction of *bcl-2* mRNA. The features of the major P1 promoter are typical for a constitutively expressed "housekeeping" gene. Only limited information is available with respect to the promoter elements and transcription factors responsible for modulating the *bcl-2* gene expression observed under different conditions. One example of such modulation is the induction of *bcl-2* mRNA upon activation of mature B-cells, which is mediated by a cyclic AMP-responsive element (CRE) located in the 5'-flanking region (Wilson *et al.*, 1996). This site binds to CREB and ATF family members following B-cell activation and may also mediate the induction of *bcl-2* mRNA in response to other stimuli that activate these transcription factors.

Several elements in the 5' flanking region function to negatively regulate *bcl-2* expression. *Bcl-2* expression is repressed by the presence of a short open reading frame located in the mRNA just upstream of the authentic *bcl-2* initiating methionine, which interferes with efficient translation of *bcl-2*

(Harigai *et al*., 1996). A negative regulatory element located upstream of the P2 promoter represses transcription initiated by P1 and heterologous promoters in transient transfection assays (Young and Korsmeyer, 1993). Although cellular factors that interact with this element were not identified, other studies have demonstrated that the p53 tumour suppressor protein can suppress transcription of *bcl-2* through a negative regulatory element localised in the same vicinity (Miyashita *et al*., 1994a). Repression of *bcl-2* expression by p53 is probably indirect since there are no sequences within the responsive region that resemble the consensus p53 binding site. A second tumour suppressor protein, the Wilms' tumor suppressor WT1, represses transcription of the *bcl-2* gene by directly binding to a recognition sequence in the 5' untranslated region (Hewitt *et al*., 1995; Heckman *et al*., 1997). Additionally, in pre-B-cells, but not mature B-cells, transcription of *bcl-2* is repressed through the direct binding of Ets family proteins to several π1 sites in the 5' untranslated region (Chen and Boxer, 1995). The fact that expression of *bcl-2* mRNA is subject to negative control by multiple cellular factors, including tumour suppressor gene products, has important implications for how expression of Bcl-2 may be deregulated in cancer cells (see below).

By contrast to Bcl-2, the expression of the pro-apoptotic homologue Bax is activated by wild type p53. Levels of *bax* mRNA and protein are induced by ionising radiation, depending on the tissue or cell line and p53 status (Miyashita *et al*., 1994b; Zhan *et al*., 1994; Kitada *et al*., 1996). Irradiation induces the activity of p53, which in turn binds to a consensus site in the *bax* promoter in the 5' untranslated region and transactivates *bax* gene expression (Miyashita and Reed, 1995). Mice deficient in p53 exhibit reduced levels of Bax protein and elevated levels of Bcl-2 protein in certain tissues, validating a role for p53 in modulating expression of these proteins in a physiological context (Miyashita *et al*., 1994b). Moreover, the ability to alter Bcl-2/Bax ratios may play a role in the p53 pro-apoptotic and tumour suppressor function (see Chapter 5). Expression of *bax* is also subject to negative control through the action of Gfi-1, a zinc-finger oncoprotein that binds to several sites in the *bax* promoter and functions as a transcriptional repressor (Grimes *et al*., 1996). Pro-apoptotic Bak is also inhibited by Gfi-1 although it is not yet clear whether this is a direct effect. Gfi-1 exhibits both anti-apoptotic and oncogenic activities which may be related to its ability to repress the expression of the pro-apoptotic genes. The promoter elements that mediate the induction of *bax* mRNA by other stimuli, including ischaemic injury in neurons and surface IgM cross-linking in B-cell lymphoma cells, have not been characterised (Bargou *et al*., 1995a; Krajewski *et al*., 1995c). In addition to the p53 and Gfi-1 binding sites, the *bax* promoter contains a TATA element and four CACGTG sequences which are potential recognition sites for c-Myc and related transcription factors.

The promoter region of the *bcl-x* gene has been characterised and like *bcl-2*, contains two distinct promoter regions (Grillot *et al.*, 1997). The more upstream promoter is the most active in brain and thymus and resembles the P1 promoter of Bcl-2 in that it is GC rich, contains multiple Sp1 binding sites and lacks a TATA element. Sequence motifs in the 5' untranslated region match the consensus binding sites for a number of transcription factors including GATA-1, Ets-1 and Oct-1, although whether these sites contribute to regulation of *bcl-x* transcription is not yet clear. Bcl-xL is induced by erythropoeitin in erythroid progenitor cells and contributes to the survival function of this cytokine (Silva *et al.*, 1996). Potentially, induction of Bcl-xL in these cells may be mediated by GATA-1 since the activity of this transcription factor is induced by erythropoeitin. Other cytokine survival factors, including IGF-I, IL-2 and IL-13, upregulate expression of Bcl-xL through as yet undefined mechanisms (Akbar *et al.*, 1996; Singleton *et al.*, 1996; Lomo *et al.*, 1997; Párrizas and Leroith, 1997). Little is known about the transcriptional control of other, more recently defined Bcl-2 family members, apart from their differential mRNA expression patterns or response to cytokines.

4.2 Phosphorylation

At a post-transcriptional level, the function of Bcl-2 and related cell death suppressors is modulated through the interaction with BH3-containing pro-apoptotic proteins and their relative abundance. It is now clear, however, that phosphorylation provides an additional level of post-translational control over both heterodimerisation and the anti-apoptotic activity of Bcl-2 and its relatives. At present, the phosphorylation of the pro-apoptotic protein Bad provides the best understood example of how phosphorylation modulates the function of Bcl-2 family members. Bad is a BH3-containing protein that promotes apoptosis principally by binding and antagonising Bcl-2/Bcl-xL (and presumably other anti-apoptotic Bcl-2 homologues). Cytokines that function as survival factors, such as insulin-like growth factor I (IGF-I) and IL-3, induce the rapid phosphorylation of Bad on two serine residues (Ser 112 and Ser 136) (Datta *et al.*, 1997; Peso *et al.*, 1997). The phosphorylation of either of these serine residues has important functional consequences, since phosphorylated Bad no longer heterodimerises with the anti-apoptotic proteins Bcl-2 and Bcl-xL and instead forms a complex in the cytosol with 14-3-3, a protein that interacts with many signalling molecules (Zha *et al.*, 1996c). Bcl-2 and Bcl-xL appear to bind exclusively to the unphosphorylated form of Bad. Thus, serine phosphorylation inactivates Bad, and indirectly activates cell death suppressors by relieving them from the repression imparted by heterodimerisation with Bad.

Cytokine-mediated phosphorylation of Bad also establishes an important connection between extracellular survival factors and the control of Bcl-2 activity. Phosphorylation of Bad induced by IGF-I or IL-3 results in the post-translational activation of resident anti-apoptotic Bcl-2 homologues and explains the observation that the anti-apoptotic effect of IGF-1 resembles Bcl-2 (in that it suppresses cell death induced by broad spectrum of stimuli including *myc*, chemotherapeutic drugs, and cycloheximide; Harrington *et al.*, 1994). A signal transduction pathway that leads to phosphorylation of Bad has been recently delineated. Bad is phosphorylated, at least on serine 136, by the serine/threonine kinase Akt (also known as protein kinase B) (Datta *et al.*, 1997; del 0Peso *et al.*, 1997). The kinase function of Akt is activated in response to IGF-I or IL-3 by a phosphotidylinositide-3' kinase (PI-3 kinase) dependent mechanism. In keeping with their proposed role in mediating the phosphorylation of Bad, both Akt and PI-3 kinase exhibit potent anti-apoptotic activity when expressed in constitutively active forms, and dominant negative mutants or inhibitors of these kinases interfere with IGF-I mediated phosphorylation of Bad (Datta *et al.*, 1997). It is not yet clear if Akt is the sole Bad kinase or whether Akt phosphorylates other proteins involved in regulating cell death. Exactly how phosphorylation of Bad prevents binding to Bcl-2/Bcl-xL is uncertain, however, it may be significant that serine 136 is located immediately adjacent to the BH3 domain of Bad that mediates binding to Bcl-xL. Nevertheless, these findings demonstrate that the activity of Bcl-2 (and related death suppressors) is regulated by cytokines through a signalling pathway which results in the phosphorylation of Bad and loss of its heterodimerisation function.

Bcl-2 itself is a phosphoprotein, although, the functional consequences of phosphorylation are controversial. IL-3 and bryostatin-1, an activator of protein kinase C (PKC), induce the serine phosphorylation of endogenous or transfected Bcl-2 in myeloid cell lines, correlating with the anti-apoptotic activity of these factors (May *et al.*, 1994). Such phosphorylation is blocked by PKC inhibitors, and PKC can phosphorylate Bcl-2 *in vitro*, implicating one or more PKC isoforms as potential Bcl-2 kinases. Each of seven potential PKC serine phosphorylation sites in Bcl-2 were mutated to examine the effect on Bcl-2 phosphorylation and its anti-apoptotic function (Ito *et al.*, 1997). Mutation of six of these serine residues had no effect. Mutation of serine 70 to alanine eliminated phosphorylation of Bcl-2 in response to IL-3 or bryostatin-1 and significantly reduced its anti-apoptotic function. Conversely, substitution of serine 70 with glutamic acid (which may mimic a phosphate charge) resulted in an enhanced ability of Bcl-2 to suppress apoptosis. Together, these data indicate that the phosphorylation of Bcl-2 at serine 70 is required to facilitate its anti-apoptotic function. Although PKC phosphorylates serine 70 (and other serine residues) *in vitro*, a different kinase may be responsible *in vivo* since Bcl-2 phosphorylation and anti-apoptotic function are still observed following treatment with high concentrations of staurosporine that block PKC activity in cells (Ito *et al.*, 1997). Bcl-2 phosphorylation in PC12 cells is stimulated by treatment with the survival factor, NGF, supporting the possibility that phosphorylation potentiates the anti-apoptotic function of Bcl-2 (Horiuchi *et al.*, 1997). The activation of MAP kinase correlates with the phosphorylation of Bcl-2 induced by NGF and may be responsible for phosphorylating Bcl-2 at serine 70 since this site conforms to a potential MAP kinase phosphorylation site.

Other studies have concluded, however, that the hyper-phosphorylation of Bcl-2 results in the inactivation of its anti-apoptotic function. Treatment of cells with the phosphatase inhibitor okadaic acid or the chemotherapeutic drug taxol induces the serine phosphorylation of Bcl-2 on as yet undetermined residues, which is manifested by a prominent mobility shift on SDS polyacrylamide gels (Holder *et al.*, 1995,1996). Hyperphosphorylation of Bcl-2 correlates with the induction of apoptosis by these agents, suggesting that it inactivates the protective function of Bcl-2. Other cytotoxic agents that affect the integrity of microtubules also trigger this hyperphosphorylation (Blagosklonny *et al.*, 1997; Haldar *et al.*, 1997). The hyper-phosphorylation of Bcl-2 induced by taxol requires the function of Raf-1, which itself is activated by diverse microtubule-damaging treatments (Blagosklonny *et al.*, 1996; Blagosklonny *et al.*, 1997). The role of Raf-1 is presumably indirect, since Bcl-2 does not appear to be a substrate for Raf-1 (Wang *et al.*, 1994). Anti-cancer drugs that operate through DNA damage pathways do not induce Bcl-2 phosphorylation (Blagosklonny *et al.*, 1997; Haldar *et al.*, 1997), demonstrating that hyperphosphorylation of Bcl-2 is not universally associated with apoptosis but rather appears to be restricted to micotubule-disrupting agents active in the G2/M cell cycle phase. Serine-phosphorylated forms of Bcl-2 with altered gel mobilities can be detected in some cell lines in the absence of taxol treatment (Guan *et al.*, 1996), however, it remains possible that the phosphorylation induced by taxol is a pharmacological effect that does not reflect how Bcl-2 is typically phosphorylated and regulated in cells. It appears that this phosphorylation is biochemically distinct from that observed following IL-3 or byrostatin treatment, in that the latter does not cause a detectable gel mobility shift of Bcl-2. These phosphorylation events may also have different functional consequences, perhaps accounting for the apparent conflicting results regarding the effect of phosphorylation on the activity of Bcl-2.

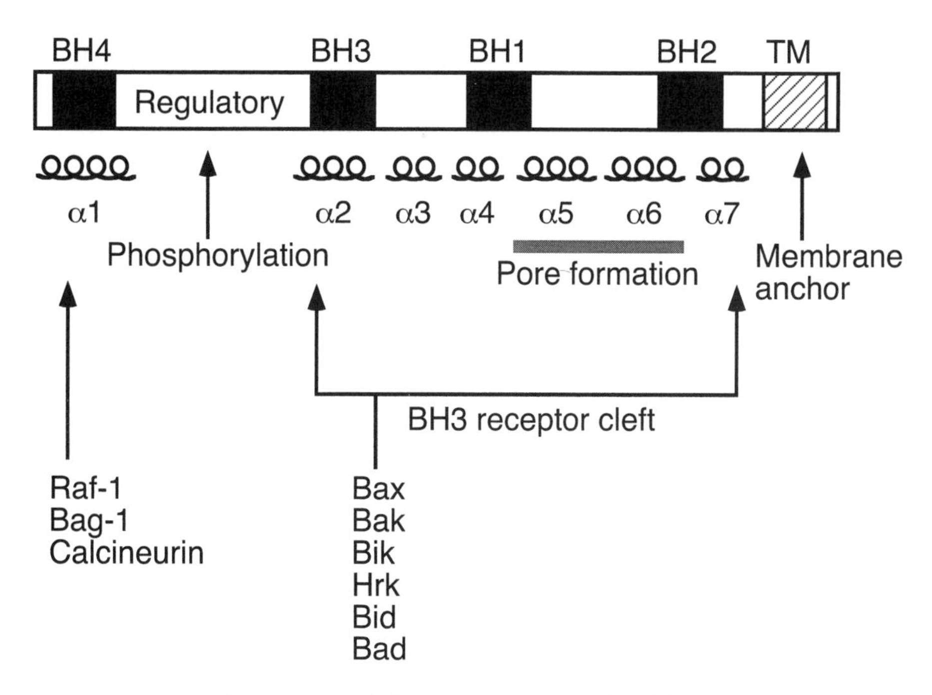

Figure 4. Multi-functional domain structure of Bcl-2.
The predicted structure of Bcl-2 includes seven α-helical domains, with α5 and α6 comprising the likely membrane pore-forming helices. The BH3, BH1 and BH2 domains contribute to the formation of a hydrophobic cleft which acts as a receptor for BH3 domain ligands presented by multiple pro-apoptotic proteins. BH4 mediates the interaction of Bcl-2 with both a protein kinase and a phosphatase (Raf-1 and calcineurin). A regulatory loop region is dispensable for anti-apoptotic activity, but phosphorylation of one or more residues within this domain appears to modulate the function of Bcl-2.

Mutational studies of Bcl-2 have also implicated a role for phosphorylation in the negative regulation of its activity. The three-dimensional structure of Bcl-xL revealed a 60 amino acid region of undefined structure between the BH4 and BH3 domains (Muchmore *et al.*, 1996), which is poorly conserved in Bcl-2 and largely absent from other homologues (depicted along with other Bcl-2 functional domains in Figure 4). Deletion of this flexible loop element in both Bcl-2 and Bcl-xL results in enhanced anti-apoptotic functions (Chang *et al.*, 1997). Most dramatically, wild-type Bcl-2 fails to protect the WEHI-231 B cell line from apoptosis induced by anti-IgM treatment, whereas the loop deletion mutant of Bcl-2 protects these cells from death in equivalent assays (Chang *et al.*, 1997). The gain of function effect of the loop deletion in Bcl-2 is associated with a significantly impaired ability of Bcl-2 to be phosphorylated in metabolic labelling experiments, indirectly suggesting that phosphorylation within the loop region negatively regulates the anti-apoptotic function of Bcl-2. The loop region encompasses serine 70 which, paradoxically, was implicated as a phosphorylation site important to anti-apoptotic function. Serine and threonine residues within this loop region of Bcl-2 can be phosphorylated by c-Jun N-terminal kinase/stress-activated protein kinases, at least *in vitro* and in transiently co-transfected cells (Maundrell *et al.*, 1997) although direct evidence for the phosphorylation of endogenous Bcl-2 on these residues is still lacking.

Many fundamental questions regarding the phosphorylation of Bcl-2 remain unanswered, including the precise definition of the relevant sites of phosphorylation in cells, and the kinases involved. It is also unclear how phosphorylation translates into an altered function of Bcl-2. Taxol-induced hyper-phosphorylation of Bcl-2 is associated with a decreased ability to co-precipitate Bax, suggesting that phosphorylation impairs the ability of Bcl-2 to heterodimerise with pro-apoptotic homologues or perhaps other proteins (Haldar *et al.*, 1996). By contrast, mutation of serine 70 to alanine and consequent loss of IL-3 induced phosphorylation has no discernible effect on the ability of Bcl-2 to interact with Bax (Ito *et al.*, 1997). Similarly, deletion of the unstructured loop elements of either Bcl-xL or Bcl-2 has no impact on their ability to interact with Bax and Bak (Chang *et al.*, 1997). Whether the phosphorylation of Bcl-2 affects its interaction with Ced-4 or its channel forming activity has not yet been examined.

5.0 ROLE OF BCL-2 FAMILY MEMBERS IN CANCER

Since the discovery of the role of Bcl-2 in lymphoma, a large number of studies have sought to examine whether it is more broadly involved in other human cancers. These efforts have typically involved surveying the levels of Bcl-2 in cancer cells and determining whether its expression is related to the survival or resistance of tumour cells to chemotherapy, or clinical outcome. The first clear link between Bcl-2 and malignancy was established in lymphoid tumours where the t(14;18) translocation and activation of Bcl-2 occurs in approximately 80% of non-Hodgkin's follicular lymphomas and about 20% of aggressive diffuse B-cell lymphomas. The pivotal role that this translocation plays in the development of lymphoma was directly demonstrated by experiments with mice that express a *bcl-2/Ig* transgene which mimics the t(14;18) fusion (McDonnell and Korsmeyer, 1991). These transgenic mice develop polyclonal follicular hyperplasia which, after a period of latency, progresses to monoclonal high grade large cell lymphomas. The latency period and evolution to lymphoma indicates that Bcl-2 hyper-expression *per se* does not cause malignancy; it is clear that secondary mutations such as the rearrangement and activation of *myc* are required for lymphomagenesis (McDonnell and Korsmeyer, 1991). The expansion of Bcl-2 expressing cells may increase the likelihood of sustaining such a secondary mutation, or, Bcl-2 may facilitate the acquisition of secondary hits since Bcl-2 over-expressing cells display enhanced susceptibility to mutation (Cherbonnel-Lasserre *et al.*, 1996) possibly because they are less likely to undergo apoptosis in response to DNA damage or genomic instability.

Elevated expression of Bcl-2 has been detected in other lymphoid malignancies including diffuse large B-cell lymphoma and chronic lymphocytic leukaemia, correlating with more frequent relapse and poorer disease-free survival (Hanada *et al.*, 1993; Hermine *et al.*, 1996; Hill *et al.*, 1996; Kramer *et al.*, 1996; Robertson *et al.*, 1996). Additionally, high-level expression of Bcl-2 has been detected in acute lymphoblastic leukaemia (Campana *et al.*, 1993) and in acute myeloid leukaemia, where it is predictive of poor response to chemotherapy and reduced survival (Campos *et al.*, 1993; Maung *et al.*, 1994; Stoetzer *et al.*, 1996). The relative ratio of Bcl-2 and Bax proteins in leukemic cells correlates with their sensitivity to chemotherapeutic drugs and is consistent with the proposed interplay between these two proteins in determining susceptibility to apoptosis (McConkey *et al.*, 1996; Pepper *et al.*, 1996; Thomas *et al.*, 1996; Salomons *et al.*, 1997).

There is accumulating evidence for the involvement of Bcl-2 in cancers of epithelial origin from solid tissues (reviewed in Lu *et al.*, 1996). Although highly variable, elevated expression of Bcl-2 has been documented in thyroid, nasopharyngeal, gastrointestinal, colorectal, renal, ovarian, cervical and endometrial carcinomas. In metastatic prostate cancer, the emergence of androgen-independent cancer is associated with high-level expression of Bcl-2 (McDonnell *et al.*, 1992; Colombel *et al.*, 1993; Apakama *et al.*, 1996). Prostate cancer cells initially undergo apoptosis in response to androgen ablation treatments, however, high level Bcl-2 expression in a fraction of the cancer cells confers an androgen-independent phenotype and allows them to escape hormone ablation therapy (Raffo *et al.*, 1995). Bcl-2 is abnormally expressed in a fraction of non-small cell lung carcinomas, but in this case its expression correlates with a more favourable prognosis (Pezzella *et al.*, 1993). Likewise, aberrant expression of Bcl-2 in breast carcinoma is associated with low histological grade and improved response to chemotherapy. This may be explained by the positive correlation of Bcl-2 expression with oestrogen receptor and wild-type p53 status, markers which independently predict a less aggressive cancer and more favourable outcome (Joensuu *et al.*, 1994; Leek *et al.*, 1994; Silvestrini *et al.*, 1994).

Elevated expression of Bcl-2 does not, by itself, establish whether it plays an essential role in the evolution or maintenance of the tumour in question. A number of studies have employed antisense strategies to inhibit Bcl-2 expression in order to examine its relevance to the survival and/or chemoresistance of cancer cells. Anti-sense oligonucleotides directed against Bcl-2 suppress the development of lymphoma in a SCID mouse model (Cotter *et al.*, 1994), reduce the viability of leukemic cells in culture (Reed *et al.*, 1990; Campos *et al.*, 1994) and sensitise them to chemotherapeutic drugs (Kitada *et al.*, 1994; Keith *et al.*, 1995). Inhibition of Bcl-2 expression by either anti-sense or ribozyme approaches restores sensitivity of prostate cancer cell lines to apoptotic signals (Dorai *et al.*, 1997) and reverses the protective effect of androgens (Berchem *et al.*, 1995). Furthermore, inhibition of Bcl-2 expression in small-cell lung cancer cell lines by anti-sense treatment significantly reduces viability of these tumour cells *in vitro* (Ziegler *et al.*, 1997). Taken together, the results of antisense Bcl-2 experiments suggest a widespread role for Bcl-2 in contributing to tumour cell survival and resistance to chemotherapy. In support of this possibility, a recombinant adenovirus that expresses Bcl-xS, induces apoptosis in carcinoma cells from solid tissues including breast and colon (Clarke *et al.*, 1995). This implicates the involvement of one or more Bcl-2 homologues in the survival of these tumour cells since Bcl-xS likely functions as a dominant inhibitor of Bcl-2 and related anti-apoptotic proteins.

There is already reason to suspect that the deregulation of other Bcl-2 family members, which function as equally potent suppressors of apoptosis, may likewise contribute to malignancy. This possibility has been demonstrated experimentally in a transgenic mouse model of multi-step tumourigenesis driven by oncogene expression in pancreatic islet ß-cells (Naik *et al.*, 1996). Bcl-xL is expressed at low levels in normal, hyperplastic, and angiogenic islets, but is significantly upregulated in tumours, operating to suppress oncogene-induced apoptosis and enhancing progression to the end-stage solid tumour. In human cancers, elevated expression of Bcl-xL has been detected in neuroblastoma cell lines (Dole *et al.*, 1995), Kaposi's sarcoma tumour cells (Foreman *et al.*, 1996), in primary gastric and colorectal carcinomas (Kondo *et al.*, 1996; Krajewska *et al.*, 1996), and in a significant fraction of primary breast cancers (Olopade *et al.*, 1997). Unlike Bcl-2, however, over-expression of Bcl-xL in breast cancer is associated with increased lymph node metastasis and high tumour grade, and does not correlate with estrogen receptor or p53 status. At least one other anti-apoptotic Bcl-2 homologue, A1, is expressed at high levels in haematopoeitic malignancies, melanomas, and in stomach cancers (Choi *et al.*, 1995; Kenny *et al.*, 1997).

In principle, the oncogenic anti-apoptotic function contributed by over-expression of Bcl-2 in tumour cells, might be mimicked by the loss of expression of Bcl-2 homologues that normally antagonise Bcl-2 and promote apoptosis. Loss of Bax expression in tumour cells, for example, may cause a defect in the regulation of apoptosis functionally equivalent to the hyperexpression of Bcl-2. In support of this scenario, the absence of Bax attenuates p53-dependent apoptosis, promotes oncogenic transformation of primary fibroblasts by E1a and Ras, and diminishes sensitivity to chemotherapeutic drugs (McCurrach *et al.*, 1997). Bax deficiency suppresses apoptosis and markedly accelerates oncogene-induced tumourigenesis in a transgenic mouse model, demonstrating that *bax* can act as a tumour suppressor gene *in vivo* (Yin *et al.*, 1997). Down regulation of Bax expression has been observed in several human cancers including certain colon and gastric carcinomas (Rampino *et al.*, 1997; Yamamoto *et al.*, 1997). Reduced Bax expression has also been noted in breast cancers and is associated with shorter survival and resistance to chemotherapy (Bargou *et al.*, 1995b; Krajewski *et al.*, 1995a). The expression of a second pro-apoptotic Bcl-2 homologue, Bak, appears to be decreased in colorectal cancers (Krajewska *et al.*, 1996), however, the potential role of other recently identified Bcl-2 homologues in cancer has not yet been investigated.

5.1 Mechanisms of deregulation

Elevated Bcl-2 expression in follicular lymphoma occurs through the transcriptional activation of the *bcl-2* promoter. The t(14;18) translocation, which may be caused by an aberrant Ig V/D/J splicing event, places the *bcl-2* proto-oncogene which normally resides at 18q21, adjacent to the JH region of the immunoglobulin heavy chain locus at chromosomal segment 14q32. The translocation results in a hybrid *bcl-2/IgH* gene fusion which encodes a wild-type Bcl-2 open reading frame but is expressed at inappropriately high levels due to a significantly increased rate of transcription (Seto *et al.*, 1988). The elevated expression of the translocated *bcl-2* gene is probably due to the presence of strong transcriptional enhancer elements within the IgH region that are active in lymphoid cells and de-regulate the normal control of the *bcl-2* promoter. *In vivo* footprinting experiments demonstrate, for example, that the WT1 binding site responsible for negatively regulating expression of the normal *bcl-2* allele, is unoccupied in the translocated allele in lymphoma cells (Heckman *et al.*, 1997).

The t(14;18) translocation appears to be unique to the development of lymphoma and has generally not been detected in other malignancies where expression of Bcl-2 is inappropriately elevated (e.g. see Hanada *et al.*, 1993; Kallakury *et al.*, 1996). Other mechanisms for activating the expression of Bcl-2 in cancer cells must exist but remain poorly defined. Changes in the activity of transcription factors that either negatively or positively regulate *bcl-2* gene expression may play a role. In particular, the function of the p53 tumour suppresser is frequently lost in human cancers and its absence may enhance expression of Bcl-2, by relieving transcriptional repression mediated by p53 (Miyashita *et al.*, 1994b). Loss of p53 function may also be responsible in part for reduced expression of *bax*, which is a direct target for p53 transactivation (Miyashita and Reed, 1995). The expression profiles of Bcl-2 and Bax in tumour cells, however, do not always correlate with p53 status as predicted by these possibilities (Krajewski *et al.*, 1997). Potentially, elevated activity of CREB proteins or other transcription factors that positively regulate *bcl-2* gene expression may contribute to increased Bcl-2 levels in cancer cells.

Somatic mutation of Bcl-2 family members may have an impact on the regulation of apoptosis in tumour cells. Point mutations within the Bcl-2 open reading frame have been detected in non-Hodgkin's lymphomas bearing the t(14;18) translocation (Seto *et al.*, 1988; Tanaka *et al.*, 1992; Reed and Tanaka, 1993; Matolcsy *et al.*, 1996) and may be the consequence of somatic hypermutation activity associated with the Ig locus. The effect, if any, of these amino acid substitutions on the biological activity of Bcl-2 remains uncertain. Conceivably, point mutations in Bcl-2 might enhance its potency by affecting its phosphorylation or protein binding functions, or may abrogate its anti-proliferative activity, and confer a selective advantage for tumour cells harbouring the mutant Bcl-2. Somatic mutations of *bax* have also been detected in tumour cells and in this case, the functional consequences are more clear. Approximately 50% of colon adenocarcinomas of the microsatellite mutator phenotype (MMP), which results in DNA replication errors (single base insertion or deletions) within short repeats, have frameshift mutations in the *bax* gene (Rampino *et al.*, 1997). These mutations fall within a polydeoxyguanosine tract spanning codons 38 to 41 of *bax*, and inactivate the protein by causing a reading frame shift early in the coding region prior to the BH3 domain. About one fourth of gastrointestinal cancers of the MMP contain *bax* frameshift mutations at this hotspot, and Bax mutations are also detected in cancers without the MMP (Yamamoto *et al.*, 1997). Interestingly, point mutations in *bax* identified in a colon cancer, Asp 68 to Val, and in a T-cell lymphoma cell line, Gly 67 to Arg, alter key residues within the BH3 domain of *bax* and abrogate its heterodimerisation and pro-apoptotic functions (Meijerink *et al.*, 1995; Zha *et al.*, 1996b; Yamamoto *et al.*, 1997).

The emerging role of phosphorylation in regulating the activity of Bcl-2 homologues represents another mechanism by which cell death may be aberrantly suppressed in tumour cells. Serine phosphorylation of Bad blocks its ability to bind and inhibit the function of Bcl-2 and Bcl-xL, resulting in the activation the anti-apoptotic function of these cell death suppressors (Zha *et al.*, 1996c). Bad phosphorylation is induced by anti-apoptotic cytokines such as IGF-I, through a PI-3 kinase/Akt pathway (Datta *et al.*, 1997). It is well known that many human tumours, particularly cancers of the breast and lung, express high levels of the IGF-I receptor which is inappropriately activated through autocrine/paracrine mechanisms and contributes to the transformed phenotype (see reviews by Baserga, 1995; Werner and LeRoith, 1996). It might be anticipated therefore that signalling by elevated IGF-I receptors in tumour cells may result in enhanced, or constitutive phosphorylation of Bad and activation of Bcl-2/Bcl-xL through the loss of heterodimerisation with Bad. There is at present no evidence for the hyperphosphorylation of Bad in cancer cells, however, this may ultimately prove to be a common way in which Bcl-2 activity is enhanced in tumour cells. Activation of Bcl-2 at this level would be missed in the numerous studies that have surveyed cancer cells for overt changes in the levels of either anti- or pro-apoptotic Bcl-2 homologues.

5.2 Potential therapeutic application in cancer

A major challenge facing cell-death researchers is to translate basic discoveries regarding the molecular biology of apoptosis into improved therapies for cancer. The widespread involvement of Bcl-2 in cancer, coupled with an understanding of how its function is aberrantly activated, has motivated efforts to develop drugs that inhibit the function of Bcl-2 or counteract its deregulation, thus restoring susceptibility to apoptosis in tumour cells. These agents may have therapeutic value by either triggering tumour cells to undergo apoptosis, or by enhancing their sensitivity to other traditional chemotherapeutic drugs. Ultimately, the success of such approaches may depend on a more refined understanding of which Bcl-2 homologues are relevant to specific cancers and their relative contribution to the survival of normal cells versus tumour cells. Part of the optimism for a good therapeutic index is based on the evidence that tumour cells are inherently more dependent on anti-apoptotic signals, due to strong intrinsic pro-apoptotic signals linked to their uncontrolled growth (e.g. activation of oncogenes).

Strategies for inhibiting Bcl-2 in cancer cells could exploit known mechanisms that regulate its function, even without a complete understanding of the molecular basis for suppression of cell death by Bcl-2. It may be possible to develop agents that inhibit the hyper-phosphorylation of Bad that may occur in tumour cells, or molecules that mimic the antagonistic effect of heterodimerisation with BH3-containing, pro-apoptotic Bcl-2 homologues. Such BH3 mimics could bind the BH3-receptor cleft in Bcl-2 and antagonise its function by preventing interaction with other cell death regulatory proteins. A more direct approach of inhibiting the expression of Bcl-2 with anti-sense oligonucleotides is already being tested in clinical trials for treatment of lymphoma, with initial data suggestive of an anti-tumour response in some patients (Webb *et al.*, 1997). Other approaches for developing inhibitors of Bcl-2 and related proteins will emerge as the mechanism of action of these proteins is better defined. There are some promising leads for finally understanding the anti-apoptotic mechanism of Bcl-2 at a molecular level, with the discovery of its membrane pore-forming activity and potential connection to Ced-4. Both of these activities provide defined biochemical handles for addressing mechanistic questions, something that has been sorely lacking in the field, and research in these areas will no doubt progress rapidly.

6.0 REFERENCES

Akbar,A.N., Borthwick,N.J., Wickremasinghe,R.G *et al.* (1996) Interleukin-2 receptor common γ-chain signaling cytokines regulate activated T cell apoptosis in response to growth factor withdrawal: selective induction of anti-apoptotic (bcl-2, bcl-xL) but not pro-apoptotic (bax, bcl-xS) gene expression. *Eur. J. Immunol.* **26**: 294-299.

Antonsson,B., Conti,F., Ciavatta,A., *et al.* (1997) Inhibition of Bax channel-forming activity by Bcl-2. *Science* **277**: 370-372.

Apakama,I., Robinson,M.C., Walter,N.M. *et al.* (1996) bcl-2 expression combined with p53 protein accumulation correlates with hormone-refractory prostate cancer. *Br. J. Cancer* **74**: 1258-1262.

Baffy,G., Miyashita,T., Williamson,J.R. and Reed,J.C. (1993) Apoptosis induced by withdrawal of interleukin-3 (IL-3) from an IL-3-dependent hematopoietic cell line is associated with repartitioning of intracellular calcium and is blocked by enforced Bcl-2 oncoprotein production. *J. Biol. Chem.* **268**: 6511-6519.

Bargou,R.C., Bommert,K., Weinmann,P. *et al.* (1995a) Induction of Bax-α precedes apoptosis in a human B lymphoma cell line: potential role of the bcl-2 gene family in surface IgM-mediated apoptosis. *Eur. J. Immunol.* **25**: 770-775.

Bargou,R.C., Daniel,P.T., Mapara,M.Y. *et al.* (1995b) Expression of the bcl-2 gene family in normal and malignant breast tissue: low bax-alpha expression in tumor cells correlates with resistance towards apoptosis. *Int. J. Cancer* **60**: 854-859.

Baserga,R. (1995) The insulin-like growth factor I receptor: a key to tumor growth? *Cancer Res.* **55**: 249-252.

Beham,A., Marin,M.C., Fernandez,A. *et al.* (1997) Bcl-2 inhibits p53 nuclear import following DNA damage. *Oncogene* **15**: 2767-2772.

Berchem,G.J., Bosseler,M., Sugars,L.Y. *et al.* (1995) Androgens induce resistance to bcl-2-mediated apoptosis in LNCaP prostate cancer cells. *Cancer Res.* **55**: 735-738.

Blagosklonny,M.V., Giannakakou,P., El-Deiry,W.S. *et al.* (1997) Raf-1/bcl-2 phosphorylation: a step from microtubule damage to cell death. *Cancer Res.* **57**: 130-135.

Blagosklonny,M.V., Schulte,T., Nguyen,P. *et al.* (1996) Taxol-induced apoptosis and phosphorylation of Bcl-2 protein involves c-Raf-1 and represents a novel c-Raf-1 signal transduction pathway. *Cancer Res.* **56**: 1851-1854.

Boise,L.H., Gonzalez-Garcia,M., Postema,C.E. *et al.* (1993) Bcl-x, a bcl-2-related gene that functions as a dominant regulator of apoptotic cell death. *Cell* **74**: 597-608.

Boise,L.H., Minn,A.J., Noel,P.J. *et al.* (1995) CD28 co-stimulation can promote T cell survival by enhancing the expression of Bcl-xL. *Immunity* **3**: 87-98.
Borner,C. (1996) Diminished cell proliferation associated with the death-protective activity of Bcl-2. *J. Biol. Chem.* **271**: 12695-12698.
Borner,C., Martinou,I., Mattmann,C. *et al.* (1994) The protein bcl-2α does not require membrane attachment, but two conserved domains to suppress apoptosis. *J. Cell Biol.* **126**: 1059-1068.
Boyd,J.M., Gallo,G.J., Elangovan,B., *et al.* (1995) Bik, a novel death-inducing protein shares a distinct sequence motif with Bcl-2 family proteins and interacts with viral and cellular survival-promoting proteins. *Oncogene* **11**: 1921-1928.
Boyd,J.M., Malstrom,S., Subramanian,T. *et al.* (1994) Adenovirus E1B 19kDa and Bcl-2 proteins interact with a common set of cellular proteins. *Cell* **79**: 341-351.
Brady,H.J., Gil-Gomez,G., Kirberg,J. and Berns,A.J. (1996) Bax alpha perturbs T cell development and affects cell cycle entry of T cells. *EMBO J.* **15**: 6991-7001.
Campana,D., Coustan-Smith,E., Manabe,A. *et al.* (1993) Prolonged survival of B-lineage acute lymphoblastic leukemia cells is accompanied by overexpression of Bcl-2 protein. *Blood* **81**: 1025-1031.
Campos,L., Rouault,J.-P., Sabido,O *et al.* (1993) High expression of bcl-2 protein in acute myeloid leukemia cells is associated with poor response to chemotherapy. *Blood* **81**: 3091-3096.
Campos,L., Sabido,O., Rouault,J.-P., and Guyotat,D. (1994) Effects of Bcl-2 antisense oligodeoxynucleotides on *in vitro* proliferation and survival of normal marrow progenitors and leukemic cells. *Blood* **84**: 595-600.
Carrió,R., Lopez-Hoyos,M., Jimeno,J. *et al.* (1996b) A1 demonstrates restricted tissue distribution during embryonic development and functions to protect against cell death. *Am. J. Pathol.* **149**: 2133-2142.
Chang,B.S., Minn,A.J., Muchmore,S.W. *et al.* (1997) Identification of a novel regulatory domain in Bcl-x_L and Bcl-2. *EMBO J.* **16**: 968-977.
Chao,D.T., Linette,G.P., Boise,L.H. *et al.* (1995) Bcl-xL and Bcl-2 repress a common pathway of cell death. *J. Exp. Med.* **182**: 821-828.
Chen,C.-Y. and Faller,D.V. (1996) Phosphorylation of Bcl-2 protein and association with p21ras in ras-induced apoptosis. *J. Biol. Chem.* **271**: 2376-2379.
Chen,G., Ray,R., Dubik,D. *et al.* (1997) The E1B 19K/Bcl-2 binding protein Nip3 is a dimeric mitochondrial protein that activates apoptosis. *J. Exp. Med.* **186**: 1975-1983.
Chen,H.-M. and Boxer, L.M. (1995) II-1 binding sites are negative regulators of bcl-2 expression in pre-B cells. *Mol. Cell. Biol.* **15**: 3840-3847.
Chen-Levy,Z., Nourse,J., and Cleary,M.L. (1989) The Bcl-2 candidate proto-oncogene product is a 24-kilodalton integral membrane protein highly expressed in lymphoid cell lines and lymphomas carrying the t(14:18). *Mol. Cell. Biol.* **9**: 701-710.
Cheng,E.H.-Y., Levine,B., Boise,L.H. *et al.* (1996) Bax-independent inhibition of apoptosis by Bcl-xL. *Nature* **379**: 554-556.
Cheng,E.H.-Y., Nicholas,J., Bellows,D.S. *et al.* (1997) A Bcl-2 homologue encoded by Kaposi sarcoma-associated virus, human herpesvirus 8, inhibits apoptosis but does not heterodimerize with Bax or Bak. *Proc. Natl. Acad. Sci.USA* **94**: 690-694.
Cherbonnel-Lasserre,C., Gauny,S. and Kronenberg,A. (1996) Suppression of apoptosis by Bcl-2 or Bcl-xL promotes susceptibility to mutagenesis. *Oncogene* **13**:1489-1497.
Chinnaiyan,A.M., Chaudhary,D., O'Rourke,K. *et al.* (1997a) Role of Ced-4 in the activation of Ced-3. *Nature* **388**: 728-729.
Chinnaiyan,A.M., O'Rourke,K., Lane,B.R. and Dixit,V.M. (1997b) Interaction of Ced-4 with Ced-3 and Ced-9: a molecular framework for cell death. *Science* **275**: 1122-1126.
Chinnaiyan,A.M., Orth,K., O'Rourke,K. *et al.* (1996) Molecular ordering of the cell death pathway. Bcl-2 and Bcl-xL function upstream of the Ced-3-like apoptotic proteases. *J. Biol. Chem.* **271**: 4573-4576.
Chiou,S.-K., Tseng,C.-C., Rao,L. and White, E. (1994) Functional complementation of the adenovirus E1B 19-Kilodalton protein with Bcl-2 in the inhibition of apoptosis in infected cells. *J. Virol.* **68**: 6553-6566.
Chittenden,T., Flemington,C., Houghton,A.B. *et al.* (1995a) A conserved domain in Bak, distinct from BH1 and BH2, mediates cell death and protein binding functions. *EMBO J.* **14**: 5589-5596.
Chittenden,T., Harrington,E.A., O'Connor,R. *et al.* (1995b) Induction of apoptosis by the Bcl-2 homologue Bak. *Nature* **374**: 733-736.
Choi,S.S., Park,I.-C., Yun,J.W. *et al.* (1995) A novel Bcl-2 related gene, Bfl-1, is overexpressed in stomach cancer and preferentially expressed in bone marrow. *Oncogene* **11**: 1693-1698.
Clarke,M.F., Apel,I.J., Benedict,M.A. *et al.* (1995) A recombinant bcl-xs adenovirus selectively induces apoptosis in cancer cells but not in normal bone marrow cells. *Proc. Natl. Acad. Sci. USA* **92**: 11024-11028.
Cleary,M. L., Smith,S.D. and Sklar,J. (1986) Cloning and structural analysis of cDNA's from bcl-2 and a hybrid bcl-2/immunoglobulin transcript resulting from the t(14:18) translocations. *Cell* **47**: 19-28.
Colombel,M., Symmans,F., Gil,S. *et al.* (1993) Detection of the apoptosis-suppressing oncoprotein bcl-2 in hormone-refractory human prostate cancers. *Am. J. Pathol.* **143**: 390-400.
Cory,S. (1995) Regulation of lymphocyte survival by the Bcl-2 gene family. *Ann. Rev. Immunol.* **13**: 513-543.

Cotter,F.E., Johnson,P., Hall,P. *et al.* (1994) Antisense oligonucleotides suppress B-cell lymphoma growth in a SCID-hu mouse model. *Oncogene* **9**: 3049-3055.
Cruz-Reyes,J. and Tata,J.R. (1995) Cloning, characterization and expression of two Xenopus bcl-2-like cell-survival genes. *Gene* **158**: 171-179.
Datta,S., Dudek,H., Tao,X. *et al.* (1997) Akt phosphorylation of Bad couples survival signals to the cell-intrinsic death machinery. *Cell* **91**: 231-241.
del Peso,L., Gonzalez-Garcia,M., Page,C. *et al.* (1997) Interleukin-3-induced phosphorylation of Bad through the protein kinase Akt. *Science* **278**: 687-689.
Diaz,J.-L., Oltersdorf,T., Haorne,W. *et al.* (1997) A common binding site mediates heterodimerization and homodimerization of Bcl-2 family members. *J. Biol. Chem.* **272**: 11350-11355.
Dole,M.G., Jasty,R., Cooper,M.J. *et al.* (1995) Bcl-xL is expressed in neuroblastoma cells and modulates chemotherapy-induced apoptosis. *Cancer Res.* **55**: 2576-2582.
Dorai,T., Olsson,C.A., Katz,A.E. and Buttyan,R. (1997) Development of a hammerhead ribozyme against bcl-2. I. Preliminary evaluation of a potential gene therapeutic agent for hormone-refractory human prostate cancer. *Prostate* **32**: 246-258.
Elangovan,B. and Chinnadurai,G. (1997) Functional dissection of the pro-apoptotic protein Bik. *J. Biol. Chem.* **272**: 24494-24498.
Evan,G.I., Wyllie,A.H., Gilbert,C.S. *et al.* (1992) Induction of apoptosis in fibroblasts by c-myc protein. *Cell* **69**: 119-128.
Fanidi,A., Harrington,E.A. and Evan,G. (1992) Cooperative interaction between c-myc and bcl-2 proto-oncogenes. *Nature* **359**: 554-556.
Farrow,S.N., White,J.H.M., Martinou,I. *et al.* (1995) Cloning of a *bcl-2* homologue by interaction with adenovirus E1B 19K. *Nature* **374**: 731-733.
Fernanadez-Sarabia,M.J. and Bischoff,J.R. (1993) Bcl-2 associates with the ras-related protein R-ras p23. *Nature* **366**: 274-275.
Foreman,K.E., Wrone-Smith,T., Boise,L.H. *et al.* (1996) Kaposi's sarcoma tumor cells preferentially express Bcl-xL. *Am. J. Pathol.* **149**: 795-803.
Frisch,S.M. and Francis,H. (1994) Disruption of epithelial cell-matrix interactions induce apoptosis. *J. Cell Biol.* **124**: 619-626.
Gibson,L., Holmgreen,S.P., Huang,D.C. *et al.* (1996) *bcl-w*, a novel member of the *bcl-2* family, promotes cell survival. *Oncogene* **13**: 665-675.
Gillet,G., Guerin,M., Trembleau,A., and Brun,G. (1995) A Bcl-2-related gene is activated in avian cells transformed by the Rous sarcoma virus. *EMBO J.* **14**: 1372-1381.
Gonzalez-Garcia,M., Perez-Ballestero,R., Ding,L. *et al.* (1994) Bcl-xL is the major bcl-x mRNA form expressed during murine development and its product localizes to mitochondria. *Development* **120**: 3033-3042.
Grillot,D.A.M., González-Garcia,M., Ekhterae,D. *et al.* (1997) Genomic organization, promoter region analysis, and chromosome localization of the mouse bcl-x gene. *J. Immunol.* **158**: 4750-4757.
Grimes,H.L., Gilks,C.B., Chan,T.O. *et al.* (1996) The Gfi-1 protooncoprotein represses Bax expression and inhibits T-cell death. *Proc. Natl. Acad. Sci. USA* **93**: 14569-14573.
Guan,R.J., Moss,S.F., Arber,N. *et al.* (1996) 30 KDa phosphorylated form of Bcl-2 protein in human colon. *Oncogene* **12**: 2605-2609.
Haldar,S., Basu,A. and Croce,C.M. (1997) Bcl-2 is the guardian of microtubule integrity. *Cancer Res.* **57**: 229-233.
Haldar,S., Chintapalli,J. and Croce,C.M. (1996) Taxol induces bcl-2 phosphorylation and death of prostate cancer cells. *Cancer Res.* **56**: 1253-1255.
Haldar,S., Jena,N. and Croce,C.M. (1995) Inactivation of Bcl-2 by phosphorylation. *Proc. Natl. Acad. Sci. USA.* **92**: 4507-4511.
Han,J., Sabbatini,P., Perez,D. *et al.* (1996a) The E1b 19K protein blocks apoptosis by interacting with and inhibiting the p53-inducible and death-promoting Bax protein. *Gen. Dev.* **10**: 461-477.
Han,J., Sabbatini,P. and White,E. (1996b) Induction of apoptosis by human Nbk/Bik, a BH3-containing protein that interacts with E1b 19kD. *Mol. Cell. Biol.*, **16**: 5857-5864.
Hanada,M., Aime-Sempe,C., Sato,T. and Reed,J.C. (1995) Structure-function analysis of Bcl-2 protein. *J. Biol. Chem.* **270**: 11962-11969.
Hanada,M., Delia,D., Aiello,A. *et al.* (1993) Bcl-2 gene hypomethylation and high-level expression in B-cell chronic lymphocytic leukemia. *Blood* **82**: 1820-1828.
Harigai,M., Miyashita,T., Hanada,M. and Reed,J.C. (1996) A cis-acting element in the bcl-2 gene controls expression through translational mechanisms. *Oncogene* **12**: 369-374.
Harrington,E.A., Bennett,M.R., Fanidi,A., and Evan,G.I. (1994) c-Myc-induced apoptosis in fibroblasts is inhibited by specific cytokines. *EMBO J.* **13**: 3286-3295.
Hawkins,C.J. and Vaux,D.L. (1997) The role of the Bcl-2 family of apoptosis regulatory proteins in the immune system. *Semin. Immunol.* **9**: 25-33.
Heckman,C., Mochon,E., Arcinas,M. and Boxer,L.M. (1997) The WT1 protein is a negative regulator of the normal bcl-2 allele in t(14;18) lymphomas. *J. Biol.Chem.* **272**: 19609-19614.
Heiden,M.G.V., Chandel,N.S., Williamson,E.K. *et al.* (1997) Bcl-xL regulates the membrane potential and volume homeostasis of mitochondria. *Cell* **91**: 627-637.
Henderson,S., Huen,D., Rowe,M. *et al.* (1993) Epstein-Barr virus BHRF1 protein, a viral homologue of Bcl-2, protects human B cells from programmed cell death. *Proc. Natl. Acad. Sci. USA.* **90**: 8479-8488.

Hengartner,M.O. and Horvitz,H.R. (1994) C. elegans cell survival gene ced-9 encodes a functional homologue of the mammalian proto-oncogene bcl-2. *Cell* **76**: 665-676.
Hermine,O., Haioun,C., Lepage,E. *et al.* (1996) Prognostic significance of bcl-2 protein expression in aggressive non-Hodgkin's lymphoma. *Blood* **87**: 265-272.
Hewitt,S.M., Hamada,S., McDonnell,T.J. *et al.* (1995) Regulation of the proto-oncogenes bcl-2 and c-myc by the Wilms's tumor suppressor gene WT1. *Cancer Res.* **55**: 5386-5389.
Hill,M.E., MacLennan,K.A., Cunningham,D.C. *et al.* (1996) Prognostic significance of Bcl-2 expression and bcl-2 major breakpoint region rearrangement in diffuse large cell non-Hodgkin's lymphoma: a British National lymphoma investigation study. *Blood* **88**: 1046-1051.
Hockenbery,D., Nunez,G., Milliman,C. *et al.* (1990) Bcl-2 is an inner mitochondrial membrane protein that blocks programmed cell death. *Nature* **348**: 334-336.
Hockenbery,D.M., Oltvai,Z.N., Yin,X. *et al.* (1993) Bcl-2 functions in an antioxidant pathway to prevent apoptosis. *Cell* **75**: 241-251.
Hohfeld,J. and Jentsch,S. (1997) GrpE-like regulation of the hsc70 chaperone by the anti-apoptotic protein Bag-1. *EMBO J.* **16**: 6209-6216.
Horiuchi,M., Hayashida,W., KambeT. *et al.* (1997) Angiotensin type 2 receptor dephosphorylates Bcl-2 by activating mitogen-activated protein kinase phosphatase-1 and induces apoptosis. *J. Biol. Chem.* **272**: 19022-19026.
Hsu,S.Y., Kaipia,A., McGee,E. *et al.* (1997) Bok is a pro-apoptotic Bcl-2 protein with restricted expression in reproductive tissues and heterodimerizes with selective anti-apoptotic Bcl-2 family members. *Proc. Natl. Acad. Sci. USA* **94**: 12401-12406.
Huang,D.C.S., Cory,S. and Strasser,A. (1997a) Bcl-2, Bcl-xL and adenovirus E1B19kD are functionally equivalent in their ability to inhibit cell death. *Oncogene* **14**: 405-414.
Huang,D.C.S., O'Reilly,L.A., Strasser,A. and Cory,S. (1997b) The anti-apoptotic function of Bcl-2 can be genetically separated from its inhibitory effect on cell cycle entry. *EMBO J.* **16**: 4628-4638.
Hunter,J.J. (1996) A peptide sequence from Bax that converts Bcl-2 into an activator of apoptosis. *J. Biol. Chem.* **271**: 8521-8524.
Hunter,J.J., Bond,B.L. and Parslow,T.G. (1996) Functional dissection of the human Bcl2 protein: sequence requirements for inhibition of apoptosis. *Mol. Cell. Biol.* **16**: 877-883.
Ink,B., Zörnig,M., Baum,B. *et al.* (1997) Human Bak induces cell death in Schizosaccharomyces pombe with morphological changes similar to those with apoptosis in mammalian cells. *Mol. Cell. Biol.* **17**: 2468-2474.
Inohara,N., Ding,L., Chen,S. and Nuñez,G. (1997) harakiri, a novel regulator of cell death, encodes a protein that activates apoptosis and interacts selectively with survival-promoting proteins Bcl-2 and Bcl-XL. *EMBO J.* **16**: 1686-1694.
Ito,T., Deng,X., Carr,B. and May,W.S. (1997) Bcl-2 phosphorylation required for anti-apoptosis function. *J. Biol. Chem.* **272**: 11671-11673.
Iwahashi,H., Eguchi,Y., Yasuhara,N. *et al.* (1997) Synergistic anti-apoptotic activity between Bcl-2 and SMN implicated in spinal muscular atrophy. *Nature* **390**: 413-417.
Jacobson,M.D. and Raff,M.C. (1995) Programmed cell death and Bcl-2 protection in very low oxygen. *Nature* **374**: 814-816.
James,C., Gschmeissner,S., Fraser,A. and Evan,G.I. (1997) CED-4 induces chromatin condensation in Schizosaccharomyces pombe and is inhibited by direct physical association with CED-9. *Curr. Biol.* **7**: 246-252.
Joensuu,H., Pylkkanen,L. and Toikkanen,S. (1994) Bcl-2 protein expression and long-term survival in breast cancer. *Am. J. Pathol.* **145**: 1191-1198.
Jürgensmeier,J.M., Krajewski,S., Armstrong,R.C. *et al.* (1997) Bax- and Bak-induced cell death in the fission yeast Schizosaccharomyces pombe. *Mol. Biol. Cell* **8**: 325-339.
Kallakury,B.V.S., Figge,J., Leibovich,B. *et al.* (1996) Increased bcl-2 protein levels in prostatic adenocarcinomas are not associated with rearrangements in the 2.8 kb major breakpoint region or with p53 protein accumulation. *Mod. Pathol.* **9**: 41-47.
Kane,D.J., Sarafian,T.A., Anton,R. *et al.* (1993) Bcl-2 inhibition of neural death: decreased generation of reactive oxygen species. *Science* **262**: 1274-1277.
Karsan,A., Yee,E. and Harlan,J.M. (1996a) Endothelial cell death induced by Tumor Necrosis Factor-alpha is inhibited by the Bcl-2 family member, A1. *J. Biol. Chem.* **271**: 27201-27204.
Karsan,A., Yee,E., Kaushansky,K. and Harlan,J.M. (1996b) Cloning of a human Bcl-2 homologue: Inflammatory cytokines induce human A1 in cultured endothelial cells. *Blood* **87**: 3089-3096.
Keith,F.J., Bradbury,D.A., Zhu,Y.-M., and Russell,N.H. (1995) Inhibition of bcl-2 with antisense oligonucleotides induces apoptosis and increases the sensitivity of AML blasts to Ara-C. *Leukemia* **9**: 131-138.
Kelekar,A., Chang,B.S., Harlan,J E. *et al.* (1997) Bad is a BH3 domain-containing protein that forms an inactivating dimer with Bcl-xL. *Mol. Cell. Biol.***17**: 7040-7046.
Kenny,J.J., Knobloch,T.J., Augustus,M. *et al.* (1997) GRS, a novel member of the Bcl-2 gene family, is highly expressed in multiple cancer cell lines and in normal leukocytes. *Oncogene* **14**: 997-1001.
Kharbanda,S., Pandey,P., Schofield,L. *et al.* (1997) Role for Bcl-xL as an inhibitor of cytosolic cytochrome c accumulation in DNA damage-induced apoptosis. *Proc. Natl. Acad. Sci. USA* **94**: 6939-6942.
Kiefer,M.C., Brauer,M.J., Powers,V.C. *et al.* (1995) Modulation of apoptosis by the widely distributed Bcl-2 homologue Bak. *Nature* **374**: 736-739.

Kitada,S., Krajewski,S., Miyashita,T. *et al.* (1996) γ-Radiation induces upregulation of Bax protein and apoptosis in radiosensitive cells *in vivo*. *Oncogene* **12**: 187-192.
Kitada,S., Takayama,S., Riel,K.D. *et al.* (1994) Reversal of chemoresistance of lymphoma cells by antisense-mediated reduction of bcl-2 gene expression. *Antisense Res. Dev.* **4**: 71-79.
Kluck,R.M., Bossy-Wetzel,E.B., Green,D.R. and Newmeyer,D.D. (1997) The release of cytochrome c from mitochondria: a primary site for Bcl-2 regulation of apoptosis. *Science* **275**: 1132-1136.
Knudson,C.M., Tung, K.S.K., Tourtellotte,W.G. *et al.* (1995) Bax-deficient mice with lymphoid hyperplasia and male germ cell death. *Science* **270**: 96-99.
Knudson,G.M. and Korsmeyer,S.J. (1997) Bcl-2 and Bax function independently to regulate cell death. *Nat. Genet.* **16**: 358-363.
Kondo,S., Shinomura,Y., Kanayama,S. *et al.* (1996) Over-expression of bcl-xL gene in human gastric adenomas and carcinomas. *Int. J. Cancer* **68**: 727-730.
Kozopas,K.M., Yang,T., Buchan,H.L. *et al.* (1993) MCL1, a gene expressed in programmed myeloid cell differentiation, has sequence similarity to BCL2. *Proc. Natl. Acad. Sci. USA.* **90**: 3516-3520.
Krajewska,M., Moss,S.F., Krajewski,S. *et al.* (1996) Elevated expression of Bcl-x and reduced Bak in primary colorectal adenocarcinomas. *Cancer Res.* **56**: 2422-2427.
Krajewski,S., Blomqvist,C., Franssila,K. *et al.* (1995a) Reduced expression of proapoptotic gene Bax is associated with poor response rates to combination chemotherapy and shorter survival in women with metastatic breast adenocarcinoma. *Cancer Res.* **55**: 4471-4478.
Krajewski,S., Bodrug,S., Krajewska,M. *et al.* (1995b) Immunohistochemical analysis of Mcl-1 protein in human tissues. *Am. J. Pathol.* **146**: 1309-1319.
Krajewski,S., Krajewska,M. and Reed,J.C. (1996) Immunohistochemical analysis of *in vivo* patterns of Bak expression, a proapoptotic member of the Bcl-2 protein family. *Cancer Res.* **56**: 2849-2855.
Krajewski,S., Krajewska,M., Shabaik,A. *et al.* (1994) Immunohistochemical analysis of *in vivo* patterns of Bcl-x expression. *Cancer Res.* **54**: 5501-5507.
Krajewski,S., Mai,J.K., Krajewska,M. *et al.* (1995c) Upregulation of Bax protein levels in neurons following cerebral ischemia. *J. Neurosci.* **15**: 6364-6376.
Krajewski,S., Tanaka,S., Takayama,S. *et al.* (1993) Investigation of the subcellular distribution of the bcl-2 oncoprotein: residence in the nuclear envelope, endoplasmic reticulum, and outer mitochondrial membranes. *Cancer Res.* **53**: 4701-4714.
Krajewski,S., Thor,A.D., Edgerton,S.M. *et al.* (1997) Analysis of Bax and Bcl-2 expression in p53-immunopositive breast cancers. *Clin. Cancer Res.* **3**: 199-208.
Kramer,M.H.H., Hermans,J., Parker,J. *et al.* (1996) Clinical significance of bcl-2 and p53 protein expression in diffuse large B-cell lymphoma: a population-based study. *J. Clin. Oncol.* **14**: 2131-2138.
Kroemer,G., Zamzami,N. and Susin,S.A. (1997) Mitochondrial control of apoptosis. *Immunol. Today* **18**: 44-51.
Lam,M., Dubyak,G., Chen,L. *et al.* (1994) Evidence that Bcl-2 represses apoptosis by regulating endoplasmic reticulum-associated Ca2+ fluxes. *Proc. Natl. Acad. Sci. USA* **91**: 6569-6573.
Leek,R.D., Kaklamanis,L., Pezzella,F. *et al.* (1994) bcl-2 in normal human breast and carcinoma, association with estrogen receptor-positive, epidermal growth factor receptor-negative tumours and in situ cancer. *Br. J. Cancer* **69**: 135-139.
Li,P., Nijhawan,D., Budihardjo,I. *et al.* (1997) Cytochrome c and dATP-dependent formation of Apaf-1/Caspase-9 complex initiates an apoptotic protease cascade. *Cell* **91**: 479-489.
Lin,E.Y., Orlofsky,A., Berger,M.S. and Prystowsky,M.B. (1993) Characterization of A1, a novel hemopoietic-specific early-response gene with sequence similarity to bcl-2. *J. Immunol.* **151**: 1979-1988.
Lin,E.Y., Orlofsky,A., Wang,H.-G. *et al.* (1996) A1, a *bcl-2* family member, prolongs cell survival and permits myeloid differentiation. *Blood* **87**: 983-992.
Linette,G.P., Li,Y., Roth,K. and Korsmeyer, S. J. (1996) Cross talk between cell death and cell cycle progression: Bcl-2 regulates NFAT-mediated activation. *Proc. Natl. Acad. Sci. USA* **93**: 9545-9552.
Liu,X., Kim,C.N., Yang,J. *et al.* (1996) Induction of apoptotic program in cell-free extracts: requirement for dATP and cytochrome c. *Cell* **86**: 147-157.
Lomo,J., Blomhoff, H.K., Jacobsen, S.E. *et al.* (1997) Interleukin-13 in combination with CD40 ligand potently inhibits apoptosis in human B lymphocytes: upregulation of Bcl-xL and Mcl-1. *Blood* **89**: 4415-4424.
Longo,V.D., Ellerby, L.M., Bredesen,D.E. *et al.* (1997) Human Bcl-2 reverses survival defects in yeast lacking superoxide dismutase and delays death of wild-type yeast. *J. Cell Biol.* **137**: 1581-1588.
Lu,Q-L., Abel,P., Foster,C.S., and Lalani,E-N. (1996) bcl-2: role in epithelial differentiation and oncogenesis. *Hum. Pathol.* **27**: 102-110.
Mangeney,M., Schmitt,J.-R., Leverrier,Y.*et al.* (1996) The product of the v-src-inducible gene nr-13 is a potent anti-apoptotic factor. *Oncogene* **13**: 1441-1446.
Manon,S., Chaudhuri,B. and Guérin,M. (1997) Release of cytochrome c and decrease of cytochrome c oxidase in Bax-expressing yeast cells, and prevention of these effects by co-expression of Bcl-xL. *FEBS Lett.* **415**: 29-32.
Marchetti,P., Castedo,M., Susin,S.A. *et al.* (1996) Mitochondrial permeability transition is a central coordinating event of apoptosis. *J. Exp. Med.* **184**: 1155-1160.
Marin,M.C., Fernandez,A., Bick,R.J. *et al.* (1996) Apoptosis suppression by bcl-2 is correlated with the regulation of nuclear and cytosolic $Ca2^{+}$. *Oncogene* **12**: 2259-2266.

Marvel,J., Perkins,G.R., Rivas,A.L. and Collins,M.K.L. (1994) Growth factor starvation of bcl-2 overexpressing murine bone marrow cells induced refractoriness to IL-3 stimulation of proliferation. *Oncogene* **9**: 1117-1122.

Matolcsy,A., Casali,P., Warnke,R.A. and Knowles,D.M. (1996) Morphologic transformation of follicular lymphoma is associated with somatic mutation of the translocated Bcl-2 gene. *Blood* **88**: 3937-3944.

Maundrell,K., Antonsson,B., Magnenat,E. *et al.* (1997) Bcl-2 undergoes phosphorylation by c-Jun N-terminal kinase/stress-activated protein kinases in the presence of the constitutively active GTP-binding protein Rac1. *J. Biol. Chem.* **272**: 25238-25242.

Maung,Z.T., MacLean,F.R., Reid,M.M. *et al.* (1994) The relationship between bcl-2 expression and response to chemotherapy in acute leukemia. *Br. J. Heamatol.* **88**: 105-109.

May,W.S., Tyler,P.G., Ito,T. *et al.* (1994) Interleukin-3 and Bryostatin-1 mediate hyperphosphorylation of Bcl2α in association with suppression of apoptosis. *J. Biol. Chem.* **269**: 26865-26870.

Mazel,S., Burtrum,D. and Petrie,H.T. (1996) Regulation of cell division cycle progression by bcl-2 expression: a potential mechanism for inhibition of programmed cell death. *J. Exp. Med.* **183**: 2219-2226.

McConkey,D.J., Chandra,J., Wright,S. *et al.* (1996) Apoptosis sensitivity in chronic lymphocytic leukemia is determined by endogenous endonuclease content and relative expression of Bcl-2 and Bax. *J. Immunol.* **156**: 2624-2630.

McCurrach,M.E., Connor,T.M.F., Knudson,C.M. *et al.* (1997) Bax-deficiency promotes drug resistance and oncogenic transformation by attenuating p53-dependent apoptosis. *Proc. Natl. Acad. Sci. USA* **94**: 2345-2349.

McDonnell,T.J. and Korsmeyer,S.J. (1991) Progression from lymphoid hyperplasia to high-grade malignant lymphoma in mice transgenic for the t(14;18). *Nature* **349**: 254-256.

McDonnell,T.J., Troncoso,P., Brisbay,S.M. *et al.* (1992) Expression of the protooncogene bcl-2 in the prostate and its association with emergence of androgen-independent prostate cancer. *Cancer Research* **52**: 6940-6944.

Meijerink,J.P., Smetsers,T.F., Sloetjes,A.W. *et al.* (1995) Bax mutations in cell lines derived from hematological malignancies. *Leukemia* **9**: 1828-1832.

Merry,D.E. and Korsmeyer,S.J. (1997) Bcl-2 gene family in the nervous system. *Annu. Rev. Neurosci.* **20**: 245-267.

Middleton,G., Nunez,G. and Davies,A.M. (1996) Bax promotes neuronal survival and antagonises the survival effects of neurotrophic factors. *Development* **122**: 695-701.

Minn,A.J., Vélez,P., Schendel,S.L. *et al.* (1997) Bcl-xL forms an ion channel in synthetic lipid membranes. *Nature* **385**: 353-357.

Miyashita,T., Harigai,M., Hanada,M. and Reed,J.C. (1994a) Identification of a p53-dependent negative response element in the bcl-2 gene. *Cancer Res.* **54**: 3131-3135.

Miyashita,T., Krajewski,S., Krajewska,M. *et al.* (1994b) Tumor suppressor p53 is a regulator of bcl-2 and bax gene expression *in vitro* and *in vivo*. *Oncogene* **9**: 1799-1805.

Miyashita,T. and Reed,J.C. (1995) Tumor suppressor p53 is a direct transcriptional activator of the human Bax gene. *Cell* **80**: 293-299.

Motoyama,N., Wang,F., Roth,K.A. *et al.* (1995) Massive cell death of immature hematopoietic cells and neurons in Bcl-x-deficient mice. *Science* **267**: 1506-1510.

Muchmore,S.W., Sattler,M., Liang,H. *et al.* (1996) X-ray and NMR structure of human Bcl-x_L, an inhibitor of programmed cell death. *Nature* **381**: 335-341.

Naik,P., Karrim,J. and Hanahan,D. (1996) The rise and fall of apoptosis during multistage tumorigenesis: down-modulation contributes to tumor progression from angiogenic progenitors. *Genes and Development* **10**: 2105-2116.

Naumovski,L. and ClearyM. L. (1996) The p53-binding protein 53BP2 also interacts with Bcl-2 and impedes cell cycle progression at G2/M. *Mol. Cell. Biol.* **16**: 3884-3892.

Nava,V.E., Cheng, E.H-Y., Veliuona,M. *et al.* (1997) Herpesvirus saimiri encodes a functional homologue of the human bcl-2 oncogene. *J. Virol.* **71**: 4118-4122.

Ng,F.W.H., Nguyen,M., Kwan,T. *et al.* (1997) p28 Bap31, a Bcl-2/Bcl-xL- and procaspase-8-associated protein in the endoplasmic reticulum. *J. Cell. Biol.* **139**: 327-338.

Nguyen,M., Branton, P.E., Walton,P.A. *et al.* (1994) Role of membrane anchor domain of Bcl-2 in suppression of apoptosis caused by E1B-defective adenovirus. *J. Biol. Chem.* **269**: 16521-16524.

Nguyen,M., Millar,D.G., Yong,V.W. *et al.* (1993) Targeting of Bcl-2 to the mitochondrial outer membrane by a COOH-terminal signal anchor sequence. *J. Biol. Chem.* **268**: 25265-25268.

O'Connor,L., Strasser,A., O'Reilly,L.A. *et al.* (1998) Bim: a novel member of the Bcl-2 family that promotes apoptosis. *EMBO J.* **17**: 384-395.

Ohr,K., Iwai,K., Kasahara,Y. *et al.* (1995) Immunoblot analysis of cellular expression of Bcl-2 family proteins, Bcl-2, Bax, Bcl-x and Mcl-1, in human peripheral blood and lymphoid tissues. *Int. Immunol.* **7**: 1817-1825.

Olivier,R., Otter,I., Monney,L. *et al.* (1997) Bcl-2 does not require Raf kinase activity for its death-protective function. *Biochem. J.* **324**: 75-83.

Olopade,O.I., Adeyanju,M.O., Safa,A.R. *et al.* (1997) Overexpression of Bcl-x protein in primary breast cancer is associated with high tumor grade and nodal metastases. *Cancer J.* **3**: 230-237.

Oltvai,Z.N. and Korsmeyer,S.J. (1994) Checkpoints of dueling dimers foil death wishes. *Cell* **79**: 189-192.

Oltvai,Z.N., Milliman,C.L. and Korsmeyer,S.J. (1993) Bcl-2 heterodimerizes *in vivo* with a conserved homologue, Bax, that accelerates programmed cell death. *Cell* **74**: 609-619.

Ottilie,S., Diaz,J-L., Chang,J. *et al.* (1997a) Structural and functional complementation of an inactive Bcl-2 mutant by Bax truncation. *J. Biol. Chem.* **272**: 16955-16961.

Ottilie,S., Diaz,J.-L., Horne,W. *et al.* (1997b) Dimerization properties of human Bad. *J. Biol. Chem.* **272**: 30866-30872.

Párrizas,M. and Leroith,D. (1997) Insulin-like growth factor-1 inhibition of apoptosis is associated with increased expression of the bcl-xL gene product. *Endocrinology* **138**: 1355-1358.

Pepper,C., Bentley,P. and Hoy,T. (1996) Regulation of clinical chemoresistance by bcl-2 and bax oncoproteins in B-cell chronic lymphocytic leukaemia. *Br. J. Hematol.* **95**: 513-517.

Petit,P.X., Susin,S-A., Zamzami,N. *et al.* (1996) Mitochondria and programmed cell death : back to the future. *FEBS Lett.* **396**: 7-13.

Pezzella,F., Turley,H., Kuzu,I. *et al.* (1993) bcl-2 protein in non-small-cell lung carcinoma. *N. Engl. J. Med.* **329**: 690-694.

Raffo,A.J., Perlman,H., Chen,M-W. *et al.* (1995) Overexpression of bcl-2 protects prostate cancer cells from apoptosis *in vitro* and confers resistance to androgen depletion *in vivo*. *Cancer Res.* **55**: 4438-4445.

Rampino,N., Yamamoto,H., Ionov,Y. *et al.* (1997) Somatic frameshift mutations in the Bax gene in colon cancers of the microsatellite mutator phenotype. *Science* **275**: 967-969.

Reed,J.C. (1994) Bcl-2 and the regulation of programmed cell death. *J. Cell Biol.* **124**: 1-6.

Reed,J.C., Miyashita,T., Takayama,S. *et al.* (1996) Bcl-2 family proteins: regulators of cell death involved in the pathogenesis of cancer and resistance to therapy. *J. Cell. Biochem.* **60**: 23-32.

Reed,J.C., Stein,C., Subasinghe,C. *et al.* (1990) Antisense inhibition of Bcl-2 protooncogene expression and leukemic cell growth and survival: comparisons of phosphodiester and phosphorothioate oligodeoxynucleotides. *Cancer Res.* **50**: 6565-6570.

Reed,J.C. and Tanaka,S. (1993) Somatic point mutations in the translocated bcl-2 genes of non-Hodgkin's lymphomas and lymphocytic leukemias: implications for mechanisms of tumor progression. *Leuk. Lymphoma* **10**: 157-163.

Reynolds,J.E., Li,J., Craig,R.W. and Eastman,A. (1996) Bcl-2 and Mcl-1 expression in Chinese hamster ovary cells inhibits intracellular acidification and apoptosis induced by staurosporine. *Exp. Cell Res.* **225**: 430-436.

Robertson,L.E., Plunkett,W., McConnell,K. *et al.* (1996) Bcl-2 expression in chronic lymphocytic leukemia and its correlation with the induction of apoptosis and clinical outcome. *Leukemia* **10**: 456-459.

Ryan,J.J., Prochownik,E., Gottlieb,C.A. *et al.* (1994) c-myc and bcl-2 modulate p53 function by altering p53 subcellular trafficking during the cell cycle. *Proc. Natl. Acad. Sci.* **91**: 5878-5882.

Salomons,G.S., Brady,H.J.M., Verwijs-Janssen,M. *et al.* (1997) The Baxα:Bcl-2 ratio modulates the response to dexamethasone in leukemic cells and is highly variable in childhood acute leukemia. *Int. J. Cancer* **71**: 959-965.

Sarid,R., Sato,T., Bohenzky,R.A. *et al.* (1997) Kaposi's sarcoma-associated herpesvirus encodes a functional Bcl-2 homologue. *Nature Med.* **3**: 293-298.

Sato,T., Hanada,M., Bodrug,S. *et al.* (1994) Interactions among members of the Bcl-2 protein family analyzed with a yeast two-hybrid system. *Proc. Natl. Acad. Sci. USA.* **91**: 9238-9242.

Sattler,M., Liang,H., Nettesheim,D. *et al.* (1997) Structure of Bcl-x_L- Bak peptide complex: recognition between regulators of apoptosis. *Science* **275**: 983-986.

Schendel,S.L., Xie,Z., Montal,M.O. *et al.* (1997) Channel formation by antiapoptotic protein Bcl-2. *Proc. Natl. Acad. Sci USA.* **94**: 5113-5118.

Schlesinger,P.H., Gross,A., Yin,X-M. *et al.* (1997) Comparison of the ion channel characteristics of proapoptotic Bax and antiapoptotic Bcl-2. *Proc. Natl. Acad. Sci. USA* **94**: 11357-11362.

Sedlak,T.W., Oltvai,Z.N., Yang,E. *et al.* (1995) Multiple Bcl-2 family members demonstrate selective dimerizations with Bax. *Proc. Natl. Acad. Sci. USA.* **92**: 7834-7838.

Sentman,C.L., Shutter,J.R., Hockenbery,D. *et al.* (1991) Bcl-2 inhibits multiple forms of apoptosis but not negative selection in thymocytes. *Cell*, **67**: 879-888.

Seshagiri,S. and Miller,L.K. (1997) Caenorhabditis elegans Ced-4 stimulates Ced-3 processing and Ced-3-induced apoptosis. *Curr. Biol.* **7**: 455-460.

Seto,M., Jaeger,U., Hockett,R.D. *et al.* (1988) Alternative promoters and exons, somatic mutation and deregulation of the Bcl-2-Ig fusion gene in lymphoma. *EMBO J.* **7**: 123-131.

Shibasaki,F., Kondo,E., Akagi,T. and McKeon,F. (1997) Suppression of signalling through transcription factor NF-AT by interactions between calcineurin and Bcl-2. *Nature* **386**: 728-731.

Shimizu,S., Eguchi,Y., Kamiike,W. *et al.* (1996) Bcl-2 expression prevents activation of the ICE protease cascade. *Oncogene* **12**: 2251-2257.

Shimizu,S., Eguchi,Y., Kosaka,H. *et al.* (1995) Prevention of hypoxia-induced cell death by Bcl-2 and Bcl-xL. *Nature* **374**: 811-813.

Silva,M., Grillot,D., Benito,A. *et al.* (1996) Erythropoietin can promote erythroid progenitor survival by repressing apoptosis through Bcl-xL and Bcl-2. *Blood* **88**: 1576-1582.

Silvestrini,R., Veneroni,S., Daidone,M.G. *et al.* (1994) The Bcl-2 protein: a prognostic indicator strongly related to p53 protein in lymph node-negative breast cancer patients. *J. Natl. Cancer Inst.* **86**: 499-504.

Simonian,P.L., Grillot,D.A.M., Merino,R. and Nunez,G. (1996) Bax can antagonize Bcl-X_L during etoposide and cisplatin-induced cell death independently of its heterodimerization with Bcl-X_L. *J. Biol. Chem.* **271**: 22764-22772.
Simonian,P.L., Grillot,D.A.M. and Nunez,G. (1997) Bak can accelerate chemotherapy-induced cell death independently of its heterodimerization with Bcl-xL and Bcl-2. *Oncogene* **15**: 1871-1875.
Singleton,J.R., Dixit,V.M. and Feldman,E. (1996) Type I insulin-like growth factor receptor activation regulates apoptotic proteins. *J. Biol. Chem.* **271**: 31791-31794.
Spector,M.S., Desnoyers,S., Hoeppner,D.J. and Hengartner,M.O. (1997) Interaction between the C. elegans cell-death regulators CED-9 and CED-4. *Nature* **385**: 653-656.
Stoetzer,O.J., Nüssler,V., Darsow,M. *et al.* (1996) Association of bcl-2, bax, bcl-xL and interleukin-1-ß-converting enzyme expression with initial response to chemotherapy in acute myeloid leukemia. *Leukemia* **10**: Suppl. 3 S18-S22.
Strasser,A., Harris,A.W., Bath,M.L. and Cory,S. (1990) Novel primitive lymphoid tumors induced in transgenic mice by cooperation between myc and bcl-2. *Nature* **348**: 331-333.
Strobel,T., Swanson,L., Korsmeyer,S. and Cannistra,S.A. (1996) Bax enhances paclitaxel-induced apoptosis through a p53-independent pathway. *Proc. Natl. Acad. Sci. USA* **93**: 14094-14099.
Susin,S.A., Zamzami,N., Castedo,M. *et al.* (1996) Bcl-2 inhibits the mitochondrial release of an apoptogenic protease. *J. Exp. Med.* **184**: 1331-1341.
Takayama,S., Bimston,D.N., Matsuzawa,S. *et al.* (1997) Bag-1 modulates the chaperone activity of Hsp70/Hsc70. *EMBO J.* **16**: 4887-4896.
Takayama,S., Sato,T., Krajewski,S.*et al.* (1995) Cloning and functional analysis of BAG-1: A novel Bcl-2-binding protein with anti-cell death activity. *Cell* **80**: 279-284.
Tanaka,S., Louie,D.C., Kant,J.A. and Reed,J.C. (1992) Frequent incidence of somatic mutations in translocated Bcl-2 oncogenes of non-Hodgkin's lymphomas. *Blood* **79**: 229-237.
Tanaka,S., Saito,K. and Reed,J.C. (1993) Structure-function analysis of the bcl-2 oncoprotein. *J. Biol. Chem.* **268**: 10920-10926.
Tao,W., Kurschner,C. and Morgan,J.I. (1997) Modulation of cell death in yeast by the Bcl-2 family of proteins. *J. Biol. Chem.* **272**: 15547-15552.
Thomas,A., Rouby,S.E., Reed,J.C.*et al.* (1996) Drug-induced apoptosis in B-cell chronic lymphocytic leukemia: relationship between p53 gene mutation and bcl-2/bax proteins in drug resistance. *Oncogene* **12**: 1055-1062.
Tsujimoto,Y. and Croce,C.M. (1986) Analysis of the structure, transcripts, and protein products of bcl-2, the gene involved in human follicular lymphoma. *Proc. Natl. Acad. Sci. USA.* **83**: 5214-5218.
Vairo,G., Innes,K.M. and Adams,J.M. (1996) Bcl-2 has a cell cycle inhibitory function separable from its enhancement of cell survival. *Oncogene* **13**: 1511-1519.
Vaux,D.L., Corey,S. and Adams,J.M. (1988) Bcl-2 promotes the survival of haemopoietic cells and cooperates with c-myc to immortalize pre-B cells. *Nature* **335**: 440-442.
Veis,D.J., Sorenson,C.M., Shutter,J.R. and Korsmeyer,S.J. (1993) Bcl-2-deficient mice demonstrate fulminant lymphoid apoptosis, polycystic kidneys, and hypopigmented hair. *Cell* **75**: 229-240.
Wagener,C., Bargou,R.C., Daniel,P.T. *et al.* (1996) Induction of the death-promoting gene bax-α sensitizes cultured breast-cancer cells to drug-induced apoptosis. *Int. J. Cancer* **67**: 138-141.
Wang,H-G.,Rapp,U.R. and Reed,J.C. (1996a) Bcl-2 targets the protein kinase Raf-1 to mitochondria. *Cell* **87**: 629-638.
Wang,H-G., Takayama,S., Rapp,U.R. and Reed, J.C. (1996b) Bcl-2 interacting protein, Bag-1, binds to and activates the kinase Raf-1. *Proc. Natl. Acad. Sci.* **93**: 7063-7068.
Wang,H-G., Miyashita,T., Takayama,S. *et al.* (1994) Apoptosis regulation by interaction of Bcl-2 protein and Raf-1 kinase. *Oncogene* **9**: 2751-2756.
Wang,K., Yin,M., Chao,D.T. *et al.* (1996c) BID: a novel BH3 domain-only death agonist. *Gen. Dev.***10**: 2859-2869.
Webb,A., Cunningham,D., Cotter,F. *et al.* (1997) BCL-2 antisense therapy in patients with non-Hodgkin lymphoma. *The Lancet* **349**: 1137-1141.
Werner,H. and LeRoith,D. (1996) The role of the insulin-like growth factor system in human cancer. *Ad. in Cancer Res.* **68**: 183-223.
Wilson,B.E., Mochon,E. and Boxer,L.M. (1996) Induction of bcl-2 expression by phophorylated CREB proteins during B-cell activation and rescue from apoptosis. *Mol. Cell. Biol.* **16**: 5546-5556.
Wu,D., Wallen,H.D., Inohara,N. and Nunez,G. (1997a) Interaction and regulation of the Caenorhabditis elegans death protease Ced-3 by Ced-4 and Ced-9. *J. Biol. Chem.* **272**: 21449-21454.
Wu,D., Wallen,H.D. and Nunez,G. (1997b) Interaction and regulation of subcellular localization of Ced-4 by Ced-9. *Science* **275**: 1126-1129.
Xiang,J., Chao,D.T. and Korsmeyer,S.J. (1996) Bax-induced cell death may not require intereukin 1ß-converting enzyme-like proteases. *Proc. Natl. Acad. Sci. USA* **93**: 14559-14563.
Yamamoto,H., Sawai,H. and Perucho,M. (1997) Frameshift somatic mutations in gastrointestinal cancer of the microsatellite mutator phenotype. *Cancer Res.* **57**: 4420-4426.
Yang,E. and Korsmeyer,S.J. (1996) Molecular thanatopsis: a discourse on the Bcl-2 family and cell death. *Blood* **88**: 386-401.
Yang, E., Zha,J., Jockel,J. *et al.* (1995) Bad, a heterodimeric partner for Bcl-xL and Bcl-2, displaces Bax and promotes cell death. *Cell* **80**: 285-291.
Yang,J., Liu,X., Bhalla,K. *et al.* (1997) Prevention of apoptosis by Bcl-2: release of cytochrome c from mitochondria blocked. *Science* **275**: 1129-1132.

Yang,T., Kozopas,K.M. and Craig,R.W. (1995b) The intracellular distribution and pattern of expression of Mcl-1 overlap with, but are not identical to, those of Bcl-2. *J. Cell Biol.* **128**: 1173-1184.
Yin,C., Knudson,C.M., Korsmeyer,S.J. and Van Dyke,T.V. (1997) Bax suppresses tumorigenesis and stimulates apoptosis *in vivo*. *Nature* **385**: 637-640.
Yin,X-M., Oltvai,Z.N. and Korsmeyer,S.J. (1994) BH1 and BH2 domains of Bcl-2 are required for inhibition of apoptosis and heterodimerization with Bax. *Nature* **369**: 321-323.
Young,R.L. and Korsmeyer,S.J. (1993) A negative regulatory element in the bcl-2 5'-untranslated region inhibits expression from an upstream promoter. *Mol. Cell. Biol.* **13**: 3686-3697.
Zha,H., Aimé-Sempé,C., Sato,T. and Reed,J.C. (1996a) Proapoptotic protein Bax heterodimerizes with and Bcl-2 and homodimerizes with Bax via a novel domain (BH3) distinct from BH1 and BH2. *J. Biol. Chem.* **271**: 7440-7444.
Zha,H., Fisk,H.A., Yaffe,M.P. *et al.* (1996b) Structure-function comparisons of the proapoptotic protein Bax in yeast and mammalian cells. *Mol. Cell. Biol.* **16**: 6494-6508.
Zha,J., Harada,H., Osipov,K. *et al.* (1997) BH3 domain of Bad is required for heterodimerization with Bcl-XL and pro-apoptotic activity. *J. Biol. Chem.* **272**: 24101-24104.
Zha,J., Harada,H., Wang,E. *et al.* (1996c) Serine phosphorylation of death agonist Bad in response to survival factor results in binding to 14-3-3 not Bcl-xL. *Cell* **87**: 619-628.
Zhan,Q., Fan,S., Bae,I. *et al.* (1994) Induction of bax by genotoxic stress in human cells correlates with normal p53 status and apoptosis. *Oncogene* **9**: 3743-3751.
Zhou,P., QianL., Kozopas,K.M. and Craig,R.W. (1997) Mcl-1, a Bcl-2 family member, delays the death of hematopoietic cells under a variety of apoptosis-inducing conditions. *Blood* **89**: 630-643.
Zhu,W., Cowie,A., Wasfy,G.W. *et al.* (1996) Bcl-2 mutants with restricted subcellular location reveal spatially distinct pathways for apoptosis in different cell types. *EMBO J.* **15**: 4130-4141.
Ziegler,A., Luedke,G.H., Fabbro,D. *et al.* (1997) Induction of apoptosis in small-cell lung cancer cells by an antisense oligonucleotide targeting the Bcl-2 coding sequence. *J. Natl. Cancer Inst.* **89**: 1027-1036.
Zou,H., Henzel,W.J., Liu,X. *et al.* (1997) Apaf-1, a human protein homologous to C. elegans Ced-4, participates in cytochrome c-dependent activation of caspase-3. *Cell* **90**: 405-413.

Chapter 4

Stress-responsive signal transduction: emerging concepts and biological significance.

Usha Kasid and Simeng Suy.
Departments of Radiation Medicine and Biochemistry and Molecular Biology, Lombardi Cancer Center, Georgetown University Medical Center, Washington D.C., 20007, USA.

1.0 INTRODUCTION

In the past few years, substantial evidence has been put forth that suggests increasing complexity of the mechanism of action of cytotoxic, mutagenic or carcinogenic agents, commonly known as stress-inducers. The developing theme in a stress-induced cellular response is stimulation of a signal transduction pathway regulated by either cell growth and proliferation-related (survival) factors or programmed cell death-related (apoptotic) factors. The balance between the survival and apoptotic signals depends on the cell type, and dictates the fate of the cell. This review summarises current knowledge of the intracellular signal transduction responses elicited by ionising radiation in rodent and mammalian cells, and focuses on the involvement of Raf-1 protein serine-threonine kinase in the signalling pathways initiated by diverse stress-inducers including ionising radiation, ultraviolet radiation, growth factor-deprivation, and cytokines.

2.0 IONISING RADIATION STIMULATES MULTIPLE SIGNAL TRANSDUCERS

Growing evidence suggests that exposure of cells to ionising radiation (IR) results in the stimulation of a series of biochemical and molecular signals, which regulate cell cycle progression, cell proliferation and survival (Table I). Various components of the IR-inducible signal transduction cascade function as either survival factors, participating in cell growth/proliferation and cellular protection, or cell death-related factors.

2.1 Protein tyrosine kinases, and tyrosine phosphorylation

Stimulation of protein tyrosine kinases (PTKs), and tyrosine phosphorylation of multiple proteins are some of the earliest biochemical events observed in a variety of cells after irradiation (Uckun *et al.* 1993; Kasid *et al.* 1996; Suy *et al.* 1997). IR-induced protein tyrosine phosphorylation occurs within 2-5 minutes after treatment, and is inhibited by PTK inhibitors. Several known and unknown proteins undergo tyrosine phosphorylation upon irradiation of cells. One of the radiation-responsive non-receptor PTKs in radiosensitive lymphoma B cells is Bruton's tyrosine kinase (BTK), a member of the Src-related TEC family of PTKs, which mediates radiation-induced apoptosis of lymphoma B cells. Cells rendered BTK-deficient failed to undergo radiation-induced apoptosis (Uckun *et al.* 1996). Introduction of the wild type human *btk* gene restored the apoptotic response, suggesting that BTK is responsible for IR-induced apoptotic death of lymphoma B cells.

TABLE I. Examples of Ionising Radiation-Responsive Signals

Gene or protein*	Nature of modification**	References
c-Abl	Activation	Kharbanda *et al.* 1995b; Liu *et al.* 1996a; Baskaran *et al.* 1997; Shafman *et al.* 1997
α-ACT	8	Hong *et al.* 1995; Patel S *et al.* 1998.
AP-1	Activation	Hallahan *et al.* 1995; Meighan-Mantha *et al.*submitted
Bax	8	Zhan Q *et al.* 1994
Bcl-xL	8	Zhan Q *et al.* 1994
BRCA1	Phosphorylation	Scully *et al.* 1997a
BTK	Activation	Burkhardt *et al.* 1991; Uckun *et al.* 1996
p34cdc2	Tyr-P	Kharbanda *et al.* 1994b
cdc25	9	Datta R. *et al.* 1992c
$p21^{WAF-1/CIP1}$	8	Kastan *et al.* 1992
CPP32	Activation	Datta R *et al.* 1997
CREB	Activation	Sahijdake *et al.* 1994
Cyclin A	9	Datta R. *et al.* 1992c
Cyclin B1	9	Muschel *et al.* 1991
egr-1	8	Hallahan *et al.* 1991b
ERK	Activation	Kharbanda *et al.* 1994a; Stevensen *et al.* 1994; Suy *et al.* 1997.
b*Fgf*	8	Haimovitz-Friedman *et al.* 1991.
*gadd*45	8	Hollander *et al.* 1993
*gadd*34	8	Hollander *et al.* 1997
JNK	Activation	Verheiji *et al.* 1996
c-*jun*	8	Sherman *et al.* 1990; Hallahan *et al.* 1991b; Chae *et al.* 1993

TABLE I. Examples of Ionising Radiation-Responsive Signals (CONTD.)

HVH1/CL100	8	Kasid *et al.* 1997
IκB	8	Kasid *et al.* 1997
p38MAPK	Activation	Pandey *et al.* 1996
NF-κB	Activation	Brach *et al.* 1991; Meighan-Mantha *et al.* submitted
OCT-1	Activation	Meighan-Mantha *et al.* submitted
PARP	Cleavage	Datta R. *et al.* 1997
p85 PI3-K	Tyr-P	Kharbanda *et al.* 1995a
PKC	Activation	Woloschak *et al.* 1990; Hallahan *et al.* 1991a
Raf-1	Tyr-P/Activation	Kasid *et al.* 1996; Suy *et al.* 1997
Ras	Activation	Suy *et al.* 1997
RNA polymerase II	Tyr-P	Liu *et al.* 1996a
SHPTP1	Tyr-P	Kharbanda *et al.* 1996
SP-1	Activation	Sahijdaka *et al.* 1994; Meighan-Mantha *et al.* submitted
Sphingomyelinase	Activation	Haimovitz-Friedman *et al.* 1994b Santana *et al.* 1996; Chmura *et al.* 1997
TNFα	8	Hallahan *et al.* 1989
p53	Activation	Kastan *et al.* 1991,1992; Kuerbitz *et al.* 1992.

* Arranged in alphabetical order.

**List does not include modifications in sucellular location, and protein-protein interactions.

Tyr-P, tyrosine phosphorylation;
8, increased gene expression;
9, decreased gene expression.

Phosphatidylinositol 3- kinase (PI3-K), a lipid and serine kinase, is stimulated by extracellular signals, especially those involving the activation of receptor PTKs. The catalytic subunit of PI3-K, p110, exhibits enzymatic activity in mammalian cells only when bound, through its amino-terminal region, to the inter-SH_2 region of p85 subunit. PI3-K activation increases the levels of 3' phosphoinositides, PI3,4,$5P_3$ and PI3,$4P_2$, in cells and these phospholipids act as second messengers to activate downstream effectors including c-Akt protein serine kinase (Franke *et al.* 1995; Datta K. *et al.* 1996). Activated Akt plays a role in protection against apoptosis, and in transformation (Franke *et al.* 1997). IR has been shown to stimulate tyrosine phosphorylation of the p85 subunit of PI3-K, and its association with c-Jun terminal kinase/stress-activated protein kinase (JNK/SAPK) via Grb2 (Kharbanda *et al.* 1995a). Whether PI3-K enzymatic activity was increased is not known, but addition of the PI3-K inhibitor, wortmannin, activated SAPK activity. These results imply that PI3-K somehow regulates SAPK; however, the significance of this association is not clear.

C-Abl, also a non-receptor protein tyrosine kinase is activated by IR in human U-937 cells (Kharbanda *et al.* 1995b), and NIH 3T3 cells (Liu *et al.* 1996a), but not in AT cells derived from ataxia telangiectasia (AT) patients (Baskaran *et al.* 1997). ATM, the gene mutated in AT, encodes a putative lipid or protein kinase, and functional ATM appears to be necessary for IR-inducible tyrosine kinase activity of c-Abl. Furthermore, functional c-Abl is required for downstream activation of JNK/SAPK, because IR has no effect on JNK/SAPK activity in *abl* $^{-/-}$ cells and this signal appears to be negatively regulated by Abl-dependent phosphorylation of the protein tyrosine phosphatase SHPTP1 in irradiated cells (Kharbanda *et al.* 1996). The JNKs have also been suggested to mediate the apoptotic response to genotoxins (Verheij *et al.* 1996). Interestingly, abrogation of Abl function in *abl*$^{-/-}$ cells does not preclude JNK activation in response to TNF-α and UV irradiation, suggesting that while Abl activation may be an IR-induced signal, JNK may be a universal effector of stress-initiated cascade .

In other studies, MCF-7 cells stably transfected with dominant negative c-Abl showed resistance to IR-induced loss of clonogenic survival and apoptosis as compared with wild type cells, and *abl* $^{-/-}$ mouse embryo fibroblasts (MEFs) also showed impaired apoptotic response to IR (Yuan *et al.* 1997). Transfection of NIH 3T3 cells with v-*abl* was demonstrated not affect clonogenic survival response to IR (Sklar *et al.* 1986; Pirollo *et al.* 1989), except under low dose conditions, where v-*abl* appeared to increase the radioresistance of NIH3T3 cells and murine haematopoietic cells (FitzGerald, 1991). The C-terminal domain of RNA polymerase is a substrate for the nuclear c-Abl tyrosine kinase; IR also causes tyrosine phosphorylation of RNA polymerase II in *abl*$^{+/+}$ 3T3 fibroblast cells (Liu *et al.* 1996a). In other studies, *abl* $^{+/+}$ and *abl*$^{-/-}$ MEFs showed similar G1/S delay, and no significant difference in long-term clonogenic survival response to IR (Baskaran *et al.* 1997). Given the fact that radiation causes modulation of transcription of several cellular genes, it has been suggested that RNA polymerase II tyrosine phosphorylation mediated by the ATM/c-Abl pathway may be involved in IR-induced transcriptional modifications.

2.2 Protein Kinase C, and the sphingomyelinase pathway

Activation of PKC occurs within seconds after irradiation of HL-60 cells. PKC activation has been linked to the transcriptional increase of c-*jun*, *egr*-1, and *TNF*-α genes. PKC inhibition by prolonged treatment with the tumour promoter, TPA, attenuates IR-inducible transcription (Sherman *et al.* 1990; Hallahan *et al.* 1991a,b). Other studies have demonstrated that a region containing six serum response or CArG motifs within the *egr*-1 promoter is responsible for IR inducible *egr*-1 gene expression (Datta R. *et al.* 1992a). PKC has also been shown to mediate basic fibroblast growth factor (bFGF) -dependent protection of bovine aortic endothelial cells (BAEC) from IR-induced apoptosis (Haimovitz-Friedman *et al.* 1994a).

In BAEC, IR has been shown to act directly on cell membrane preparations, and within seconds, stimulate sphingomyelin hydrolysis via a neutral sphingomyelinase, to generate ceramide and phosphocholine. Ceramide then acts as a second messenger turning on an apoptotic pathway (Haimovitz-Friedman *et al.* 1994b). A role for acidic sphingomyelinase in the generation of ceramide following IR has also been shown, as acidic sphingomyelinase knock out mice are resistant to IR induced apoptosis (Santana *et al.* 1996). Interestingly, TNFα, a well known activator of the sphingomyelinase pathway, is also induced by IR at the level of transcription (Hallahan *et al.* 1989). It is not known whether in intact cells, IR-induced TNFα accelerates ceramide generation. Downstream effectors of IR-induced ceramide include JNK/SAPK, but not p42/p44 mitogen-activated protein kinase (MAPK/ERK) (Verheij *et al.* 1996). This pathway seems to be regulated via protein kinase C (PKC), because PKC activation blocks both IR-induced ceramide levels and apoptosis.

2.3 The Raf-1 pathway

The c-*raf*-1 gene is a member of a family of genes (c-*raf*-1, A-*raf*, and B-*raf*) identified as the cellular counterparts of v-*raf*, the transforming gene of murine retrovirus 3611-MSV (Rapp *et al.* 1983; Bonner *et al.* 1986; Beck *et al.* 1987; Ikawa *et al.* 1988). The c-*raf*-1 gene is expressed in normal and transformed cells of epithelial, neuronal, and haematopoietic origin (Storm *et al.* 1990). Raf-1, the product of the c-*raf*-1 gene, is ~75kDa cytoplasmic phosphoprotein with intrinsic kinase activity toward serine and threonine residues (Moelling *et al.* 1984; Bonner *et al.* 1986). Sequence analysis of Raf-1 suggests two major functional domains: the amino-terminal regulatory domain, which contains two highly conserved regions, CR1 and CR2; and the carboxy-terminal domain, which represents the third highly conserved region CR3. CR1 contains a cysteine motif, CR2 contains many serine and threonine residues, and CR3 with an ATP-binding domain is the catalytic portion of Raf-1 (Daum *et al.* 1994). Several critical phosphorylation sites have been identified in Raf-1, including S^{338}, S^{339}, Y^{340}, and Y^{341} in the CR3 region which are important for regulation of the biological activity of Raf-1 (Fabian *et al.* 1993; Marais *et al.* 1995; Diaz *et al.* 1997).

We and others have shown that oncogenic activation of the Raf-1 protein kinase occurs by truncation of its regulatory amino-terminus and retention of the kinase domain, resulting in the neoplastic growth of recipient cells (Kasid *et al.* 1987; Stanton *et al.* 1989; Heidecker *et al.* 1990), and transcriptional transactivation-related modification (Wasylyk *et al.* 1989; Qureshi *et al.* 1991; Bruder *et al.* 1992; Finco and Baldwin, 1993). The mitogenic function of oncogenic Raf-1 is well established. For example, constitutively activated Raf-1 induces colony stimulating factor (CSF-1)-independent growth of macrophage cells (Buscher *et al.* 1993), and initiates DNA synthesis in serum-starved cells (Smith *et al.* 1990). A conditional construct expressing the catalytic domain of Raf-1 under the control of a mutated oestrogen receptor has been designed, and is activated by 4-hydroxytamoxifen (OHT) (Littlewood *et al.* 1995). Quiescent NIH/3T3 cells transfected with this construct expressed oncogenic Raf-1 and were induced to proliferate in the presence of OHT, implying that oncogenic Raf-1 was sufficient to substitute for growth factor dependency (Kerkhoff and Rapp, 1997). Interestingly, activation of Raf-1 in these cells coincided with the amplified expression of cyclin D1, and repression of p27 cdk (cyclin-dependent kinase) inhibitor. The p27 cdk inhibitor seems to mediate cell cycle arrest (Kato *et al.* 1994; Polyak *et al.*, 1994; Toyoshima and Hunter, 1994), and its suppression by constitutively activated Raf-1 may be important in transformation.

Other authors have reported similar findings using a conditionally active form of Raf-1 fused to the hormone binding domain of oestrogen receptor. Low levels of Raf-1 promoted cell cycle progression, via activation of cyclin D1-cdk4 and cyclin E-cdk2 complexes (Samuels *et al.* 1993 Woods *et al.* 1997). High levels of cyclin D1 are also seen in Ras transformed cells, it has been suggested that Ras and Raf pathways have overlapping downstream effectors leading to cell proliferation (Lavoie *et al.* 1996, Winston *et al.* 1996).

Contrary to its ability to promote cell proliferation, the Ras/Raf pathway has been reported to play a role in differentiation of a variety of cell types including medullary thyroid carcinoma cells, pheochromocytoma cells, and neuronal cells (Nakagawa *et al.* 1987; Wood *et al.* 1993; Cowley *et al.*

1994; Carson *et al.* 1995; Kuo *et al.* 1996). Using the oestrogen-regulated system mentioned, expression of high levels of Raf-1 activity in NIH3T3 cells induced $p21^{WAF-1/CIP1}$ expression and cell cycle arrest in a p53-independent-manner (Woods *et al.* 1997; Sewig *et al.* 1997). Elevated levels of p21 have been associated with differentiation and cellular senescence in a p53-dependent and independent fashion (Halevy *et al.* 1995; Macleod *et al.* 1995; Harper and Elledge, 1996). In addition, an oestradiol-regulated, constitutively activated form of Raf-1 has been shown to cause induction of $p27^{Kip1}$ cdk inhibitor, and p53-independent G1 and G2 block, in small cell lung carcinoma (SCLC) cells. These effects were blocked using a MEK inhibitor, suggesting that activated MAPK pathway plays a role in growth arrest of SCLC cells (Ravi *et al.* 1998). Premature cell senescence induced by the Ras pathway via induction of p16INK4a cdk inhibitor has also been reported (Serrano *et al.* 1997).

A variety of biochemical experiments show that Raf-1 functions as an effector for several growth factor receptor and non-receptor tyrosine kinases, as well as for non-tyrosine kinases (Morrison, 1990; Rapp, 1991; Daum *et al.* 1994; Marshall, 1996a and 1996b). Growth factor-dependent modifications of Raf-1 include hyperphosphorylation and reduced mobility in gels, recruitment from the cytosol to the membrane, and elevated serine/threonine kinase activity. Raf-1 is a well-recognised component of the mitogen-activated signal transduction pathway involving upstream effectors such as Ras, PI3-K, PKC, and downstream effectors, MEK and ERK (Kyriakis *et al.* 1992; Nakielny *et al.* 1992; Howe *et al.* 1992; Moodie *et al.* 1993; Kolch *et al.* 1993; King *et al.* 1997; Chen Q. *et al.* 1996; Leevers *et al.* 1994; Marais *et al.* 1995; Jelinek *et al.* 1996). Activated ERK's translocate to the nucleus, resulting in the activation of transcription factors, and eventually cell proliferation (Chen R-H *et al.* 1992; Lenormand *et al.* 1993; Marshall, 1995; Treisman, 1996). Consistent with this, Raf-1-mediated signal transduction was demonstrated to be essential for growth factor-induced cell cycle progression of fibroblasts (Kolch *et al.* 1991). Activation of 3pK, a novel member of the MAPK-family and downstream of all known members of the MAPK family, has been shown to be Raf-1-dependent, and this kinase is activated by both mitogens and stress-inducing agents in a cell type-specific and perhaps dose-dependent fashion (McLaughlin *et al.* 1996; Ludwig *et al.* 1996). Mitogen stimulation of 3pK is temporal, whereas stress stimulation is sustained (Ludwig *et al.* 1996). Difference in the kinetics of stimulation due to different stimuli has been observed in other cell types. For example, IR induced a temporal expression of a MAPK phosphatase gene *HVH1*, and the Inhibitor of Nuclear Factor κB, *IκBα* gene. TNFα, however, induced a sustained expression of these genes (Kasid *et al.* 1997).

Tyrosine phosphorylation, membrane-recruitment and activation of endogenous Raf-1 protein kinase have been demonstrated after the irradiation of the human tumour cells PCI-04A and MDA-MB231 (Kasid *et al.* 1996; Suy *et al.* 1997). Somewhat analogous to mitogen-induced signalling, IR stimulated an increase in GTP-binding and hydrolysis on Ras, and co-immunoprecipitation of endogenous Grb2 with Sos, and Raf-1 with Ras. These data suggest that endogenous Raf-1 appears to function as an effector of Ras and IR-stimulated PTKs, leading to activation, in part, of a radioresponsive ERK pathway. The physiological consequences of Raf-1 activation after IR are not fully clear, but this is likely to be a stress-induced protective response.

2.4 Mitogen-activated protein kinases

The three important members of the mitogen-activated protein (MAP) kinase family, namely, ERK (extracellular signal-regulated kinase), JNK/SAPK, and p38 MAPK, provide connecting links between membrane-to-nucleus signalling cascade. Several independent studies have demonstrated that IR exposure of cells causes activation of these MAPKs (Kharbanda *et al.* 1994a, 1995a; Stevenson *et al.* 1994; Kasid *et al.* 1996; Chen Y.R. *et al.* 1996; Pandey *et al.* 1996; Suy *et al.* 1997). Based on the significant differences in the levels and kinetics of IR-induced activation of Raf-1 and MAPK, it is likely that multiple effectors, including Raf-1, are upstream of ERK (Suy *et al.* 1997).

Ionising radiation-induced cellular and biological effects have been associated with the formation of reactive oxygen intermediates (ROIs) (Hall, 1988). It appears that ROIs play a role in the phosphorylation of ERK, because prior exposure of cells to the antioxidant N-acetyl-cysteine (NAC) inhibits ERK activation by IR, but not ERK activation by serum (Stevenson *et al.* 1994). This suggests that IR-stimulation of the ERK pathway involves the intracellular free radicals. IR has been shown to increase the expression of a dual specificity phosphatase HVH1 (CL100) gene, which dephosphorylates and inactivates MAPKs (Kwak *et al.* 1994; Liu *et al.* 1995; Kasid *et al.* 1997). One speculation is that increased levels of HVH1 may dephosphorylate and inactivate IR-induced MAPKs, thereby providing a negative feedback regulation of the MAPK pathway. It is likely, though not proven yet, that IR-inducible ERK pathway is a protective response. As mentioned earlier, depending on the cell type, IR induces JNK/SAPK activity, which in many instances precedes apoptosis. Induction of JNK activation via IR is a ceramide-dependent or independent mechanism.

Mitogen-induced activation of ERKs has been associated with enhanced AP-1 DNA binding activity, and IR has been shown to increase AP-1 DNA binding activity and *c-jun* expression (Hallahan *et al.* 1995; Meighan-Mantha *et al.* manuscript submitted, 1998).

2.5 Cell cycle control elements

Gamma irradiation of eukaryotic cells results in a delay in cell division due to arrest of cells mainly in the G_1 and G_2 phases of the cell cycle (Howard and Pelc, 1953; Elkind *et al.* 1963; Bedford and Mitchell, 1977; Hartwell and Kastan, 1994; Murname, 1995). The *p53* and *ATM* gene products play major roles in the regulation of the cell cycle following DNA damage (reviewed in Morgan and Kastan, 1997, and references therein). Specifically, G1 arrest following IR correlates with the induction of wild type (wt) p53 levels, whereas lack of G1 arrest is seen in cells overexpressing mutant p53, or in fibroblasts from *p53*-null mice (Kastan *et al.* 1991, 1992). In the presence of wt p53, IR exposure induces the expression of several genes including those coding for gadd45; the cdk inhibitor, $p21^{WAF-1/CIP1}$; the negative regulator of p53, mdm2 and a cell-death promoting protein Bax (Kastan *et al.* 1992; Hollander *et al.* 1993; Zhan Q. *et al.* 1994; Dulic *et al.* 1994; Miyashita *et al.* 1994; Selvakumaran *et al.* 1994). *p21*-null cells are partially defective in a G1 arrest following IR, suggesting that p53-regulated G1 arrest proceeds following more than one signal (Brugarolas *et al.* 1995; Deng *et al* 1995). Another downstream effector of the p53-dependent G1 arrest pathway is proliferating cell nuclear antigen (PCNA) which interacts with GADD 45 and is involved in DNA replication and repair (Smith M.L. *et al.* 1994; Chen I.T. *et al.* 1995). Interestingly, another member of the Gadd family, gadd34 is inducible in a p53-independent manner in cells undergoing IR-inducible apoptosis (Hollander *et al.* 1997). Functional significance of p53-mediated, damage-responsive G1 arrest is indicated by the observations of a correlation between loss of p53 function and decrease in DNA excision repair (Ford and Hanawalt, 1995; Wang X.W. *et al.* 1995). Cells derived from AT patients contain mutated ATM, and show multiple defects including hypersensitivity to IR, lack of p53-dependent G1 arrest following IR, and defective p53-independent S and G2-M cell cycle check points (reviewed in Morgan and Kastan, 1997). The normal ATM protein is required to optimally induce p53 levels following IR, suggesting that ATM is upstream of p53 in the IR-induced G1 arrest pathway (Kastan *et al.* 1992; Canman *et al.* 1994). The roles of ATM and p21 in G2/M delay or in modulating apoptosis are not known, but some clues are emerging. For example, *ATM*-null lymphoma cells arrest normally at G2, but *ATM/p21* double-null lymphoma cells show defective or abolished G2 arrest. Consistent with these data, *ATM/p21* double-null mice are more sensitive to ionising radiation than *ATM* single-null mice (Wang,Y.A. *et al.*, 1997).

IR-induced G2 delay correlates with the reduced levels of cyclin B1 mRNA and protein (Muschel *et al.* 1991). Cyclin B1 forms a complex with p34cdc2 kinase and this association is required for p34cdc2 activity, and exit from G2 (Solomon *et al.* 1990; Pines, 1995). Dephosphorylation and inactivation of p34cdc2 is mediated by the phosphorylated and activated cdc25C phosphatase. Interestingly, IR stimulates tyrosine phosphorylation of p34cdc2 kinase, and inhibits phosphorylation of cdc25C (Kharbanda *et al.* 1994b; Barth *et al.* 1996). More recently, it has been shown that the levels of cyclin B1 regulate the length of G2 in irradiated but not unirradiated cells, implying that other damage-inducible factors may be involved in regulation of the mitotic delay (Kao *et al.* 1997). In addition, transient down-regulation of expression of a number of cell cycle control genes (including those coding for cdc2, cyclin A , cyclin B, cdc25, histone H2A/B) has been reported (Datta R. *et al.* 1992c; Kharbanda *et al.* 1994b; Meighan-Mantha *et al.* manuscript submitted, 1998). Taken together, it appears that activities of many cell cycle controlling elements are modulated in response to IR. Earlier studies have demonstrated an association between increased G2 delay and radioresistance in primary rat embryo fibroblasts (REFs) transformed with H-*ras* and v-*myc* (Mckenna *et al.* 1991). REFs transfected with *myc* alone undergo apoptosis upon serum withdrawal and irradiation. While the presence of H-*ras* does not prevent apoptosis of REFs due to overexpression of *myc* following serum withdrawal, H-*ras* suppresses apoptosis of *myc*-transfected cells due to X-rays (McKenna *et al.* 1996). REFs containing H-*ras* show substantial G2 arrest associated with suppression of cyclin B1 mRNA expression, whereas cells with *myc* alone have minimal G2 delay and no suppression of cyclin B1 mRNA. These studies suggest that a profound G2 delay is an important anti-apoptotic signal and H-Ras is a regulator of this checkpoint in irradiated cells.

Little is known about the S phase controlling elements. *Brca*1 and *Brca*2 are tumour suppressor genes because mutations or loss of one allele is associated with a dominant predisposition to cancer (Wooster *et al.* 1994; Smith S.A. *et al.* 1992). Both genes code for nuclear phosphoproteins and are highly expressed in proliferating cells with expression peaking at the G1/S boundary of the cell cycle. More recently, BRCA1 has been shown to undergo phosphorylation in G1 and S phases within 1 h after irradiation of MCF-7 cells, and it co-localises with PCNA/replicating DNA structures, suggesting a role of BRCA1 in the processing of abnormal DNA, and in the S-phase check point (Scully *et al.* 1997a,b).

2.6 The Bcl-2 family, and caspases

The Bcl-2 family of proteins includes both anti-apoptotic (Bcl-2, Bcl-X_L, MCL-1, A1) and pro-apoptotic members (Bax, Bcl-X_S, Bad, Bak, Bid, and Bik) (Farrow and Brown, 1996; Oltvai and Korsmeyer, 1994). The family members with a hydrophobic C-terminal signal anchor sequence are intracellular membrane proteins localised to outer mitochondrial, endoplasmic reticulum, and outer nuclear membranes (Hockenbery *et al.* 1990, Krajewski *et al.* 1993, Nguyen *et al.* 1993). The Bcl-2 family members constitute important checkpoints in a distal cell death pathway with Bcl-2 functioning to prevent the activation of caspases, the terminal effectors of apoptosis, and Bax inducing a downstream programme of mitochondrial dysfunction and activation of caspases (see Chittenden, this volume and references therein).

p53 downregulates endogenous Bcl-2 expression, and upregulates Bax expression, which may then lead to apoptosis (Harper *et al.* 1993; Miyashita *et al.* 1994; Selvakumaran *et al.* 1994; Strasser *et al.* 1994). Furthermore, apoptotic effects of p53 but not $p21^{Cip1}$-mediated growth arrest are overcome by overexpressed Bcl-2. More recently, a cytoplasmic protein 53BP2 has been shown to interact with both Bcl-2 and p53. The transient transfection of 53BP2 correlates with arrest of cells at G2/M phase of the cell cycle, thereby suggesting a mechanism for the ability of wild type p53 to induce G2/M-arrest (Aloni-Grinstein *et al.* 1995; Naumovski *et al.* 1996).

Recent investigations have revealed other interesting characteristics of the anti-apoptotic members (Bcl-2, Bcl-X_L) including the pore-forming domains which would serve as channels to regulate the transport of ions or proteins across membranes to which Bcl-2 members are bound to, and interactions with other proteins (Raf-1, Bag-1, 53BP2), thereby generating additional means of cell survival (Minn *et al.* 1997; reviewed in Reed, 1997). Accordingly, it is suggested that Bcl-2 may regulate the release of protease activators, cytochrome C and apoptosis inducing factor (AIF) from mitochondria (Kluck *et al.* 1996; Yang *et al.* 1997; Li *et al.* 1997; reviewed in Salvesen and Dixit, 1997).

Bcl-2 and Bcl-X_L have been associated with the resistance of cells to cytotoxic agents (Kitada *et al.* 1994; Minn *et al.* 1995; reviewed in Rudin and Thompson, 1997). IR has been shown to induce *bax* expression and *bcl-X_L* expression, but to down regulate *bcl-2* in cells having an apoptosis-susceptible phenotype and wild type p53 (Zhan *et al.* 1994, 1996; Haldar *et al.* 1994; Selvakumaran *et al.* 1994; Miyashita *et al.* 1994). Although, downregulation of Bcl-2 and upregulation of Bax after IR are consistent with the ability of p53 to induce apoptosis, induction of Bcl-X_L, an anti-apoptotic factor seems paradoxical. One suggestion has been made that induction of Bcl-X_L serves to limit the severity of Bax-promoted cell death, and increased Bax and Bcl-X_L proteins might operate in a balance between p53-mediated apoptosis and G1-arrest which allows time to recover from IR-induced damage (Zhan *et al.* 1996).

In other studies, IR exposure of cells has been shown to increase the cytosolic cytochrome C levels, coinciding with the activation of proteases as evidenced by the presence of a cleavage products of poly(ADP-Ribose) polymerase (PARP) and PKCδ. These IR-induced signals are blocked by overexpression of Bcl-X_L, consistent with the resistance of cells that overexpress Bcl-X_L/Bcl-2 to IR-induced DNA fragmentation (Sentman *et al.* 1991; Datta R. *et al.* 1995; Emoto *et al.* 1995; Datta S.R. *et al.* 1997; Kharbanda *et al.* 1997). Furthermore, cells that overexpress the cysteine protease inhibitor, the baculovirus p35 protein show IR-induced accumulation of cytochrome C in the cytosol, but fail to show cleavage of PARP, indicating that release of cytochrome C from mitochondria precedes the activation of cysteine proteases (Datta R. *et al.* 1997).

2.7 DNA double-strand breaks, and repair pathways

Ionising radiation is an efficient producer of DNA double-strand breaks (DSBs), and the repair of DSBs is important for the survival of cells. The signal transduction mechanism of the repair of DSBs is not known. In severe combined immunodeficient (SCID) mice, increased sensitivity to IR is due to decreased efficiency of the repair of DSBs (Fulop and Philips, 1990; Biedermann *et al.* 1991). DNA-dependent protein kinase (DNA-PK) is composed of a ~460 kDa catalytic subunit, DNA-PKcs, a serine threonine protein kinase that is dependent on DNA double-strand ends for its activity, and the heterodimeric p70/p80 DNA binding subunit, Ku protein that is essential for the DNA binding activity of this complex (Gottlieb and Jackson, 1993). Several reports suggest a defect(s) in DNA-PK is an underlying mechanism of IR-sensitivity, and deficiency in V(D)J recombination associated with the SCID phenotype (Anderson, 1993; Kirchgessner *et al.* 1995; Blunt *et al.* 1995; Malynn *et al.* 1998). Based on somatic cell hybrid studies, eight x-ray cross-complementation (XRCC) groups have been suggested (Jeggo *et al.* 1991). DNA-PKcs is the product of XRCC7 gene, Ku p70 is the product of XRCC6 gene, and Ku p80 is the product of XRCC5 gene (Taccioli *et al.* 1994; Sipley *et al.* 1995, Thompson and Jeggo, 1995, Singleton *et al.* 1997; Gu *et al.* 1997). The role of DNA-PK in the SCID phenotype is not known. SCID mice-derived cells (SCID-MEFs) display normal cell cycle check point response to IR (Huang *et al.* 1996). Consistent with these data, wild-type p53 levels are elevated following irradiation of SCID-MEFs (Fried *et al.* 1996).

Mammalian Rad54 is a member of the Rad52 family of DNA repair proteins (Rad51, Rad52, and Rad54) and appears to be directly involved in the repair of DNA damage. Targeted disruption of *rad54*, results in growth retardation, x-ray sensitivity, and defects in homologous recombination (Bezzubova *et al.* 1997; Essers *et al.* 1997). In contrast, homozygous null mutants of *rad51* are not viable, suggesting Rad51 has additional important function. BRCA1/2 co-localises with Rad51 (Rajan *et al.* 1996; Scully *et al.* 1997a,b). Furthermore, mice lacking *Brca2* show embryonic lethality and radiation hypersensitivity, implying that BRCA2 may be a cofactor in the Rad51-dependent repair of double-strand breaks (Sharan *et al.* 1997a,b). These associations also suggest that BRCA1/BRCA2 and Rad51 are important for maintenance of genome integrity.

2.8 Transcription factors

Several genes are modulated at the level of transcription following irradiation of cells. Consistent with this, IR stimulates DNA binding activities of a number of transcription factors including NF-κb, AP-1, SP-1, OCT1, and CREB (Brach *et al.* 1991; Hallahan *et al.* 1993; Sahijdak *et al.* 1994; Meighan-Mantha *et al.* manuscript submitted, 1998). In certain cases, inhibition of transcription factor activity has been directly associated with cell death. For example, IR stimulates nuclear translocation and activation of NF-κB. Retention of NF-κB in the cytosol through expression of a super-repressor form of IκBα has been correlated with increased cell death (Wang C-Y *et al.* 1996). Second, the inducible expression of a dominant negative c-*jun* construct prevents X-ray inducibility of transcription through the AP-1 binding site, delays the onset of S phase, and reduces the survival of human cells exposed to ionising radiation (Hallahan *et al.* 1995). Third, the *bax* gene promoter has been shown to contain several p53 binding sites, and can be directly transcriptionally transactivated by p53. IR also induces *bax* expression in cells containing wt p53, but not in most tumour cells that lack p53 or contain mutant p53, and regulation of *bax* expression via p53 seems to be associated with *in vitro* and *in vivo* radiosensitivity (Zhan *et al.* 1994; Kitada *et al.* 1996). It is important to note here that mechanisms other than Bax-mediated cell death cannot be ruled out, because radiation has been reported to induce p53-dependent elevations in $p21^{WAF-1/CIP1}$ in a wide variety of tissues and cell types, including some which do not show elevation in Bax or apoptosis after irradiation (Macleod *et al.* 1995). Therefore, p53 may be necessary, but insufficient to cause *bax* expression and apoptosis after irradiation.

Although the precise mechanism of IR-mediated transcriptional transactivation is not known, several reports suggest that reactive oxygen species (ROS) generated by IR may be involved in the activation of transcription factors. Accordingly, N-actetyl-*l*-cysteine inhibits IR-induced NF-κB DNA binding activity in human B lymphocytes (Schieven *et al.* 1993) and IR-induced c-*jun* expression in HL-205 cells (Datta R. *et al.* 1992b). Interestingly, hypoxia also induces NF-κB activity via the Ras/Raf signalling pathway (Koong *et al.* 1994). Furthermore, tyrosine kinase inhibitor inhibits both IR and hypoxia-induced NF-κB activity, suggesting that stress-responsive NF-κB activation is mediated by prior stimulation of protein tyrosine kinase(s) (Schieven *et al.* 1993; Koong *et al.* 1994).

3.0 RAF-1 PATHWAY GENERATES A SURVIVAL SIGNAL

Several lines of evidence suggest that Raf-1 functions as a survival factor in a variety of cell types, and in response to a number of stress-inducers including ionising radiation, ultraviolet radiation, growth factor-deprivation, and cytokines.

3.1 Ionising radiation

The possibility of a dual role of Raf-1 in oncogenesis and the cell survival response was first suggested by the antisense c-*raf*-1 cDNA transfection experiments using a relatively radioresistant human laryngeal squamous carcinoma cell line (SQ-20B) (Weichselbaum *et al.* 1988; Kasid *et al.* 1989). SQ-20B cells transfected with full length antisense c-*raf*-1 cDNA show delayed tumour growth in athymic mice, and enhanced sensitivity to ionising radiation, compared with sense c-*raf*-1 cDNA transfectants (Kasid *et al.* 1989). Radiation dose response was measured using a long-term clonogenic survival assay. Recently, several independent studies have demonstrated a decrease in the clonogenic survival response of SQ-20B cells following treatment with a combination of antisense *raf* oligodeoxynucleotides and IR (Soldatenkov *et al.* 1997; Gokhale *et al.* 1997; Pirollo *et al.* 1997). In addition, antitumour activity and the radiosensitisation effect of antisense *raf* oligodeoxynucleotide have been observed *in vivo* in athymic mice model (Monia *et al.* 1996; Gokhale *et al.*, 1998). It appears that IR-stimulation of Raf-1 kinase activity discussed earlier may be a protective response (Kasid *et al.* 1996; Suy *et al.* 1997), and the antisense-mediated reduction in the levels of endogenous Raf-1 enhances the susceptibility of cells to radiation-induced toxicity. Increased radioresistance due to overexpression of *raf-1* gene has been also reported in NIH3T3 cells (Pirollo *et al.* 1989). Furthermore, immortalised human bronchial epithelial cells (BEAS-2B) transfected with c-*raf*-1 gene (BEAS-

2B- *raf*) are non-tumorigenic (Pfeifer *et al.* 1989), but show enhanced resistance to radiation-induced killing, implying that a dissociation exists between the roles of Raf-1 as a survival factor and as an oncogenic protein (Kasid *et al.* 1993; Pfeifer *et al.* manuscript in preparation).

More recently, mouse primary embryo fibroblasts (MEF's) containing targeted disruption of c-*raf-1* (*raf* -/-) showed increased radiation-induced apoptosis, compared to wild type MEF's (*raf* +/+). Interestingly, Bad and Cyclin B1 levels were enhanced following irradiation of *raf* -/- MEF's, relative to *raf* +/+ MEF's (Suy *et al.* manuscript in preparation). These data provide a genetic proof of principle and are consistent with a role for Raf-1 in the generation of a radioresponsive survival signalling pathway.

3.2 Ultraviolet radiation

Ultraviolet (UV) irradiation causes DNA damage, lipid peroxidation followed by generation of free radicals, and depletion of the intracellular pool of reduced glutathione (GSH), resulting in oxidative stress (Devary *et al.* 1992). Exposure of cells to UV radiation triggers a signal transduction pathway involving activation of Src tyrosine kinase, followed by activation of Ha-Ras and Raf-1, and increased activity of transcription factors AP-1 and NF-κB (Devary *et al.* 1992; Radler-Pohl *et al.* 1993; Devary *et al.* 1993). A free-radical scavenger, NAC inhibits activation of Src by UV and expression of c-*jun*. In addition, UV-inducible signalling is blocked by tyrosine kinase inhibitors and dominant negative mutants of v-*src*, Ha-*ras* and c-*raf*-1 (Devary *et al.* 1992, 1993), implying that the UV response is mediated by oxidative-stress and activation of protein tyrosine kinase(s). HeLa cells pretreated with tyrosine kinase inhibitors, lavendustin or tyrphostin, show increased sensitivity to UV radiation. These observations indicate that the activation of the Src/Ras/Raf-1/AP-1 signalling pathway is a protective response to UV-induced stress in HeLa cells (Devary *et al.* 1992).

3.3 Growth factor deprivation

The ability of growth factors to enhance cell survival has been demonstrated in a number of cell systems (Neta and Oppenheim, 1991; Leigh *et al.* 1993; Waddick and Uckun, 1993; Fuks *et al.* 1994; Haimovitz-Friedman *et al.* 1991). One of the well characterised model systems is the interleukin-3 (IL-3)-dependent myeloid progenitor cell line, 32D.3, which is dependent upon IL-3 for proliferation and survival (Greenberger *et al.* 1983). In the absence of IL-3, 32D.3 cells accumulate in G1 and eventually undergo apoptosis. Since overexpression of c-Myc promotes progression of these cells into S-phase but accelerates apoptosis in the absence of IL-3, it appears that cell cycle progression and cell survival signalling pathways are distinct (Askew *et al.* 1991; Evan *et al.* 1992). Taken together, these observations imply that growth factor treatment of cells triggers signals which promote cell survival and cell cycle progression.

32D.3 cells expressing v-*raf* do not undergo apoptosis in the absence of IL-3; and in the presence of IL-3, these cells show an enhanced rate of cell proliferation due to a shortened G1 phase (Troppmair *et al.* 1992; Cleveland *et al.* 1994). Other studies have shown that the IL-3 receptor-activated Ras/Raf/MAPK pathway leads to the anti-apoptotic effect of IL-3 (Sato *et al.* 1993; Kinoshita *et al.* 1995; Chao *et al.* 1996). These data support a general notion that growth-factor dependent pathways promote cell cycle progression and cell survival, and that consitutively-activated Raf provides a survival signal which balances the mitogenic signalling pathway by suppression of apoptosis (Kolch *et al.* 1990).

3.3.1 Raf-1 and Bcl-2/Bad/Bag-1

In the absence of IL-3, expression of Bcl-2 also supports survival of IL-3-dependent progenitor cells (Vaux *et al.* 1988; Nunez *et al.* 1990). A functional synergy between Bcl-2 (26kDa) and constitutively activated Raf-1 (~45kDa) in the suppression of apoptosis following IL-3 withdrawal of 32D.3 cells has been reported (Wang H.G. *et al.* 1994). Further support for the role of Raf-1 in Bcl-2-mediated inhibition of apoptosis is provided by the observation that mitochondrial targeting of constitutively-activated Raf-1 delays cell death, but maintains 32D.3 cells arrested in G1/G0 in IL-3-deficient medium (Wang H.G. *et al.* 1996a). The pro-apoptotic Bcl-2 family member, Bad, is phosphorylated *in vitro* in the presence of activated Raf-1 targeted to the mitochondria, as well as by IL-3 in IL-3-dependent haematopoietic cells (Wang H.G. *et al.*1996a; Zha *et al.* 1996). Phosphorylated Bad is sequestered in the cytosol through interaction with the 14-3-3 protein, and is unable to heterodimerise with the anti-apoptotic members Bcl-2/Bcl-X_L. It has been suggested that mitochondrial targeting of Raf-1 may bring this protein in closer vicinity to the substrates or adaptors essential for suppression of apoptosis (Wang H.G. *et al.* 1996a; Reed, 1997).

Unlike plasma membrane targeting of Raf-1 where Ras binding occurs on the amino-terminus of Raf-1, Bcl-2 interaction requires only the catalytic domain of Raf-1. Whether Bcl-2 is able to target full length or endogenous Raf-1 to the mitochondria following IL-3 withdrawal remains to be determined. It is also not known if any activating point mutation(s) of the natural pool of Raf-1 is required for its translocation to the mitochondria. It is noteworthy that a very small (approx 1%) but perhaps functionally significant pool of endogenous Raf-1 interacts with Bag-1, another anti-apoptotic member of the Bcl-2 family interacting with Bcl-2 (Wang H.G. *et al.* 1996b). Transfection of Bag-1 enhances the kinase activity of endogenous Raf-1 protein, perhaps in preparation for the mitochondrial activities of Raf-1 in conjunction with the Bcl-2 family members. Whether interaction of Raf-1 with Bag-1 is important for the suppression of apoptosis remains to be seen.

The possibility of the initiation of a factor-dependent survival signalling pathway(s) at the plasma membrane may not be completely ruled out, because *in vivo* phosphorylation of exogenous or endogenous Bad occurs after treatment of appropriate cells with a variety of other potent survival factors like PDGF and nerve growth factor (NGF), which also activate Raf-1 and MAPK signalling cascade, implying a growth factor receptor-mediated cell survival mechanism (Zha *et al.* 1996; Datta *et al.* 1997). Furthermore, the survival response is also regulated by the Ras-MAPK-p90 Rsk pathway (Xia *et al.* 1995). Inhibition of MEK, the only known physiological substrate of Raf-1 (Dent *et al.* 1992), does not significantly block Bad phosphorylation (Datta *et al.* 1997). Therefore, it is likely that Raf-1 (~75kDa) has as-yet unidentified cell survival checkpoint-related substrate(s) *in vivo.* Nonetheless, the possibility of a plasma membrane-/Ras-independent role of Raf-1 as a survival factor under growth factor-deprived conditions is very attractive. It would be important to determine if endogenous Raf-1 is recruited to the mitochondria following stress, especially under conditions where Ras activity is negated in the cells. Since overexpression of Bad does not induce cell death in all cell types (Yang E. *et al.* 1995), as-yet unknown players in the cell survival pathway may also exist upstream or downstream of the Raf-1 pathway.

3.3.2 Akt

IL-3-dependent survival of haematopoietic cells also relies on the activity of a serine/threonine kinase Akt (also known as PKB), mediated by the activation of PI3-kinase signalling pathway which involves growth factor receptor activation, recruitment of PI3-K to the membrane, and phosphorylation of phosphoinositides (Corey *et al.* 1993; Carpenter and Cantley, 1996; Songyang *et al.* 1997; Kennedy *et al.* 1997; Khwaja *et al.* 1997). In 32D cells, consitutively activated Akt promotes survival after IL-3 withdrawal. IL-3-dependent Akt signalling pathway in these cells is distinct from the pathway leading to MAPK activation, because dominant negative mutants of Akt inhibit endogenous Akt activation by IL-3 and accelerate apoptosis, without blocking MAPK activation. Akt signalling mechanism appears not to involve Bcl-2, because dominant negative Akt does not compromise the Bcl-2 dependent survival, and Bcl-2 and Akt do not synergise in protection against apoptosis (Songyang *et al.* 1997). In primary neurons, however, insulin-like growth factor 1 (IGF-1)-induced activation of the PI3-K/Akt pathway leads to Bad phosphorylation, inhibition of Bad-induced cell death, and promotion of cell survival (Dudek *et al.* 1997; Datta S.R. *et al.* 1997). Furthermore, a consitutively activated Akt substitutes for IGF-1-induced survival response, and suppresses Bad-mediated cell death (Datta S.R. *et al.* 1997).

The mechanism of Akt-mediated prevention of apoptosis is not known. Ras-induced activation of the PI3-K/Akt pathway prevents apoptosis induced by conditional overexpression of c-Myc protein in serum-deprived Rat-1 fibroblasts; and glycogen synthase kinase-3 (GSK3), the only known downstream target of Akt, is not implicated in the regulation of cell survival (Kauffmann-Zeh *et al.* 1997). Interestingly, constitutively-activated oncogenic V12 Ras mutant, the partial loss-of-function V12S35 Ras mutant, or a constitutively-activated Raf-CAAX mutant enhance c-Myc-induced apoptosis in serum-starved Rat-1 cells. Ras binds and activates PI3-K, causing activation of Akt, but the Ras-Akt pathway does not lead to activation of MAPK or p70S6kinase, implying the existence of a Ras pathway parallel to the classic Ras/Raf/MEK/MAPK pathway (Franke *et al.* 1995; Kauffmann-Zeh *et al.* 1997; reviewed in Franke *et al.* 1997). Akt may not be a general inhibitor of apoptosis as it does not effectively block etoposide or UV-induced apoptosis in the presence of serum.

3.3.3 Raf-1 and p53/p21$^{WAF-1/CIP1}$

In a non-malignant IL-3-dependent murine lymphoid cell line (Baf-3), exposure to IR in the presence of IL-3 causes cell cycle arrest, whereas irradiation causes apoptosis in the absence of IL-3 (El-Deiry *et al.* 1993). More recently, IR-induced G1 arrest or apoptosis in Baf-3 cells has been shown to depend on p53 function, suggesting that growth factor treatment generates a dominant survival signal which is downstream of p53 (Canman *et al.* 1995). Furthermore, constitutive activation of Src or Raf-1 kinase inhibits IR-induced apoptosis, and restores G1 arrest in the absence of IL-3; like in the IL-3 response, these protein kinases induce p21$^{WAF-1/CIP1}$, a previously established mediator of p53-induced G1 arrest and apoptosis (El-Deiry *et al.* 1993), and decrease the expression of gadd45 gene, previously shown to be induced by all types of DNA-damaging agents (Fornace, 1992; Canman *et al.* 1995). These data strongly support the general notion that growth factors function as survival factors (Williams G.T. *et al.* 1990, Canman *et al.* 1995) and suggest that activation of Src/Raf-1 kinase substitutes for growth factor dependencies of several cell types, thereby implying a role of activated Src or Raf-1 kinase in cell survival.

3.4 TNFα

Cell death induced by the members of the TNFα super family is mediated by a sphingomyelinase pathway involving ceramide generation, followed by activation of the SAPK/JNK pathway (Obeid *et al.* 1993). In contrast, ceramide metabolites, sphingosine and sphingosine-1-phosphate, suppress apoptosis, and induce cell proliferation via counteracting ceramide-induced JNK activation and stimulating a Raf/MEK/ERK pathway in TNFα treated U937 cells (Olivera and Spiegel, 1993; Wu *et al.* 1995; Cuvillier *et al.* 1996). Thus, Raf-1 activity appears to generate a protective response in TNFα-treated cells.

Several lines of evidence indicate that Raf-1 mediates growth factor- or stress-induced activation of NF-κB site-driven gene expression (Finco and Baldwin, 1993; Li and Sedivy, 1993; Devary *et al.* 1993; Koong *et al.* 1994; Bertrand *et al.* 1995; Zhou and Kuo, 1997). Consistent with these observations, constitutive activation of Raf-1 increases NF-κB activity (Li and Sedivy, 1993; Folgueira *et al.* 1996). Moreover, NF-κB activation following TNFα treatment requires Raf-1, and activation of this transcription factor provides protection against apoptosis (Wang C.Y *et al.* 1996; Beg and Baltimore, 1996; Van Antwerp *et al.* 1996; Arsura *et al.* 1996; Liu Z. G. *et al.* 1996b, Baeuerle and Baltimore, 1996).

TNFα treatment of cells stimulates the expression of manganese superoxide dismutase (MnSOD), an antioxidant enzyme, via activation of NF-κB, and increase in MnSOD level is suggested to be a cellular defence mechanism in response to TNFα-induced oxidative stress (Smith D.M. *et al.* 1994). In human tumour cells transfected with antisense c-*raf*-1 cDNA, a relative increase in the steady state levels of MnSOD has been observed compared to vector transfected cells, and TNFα treatment of antisense c-*raf*-1 cDNA transfectants show significant release of hydrogen peroxide. These data imply, though not prove as-yet, a role of Raf-1 in cellular protection against cytokines (Kasid *et al.* unpublished data).

3.5 Paclitaxel

When is Raf-1 not a survival factor? The cytotoxic drug paclitaxel (PTX) binds to and stabilises microtubules, preventing depolymerisation, and leading to G2/M arrest, microtubular bundling and cell death (Horwitz, 1992). PTX has been shown to induce phosphorylation of Raf-1 and Bcl-2, and apoptosis in PTX-sensitive, but not PTX-resistant cells (Blagosklonny *et al.* 1996, 1997). It appears that Raf-1 phosphorylation requires the interaction of PTX with tubulin because in resistant cells tubulin polymerisation does not occur following PTX treatment. The mechanism linking PTX-induced disruption of microtubules and Raf-1/Bcl-2 phosphorylation is not known. Earlier studies reported rapid fragmentation of microtubules when polymerised tubulin was incubated with recombinant human translational elongation factor-1α (EF-1α), a regulator of cell proliferation and transformation. Microinjection of EF-1α into fibroblasts caused rapid destruction of microtubule networks (Tatsuka, 1992; Shiina *et al.* 1994). The promoter region of the human EF-1α gene contains the AP-1 binding site (Wakabayashi-Ito and Nagata 1994), and PTX induces AP-1 activity (Lee *et al.* 1997). PTX may induce EF-1α expression via activation of AP-1 in PTX-sensitive cells.

Constitutive activation of Raf-1 protein kinase activity has been associated with a significant increase in EF-1α gene expression in human squamous carcinoma cells (Patel B. *et al.* 1997; PatelS. *et al.*1997). Activated Raf-1 may enhance PTX-induced microtubule damage via newly synthesised EF-1α in PTX-sensitive cells. Although other possibilities remain, this speculation seems to be consistent with the previous work indicating that cycloheximide or actinomycin D inhibits toxicity due to microtubule-active drugs (Donaldson *et al.* 1994).

3.6 Raf-1 vs. Ras

Several reports have identified the cyclic AMP (cAMP)-dependent protein kinase A (PKA) as a negative regulator of Raf-1 (Burgering *et al.* 1993; Cook and McCormick, 1993; Wu *et al.* 1993; Hafner *et al.* 1994). PKA does not affect the activity of Ras, MEK, or MAPK, and inhibits Raf-1 activity independent of Ras (Cook and McCormick, 1993; Wu *et al.* 1993). In v-*abl* transformed NIH3T3 cells, endogenous Raf-1 is activated and endogenous Myc is induced. In these cells, PKA inhibits Raf-1 activity and induces a pronounced G2/M phase arrest and apoptosis without affecting v-Abl kinase activity or Abl-dependent induction of c-Myc (Wong *et al.* 1995; Renshaw *et al.* 1992; Weissinger *et al.* 1997). Since overexpression of c-Myc in the absence of appropriate growth factor or a growth factor-like survival, signal triggers apoptosis (reviewed in Evan and Littlewood, 1993; Harrington *et al.* 1994), it is suggested that Raf-1 activation in v-*abl* transformed cells contributes a survival signal, and abrogation of Raf-1 activity by PKA creates a scenario whereby a high level of c-Myc expression is faced with a growth factor deprivation signal, which leads to apoptosis (Weissinger *et al.* 1997). In further support of this concept, conditional expression of constitutively-activated Raf-1 abrogates c-*myc*-induced apoptosis in a growth factor-deprived TGR-1 cell line, which is a

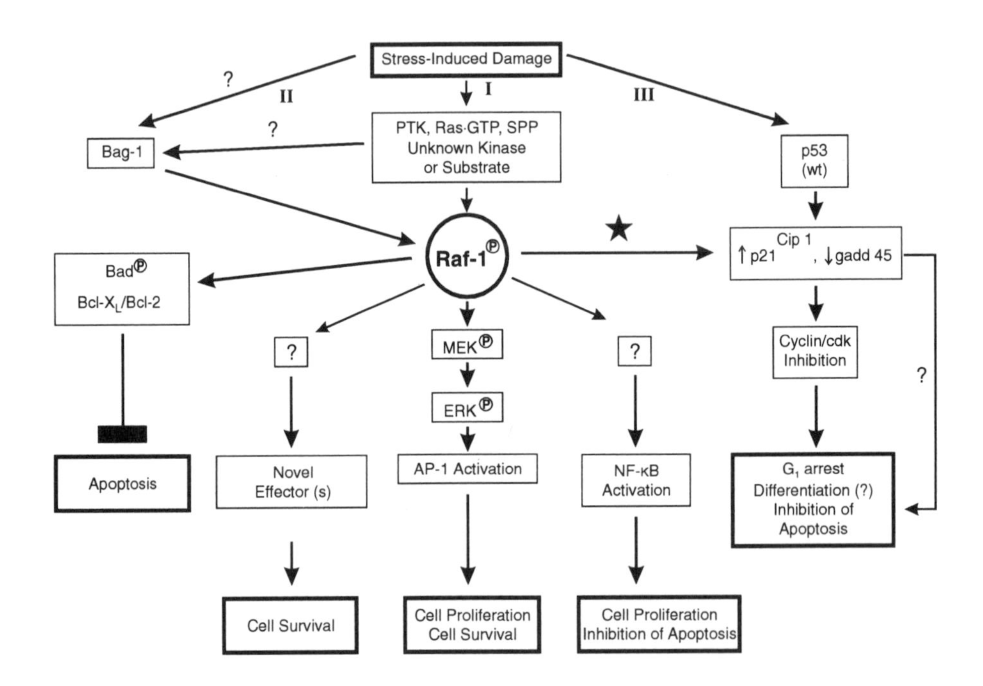

Figure 1. *Speculative model of multiple routes (I-III) of Raf-1-mediated signalling.*
See text for details. PTK, protein tyrosine kinase; SPP, sphingosine-phosphate;wt, wild type.

subclone of the Rat-1 fibroblasts (Weissinger *et al.* 1997). Interestingly, a MEK inhibitor does not induce apoptosis in v-*abl* transformed cells (Weissinger *et al.* 1997), ruling out the involvement of MEK in Raf-1-mediated inhibition of apoptosis in these cells.

4.0 SUMMARY

Genetic, biochemical, and biological studies during the past few years have redefined the functions of a large number of cellular proteins initially associated with oncogenesis, tumour suppression, cell cycle regulation, or transcription/translation-related processes. Many of these activities have been unleashed following the exposure of cells to genotoxic or non-genotoxic damage. While no one pathway can be charted for a particular stress-inducer, a functional balance between the components of different pathways appears to regulate cellular response to stress. Signal transduction pathways from plasma membrane to nucleus, and involvement of other subcellular compartments including mitochondria may have significant ramifications in the biological effects of ionising radiation. Further, tumour cells may respond to cytotoxic agents by launching a compensatory defence system, and resistance to therapy may be acquired at several points along the apoptotic signalling pathway.

Based on the current knowledge, a scheme of Raf-1-mediated stress-responsive signal transduction is shown in Figure 1. In response to stress-induced genotoxic/non-genotoxic damage, activation of a Raf-1 kinase pathway occurs in a cell type- and stimulus-specific manner. Somewhat analogous to mitogen-induced signalling, radiation stimulates protein tyrosine phosphorylation, membrane-recruitment and activation of Raf-1, followed by activation of a downstream MEK/ERK pathway leading to cell cycle progression, proliferation, and resistance to stress-induced cytotoxicity. Following TNFα treatment, constitutively-activated Raf-1 may activate NF-κB-dependent transcription of a protective/anti-apoptotic gene(s). Under growth factor-deprived conditions, constitutively-activated Raf-1 interacts with Bcl-2 in the mitochondria resulting in the block of apoptosis. In addition, in growth factor-deprived cells containing normal p53, inducible expression of constitutively-activated Raf-1 (★, Fig.1) restores G1-arrest and blocks radiation-induced apoptosis. Alternatively, Raf-1 may have a physiological substrate(s) other than MEK, which plays a role in the stress-induced cellular response. Future identification of potential stress-inducible kinase(s) or other effector(s) may provide further insight into cellular protection mechanisms mediated by the Raf-1 signalling pathway.

Acknowledgements

The authors apologise to all those whose work in this rapidly expanding field has not been referenced, and thank colleagues and collaborators for their valuable contributions. The authors' work was funded by the NIH grants CA46641, CA58984, CA68322 and CA74175.

5.0 REFERENCES

Aloni-Grinstein,R., Schwartz,D., and Rotter,V. (1995) Accumulation of wild-type p53 protein upon gamma-irradiation induces a G2 arrest-dependent immunoglobulin kappa light chain gene expression. *EMBO J.* **14**: 1392-1401.

Anderson,C.W. (1993) DNA damage and the DNA-activated protein kinase. *Trends Biochem. Sci.* **18**: 433-437.

Arsura,M., Wu,M., and Soneshein,G. (1996) TGF beta 1 inhibits NF-kappaB/Rel activity inducing apoptosis of B cells: transcriptional activation of I kappa B alpha. *Immunity* **5**: 31-40.

Askew,D.S., Ashmun,R.A., Simmons,B.C. *et al.* (1991) Constitutive c-myc expression in an IL-3-dependent myeloid cell line suppresses cell cycle arrest and accelerates apoptosis. *Oncogene* **6**: 1915-1922.

Baeuerle,P.A., and Baltimore,D. (1996) NF-kappa B: ten years after. *Cell* **87**: 13-20.

Baffy,G., Miyashita,T., Williamson,J. *et al.* (1993) Apoptosis is induced by withdrawal of interleukin-3(IL-3) from an IL-3 dependent hematopoietic cell line associated with repartitioning of intracellular calcium and is blocked by enforced Bcl-2 oncoprotein production. *J.Biol.Chem.* **268**: 6511-6519.

Baldwin,A.S.Jr. (1996) The NF-kappaB and I kappa B proteins: new discoveries and insights. *Ann.Rev.Immunol.* **14**: 649-683.

Barth,H., Hoffmann,I., Klein,S. *et al.* (1996) Role of cdc25-C phosphatase in the immediate G2 delay induced by the exogenous factors epidermal growth factor and phorbol ester. *J.Cell Physiol.* **168**: 589-599.

Baskaran,R., Wood,L.D., Whitaker,L.L. *et al.* (1997) Ataxia telangiectasia mutant protein activates c-Abl tyrosine kinase in response to ionising radiation. *Nature* **387**: 516-519.

Beck,T.W., Huleihel,M., Gunnell,M. *et al.* (1987) The complete coding sequence of human A-raf-1 oncogene and transforming activity of a human A-raf carrying retrovirus. *Nuc. Acids Res.* **15**: 595-609.

Bedford,J.S. and Mitchell,J.B. (1977) Mitotic accumulation of HeLa cells during continuous irradiation. *Radiat. Res.* **70**: 173-180.

Beg,A.A., and Baltimore,D. (1996) An essential role for NF-kappaB in preventing TNF-alpha-induced cell death. *Science* **274**: 782-784.

Bertrand,F., Philippe,C., Antoine,P.J. *et al.* (1995) Insulin activates nuclear factor κB in mammalian cells through a Raf-1-mediated pathway. *J. Biol. Chem.* **270**: 24435-24441.

Bezzubova,O., Silbergleit,A., Yamaguchi-Iwai,Y. *et al.* (1997) Reduced X-ray resistance and homologous recombination frequencies in a Rad54$^{-/-}$ mutant of the chicken DT40 cell line. *Cell* **89**: 185-193.

Biedermann,K.A., Sun,J.R., Giaccia,A.J. *et al.* (1991) Scid mutation in mice confers hypersensitivity and a defieciency in DNA double-strand break repair. *Proc.Natl.Acad.Sci. USA* **88**: 1394-1397.

Blagoskonny,M.V., Schulte,T.W., Nguyen,P. *et al.* (1996) Taxol-induced apoptosis and phosphorylation of bcl-2 protein involves c-Raf-1 and represents a novel c-raf-1 signal transduction pathway. *Cancer Res.* **56**: 1851-1854.

Blagosklonny,M.V., Giannakakou,P., El-Deiry,W.S. *et al.* (1997) Raf-1/Bcl-2 phosphorylation: A step from microtubule damage to cell death. *Cancer Res.* **57**: 130-135.

Blunt,T., Finnie,N.J., Taccioli,G.E. *et al.* (1995) Defective DNA-dependent protein kinase activity is linked to V(D)J recombination and DNA repair defects associated with the murine *Scid* mutation. *Cell* **80**: 813-823.

Bonner,T.I., Oppermann,H., Seeburg,I. *et al.* (1986) The complete sequence of the human raf oncogene and the corresponding structure of the c-raf-1 gene. *Nuc. Acids Res.* **14**: 1009-1015.

Boulakia,C., Chen,G., Ng,F.W.H. *et al.* (1996) Bcl-2 and adenovirus E1B 19 kDA protein prevent E1A-induced processing of CPP32 and cleavage of poly (ADP-ribose) polymerase. *Oncogene* **12**: 29-36.

Brach,M.A., Hass,R., Sherman,M.L. *et al.* (1991) Ionising radiation induces expression and binding activity of the nuclear factor κB. *J.Clin.Invest.* **88**: 691-695.

Bruder,J.T., Heidecker,G., Rapp,U.R. (1992) Serum-, TPA-, and Ras-induced expression from AP-1/Ets-driven promoters require Raf-1 kinase. *Gen. Dev.* **6**: 545-556.

Brugarolas,J., Chandrasekaran,C., Gordon,J.I. *et al.* (1995) Radiation-induced cell cycle arrest compromised by p21 deficiency. *Nature* **377**: 552-557.

Burgering, B.M., Pronk, G.J., van Weeren, P. *et al.* (1993) cAMP antagonizes p21ras directed activation of extracellular signal-regulated kinase 2 and phosphorylation of mSos nucleotide exchange factor.

EMBO J. **12**: 4211-4220.
Burkhardt,A.L., Brunswick,M., Bolen,J.B. *et al.* (1991) Anti-immunoglobulin stimulation of B lympohocytes activates src-related protein-tyrosine kinases. *Proc.Natl.Acad.Sci. USA* **88**: 7410-7414.
Buscher,D., Dello Sparba,P., Hipskind,R.A. *et al.* (1993) v-Raf confers CSF-1 independent growth to a macrophage cell line and leads to immediate early gene expression without MAP-kinase activation. *Oncogene* **8**: 3323-3332.
Canman,C.E., Wolff,A.C., Chen,C.Y. *et al.* (1994) The p53-dependent G1 cell cycle checkpoint pathway and ataxia-telangiectasia. *Cancer Res.* **54**: 5054-5058.
Canman,C.E., Gilmer,T.M., Coutts,S.B. *et al.* (1995) Growth factor modulation of p53-mediated growth arrest versus apoptosis. *Gen. Dev.* **9**: 600-611.
Carpenter,C.L and Cantley,L.C. (1996) Phosphoinositide kinases. *Curr.Opin.Cell Biol.* **8**: 153-158.
Carson,E.B., McMahon,M., Baylin,S. *et al.* (1995) Ret gene silencing is associated with Raf-1- induced medullary thyroid carcinoma cell differentiation. *Cancer Res.* **55**: 2048-2052.
Chae,H.P., Jarvis,L.J. and Uckun,F.M. (1993) Role of tyrosine phosphorylation in radiation-induced activation of c-jun protooncogene in human lymphohematopoietic precursor cells. *Cancer Res.* **53**: 447-451.
Chao,J.R., Chen,C. and Wang,T. (1996). Characterization of factor-independent variants derived from TF-1 hematopoietic progenitor cells: the role of the Raf/MAP kinase pathway in the anti-apoptopic effect of GM-CSF. *Oncogene* **14**: 721-728.
Chen,I.T., Smith,M.L., O'Connor,P.M. *et al.* (1995) Direct interaction of Gadd45 with PCNA and evidence for competitive interaction of Gadd45 and p21Waf1/Cip1 with PCNA. *Oncogene* **11**: 1931-1937.
Chen,Q., Lin,T.H., Der,C.J. *et al.* (1996) Integrin-mediated activation of mitogen activated protein (MAP) or extracellular signal-related kinase (MEK) and kinase is independent of Ras. *J.Biol. Chem.* **271**: 18122-18127.
Chen,R.-H, Sarnecki,C. and Blenis,J. (1992) Nuclear localization and regulation of *erk* - and *rsk* - encoded protein kinases. *Mol.Cell.Biol.* **12**: 915-927.
Chen,Y.-R., Meyer,C.F. and Tan,T.-H. (1996) Persistent activation of c-Jun N-terminal kinase (JNK1) in (- radiation-induced apoptosis. *J.Biol. Chem.* **271**: 631-634.
Chmura,S.J., Nodzenski,E., Beckett,M.A. *et al.* (1997) Loss of ceramide production confers resistance to radiation-induced apoptosis. *Cancer Res.* **57**: 1270-1275.
Cleveland,J.L., Troppmair,J., Packham,G. *et al.* (1994) v-Raf suppresses apoptosis and promotes growth of interleukin-3-dependent myeloid cells. *Oncogene* **9**: 2217-2226.
Cook,S.J. and McCormick,F. (1993) Inhibition by cAMP of Ras-dependent activation of Raf. *Science* **262**: 1069-1072.
Corey,S., Eguinoa,A., Puyana-Theall,K. *et al.* (1993) Granulocyte macrophage-colony stimulating factor stimulates both association and activation of phosphoinositide 3-OH-kinase and src-related tyrosine kinase(s) in human myeloid derived cells. *EMBO J.* **12**: 2681-2690.
Cowley,S., Paterson,H., Kemp,P. *et al.* (1994) Activation of MAP kinase kinase is necessary and sufficient for PC12 differentiation and transformation of NIH 3T3 cells. *Cell* **77**: 841-852.
Cuvillier,O., Pirianov,G., Kleuser,B. *et al.* (1996) Suppression of ceramide-mediated programmed cell death by sphingosine-1-phosphate. *Nature* **381**: 8000-8003.
Datta,K., Bellacosa,A., Chan,T.D. *et al.* (1996) Akt is a direct target of the PI 3-kinase: Activation by growth factors, v-src and v-Ha-ras in Sf9 and mammalian cells. *J. Biol.Chem.* **271**: 30835-30839.
Datta,S.R., Dudek,H., Tao,X. *et al.* (1997) Akt phosphorylation of Bad couples survival signals to the cell-intrinsic death machinery. *Cell* **91**: 231-241.
Datta,R., Rubin,E., Sukhatme,V. *et al.* (1992a) Ionising radiation activates transcription of the EGR1 gene via CArG elements. *Proc.Natl.Acad.Sci. USA* **89**: 10149-10153.
Datta,R., Hallahan,D.E., Kharbanda,S.M. *et al.* (1992b) Involvement of reactive oxygen intermediates in the induction of c-Jun gene transcription by ionising radiation. *Biochemistry* **31**:83000-83006.
Datta,R., Hass,R., Gunji,H. *et al.* (1992c) Down-Regulation of cell cycle control genes by ionising radiation. *Cell Growth Diff.* **3**: 637-644.
Datta,R., Manome,Y., Taneja.,N. *et al.* (1995) Over expression of Bcl-X_L by cytotoxic drug exposure confers resistance to ionising radiation-induced internucleosomal DNA fragmentation. *Cell Growth Diff.* **6**: 363-370.
Datta,R., Kojima,H., Banach,D. *et al.* (1997) Activation of a CrmA-insensitive , p35-sensitive pathway in ionising radiation-induced apoptosis. *J.Biol.Chem.* **272**: 1965-1969.
Daum,G., Eisenman-Tappe,I., Fries,H.W. *et al.* (1994) The ins and outs of Raf kinases. *Trends Biochem.Sci.* **19**: 474-480.
Deng,C., Zhang,P., Harper,J.W. *et al.* (1995) Mice lacking p21CIP1/WAF1 undergo normal development, but are defective in G1 checkpoint control. *Cell* **82**: 675-684.
Dent,P., Haser,W., Haystead,T.A. *et al.* (1992) Activation of mitogen-activated protein kinase kinase by v-Raf in NIH 3T3 cells and *in vitro*. *Science* **257**: 1404-1407.
Devary,Y., Gottlieb,R.A., Smeal,T. *et al.* (1992) The mammalian ultraviolet response is triggered by activation of src tyrosine kinases. *Cell* **71**: 1081-1091.
Devary,Y., Rosette,C., DiDonato,J.A. *et al.* (1993) NF-κB activation by ultraviolet light not dependent on a nuclear signal. *Science* **261**: 1442-1445.

Diaz,B., Barnard,D., Filson,A. *et al.* (1997) Phosphorylation of Raf-1 serine 338 -serine 339 is an essential regulatory event for Ras-dependent activation and biological signaling. *Mol.Cell.Biol.* **17**: 4509-4516.
Donaldson,K.L., Goolsby,G., Kiener,P.A. *et al.* (1994) Activation of p34cdc coincident with Taxol-induced apoptosis. *Cell Growth Diff.* **5**: 1041-1050.
Dudek,H., Datta,S.R., Franke,T.F. *et al.* (1997) Regulation of neuronal survival by the serine-threonine protein kinase Akt. *Science* **275**: 661-668.
Dulic,V., Kaufmann,W.K., Wilson,S.J. *et al.* (1994) p53-dependent inhibition of cyclin-dependent kinase activities in human fibroblasts during radiation-induced G1 arrest. *Cell* **76**: 1013-1023.
El-Deiry,W.S., Tokino,T., Velculescu,V.E. *et al.* (1993) WAF1, a potential mediator of p53 tumor suppression. *Cell* **75**: 817-825.
Elkind,M.M., Han,A. and Volz,K. (1963) Radiation response of mammalian cells grown in culture. IV. Dose dependence of division delay and post-irradiation growth of surviving and non-surviving Chinese hamster cells. *J.Natl.Cancer Inst.* **30**: 705-711.
Emoto,Y., Manome,Y., Meinhardt,G. *et al.* (1995) Proteolytic activation of protein kinase C delta by an ICE-like protease in apoptotic cells. *EMBO J.* **14**: 6148-6156.
Essers,J., Hendriks,R.W., Swagemakers,S.M.A. *et al.* (1997) Disruption of mouse Rad54 reduces ionising radiation resistance and homologous recombination. *Cell* **89**: 195-204.
Evan,G.I., Wyllie,A.H., Gilbest,C.S. *et al.* (1992) Induction of apoptosis in fibroblasts by c-myc protein. *Cell* **69**: 119-128.
Evan,G.I. and Littlewood,T.D. (1993) The role of c-myc in cell growth. *Curr. Opin. Genet. Dev.* **3**: 44-49.
Fabian,J.R., Daar,I.O. and Morrison,D.K. (1993) Critical tyrosine residues regulate the enzymatic and biological activity of Raf-1 kinase. *Mol.Cell.Biol.* **13**: 7170-7179.
Farrow,S.N. and Brown,R. (1996) New members of the Bcl-2 family and their protein partners. *Curr.Opin.Gen.Dev.* **6**: 45-49.
Finco,T.S. and Balwin,A.S.Jr. (1993) 6 site-dependent induction of gene expression by diverse inducers of nuclear factor κB requires Raf-1. *J.Biol.Chem.* **268**: 17676-17679.
Finco,T., Westwick,J.K., Norris,J.L. *et al.* (1997) Oncogenic Ha-Ras-induced signaling activates NF-kappaB transcriptional activity, which is required for cellular transformation. *J.Biol.Chem.* **272**: 24113-24116.
Fitzgerald,T.J., Santucci,M.A., Das,I. *et al.* (1991) The v-abl, c-fms, or v-myc oncogene induces gamma radiation resistance of hematopoietic progenitor cell line 32d cl 3 at clinical low dose rate. *Int.J. Radiat. Oncol. Biol.Phys.* **21**: 1203-1210.
Folgueira,L., Algeciras,A. and MacMorran,W.S. (1996) The Ras-Raf pathway is activated in human immunodeficiency virus-infected monocytes and participates in the activation of NF-κB. *J.Virology* **70**: 2332-2338.
Ford,J.M. and Hanawalt,P.C. (1995) Li-Fraumeni syndrome fibroblasts homozygous for p53 mutations are deficient in global DNA repair but exhibit normal transcription-coupled repair and enhanced UV resistance. *Proc.Natl.Acad.Sci. USA* **92**: 8876-8880.
Fornace,A.J.Jr., (1992) Mammalian genes induced by radiation: Activation of genes associated with growth control. *Ann. Rev. Gen.* **26**: 507-526.
Franke,T.F., Yang,S.-I, Chan,T.O. *et al.* (1995) The protein kinase encoded by the Akt proto-oncogene is a target of the PDGF-activated phosphatidylinositol-3-kinase. *Cell* **81**: 727-736.
Franke,T.F., Kaplan,D.R. and Cantley,L.C. (1997) PI3K: Downstream AKTion blocks apoptosis. *Cell* **88**: 435-437.
Fried,L.M., Koumenis,C., Peterson, S.R. *et al.* (1996) The DNA damage response in DNA-dependent protein kinase-deficient SCID mouse cells: Replication protein A hyperphosphorylation and p53 induction. *Proc.Natl.Acad.Sci. USA* **93**: 13825-13830.
Fuks,Z., Persaud,R.S., Alfieri,A. *et al.* (1994) Basic fibroblast growth factor protects endothelial cells against radiation-induced programmed cell death *in vitro* and *in vivo*. *Cancer Res.* **54**: 2582-2590.
Fulop,G.M. and Phillips,R.A. (1990) The scid mutation in mice causes a general defect in DNA repair. *Nature* **347**: 479-482.
Gokhale,P.C., Soldatenkov,V., Wang,F.-W *et al.* (1997) Antisense raf oligodeoxyribonucleotide is protected by liposomal encapsulation and inhibits raf-1 protein expression *in vitro* and *in vivo*: implication for gene therapy of radioresistant cancer. *Gene Therapy* **4**: 1289-1299.
Gokhale,P.C., McRae,D., Monia,B.P. *et al.* (1998). Antisense *raf* oligonucleotide is a tumor radiosensitizer. Cambridge Healthtech Institute's Conference on Antisense Technologies.
Gottlieb,T.M. and Jackson,S.P. (1993) The DNA-dependent protein kinase: requirement for DNA ends and association with Ku antigen. *Cell* **72**: 131-142.
Greenberger,J.S., Sakakeeny,M.A. *et al.* (1983) Demonstration of permanent factor-dependent multipotential (erythroid/neutrophil/basophil) hematopoietic progenitor cell lines. *Proc.Natl.Acad.Sci.USA* **80**: 2931-2935.
Gu,Y., Jin,S., Gao,Y. *et al.* (1997) Ku70-deficient embryonic stem cells have increased ionising radiosensitivity, defective DNA end-binding activity, and inability to support V(D)J recombination. *Proc.Natl.Acad.Sci. USA.* **94**: 8076-8081.
Hafner,S., Adler,H.S., Mischak,H. *et al.* (1994) Mechanism of inhibition of Raf-1 by protein kinase A. *Mol.Cell.Biol.* **14**: 6696-6703.
Haimovitz-Friedman,A., Vlodavsky,I., Chaudhuri,A. *et al.* (1991) Autocrine effects of fibroblast growth

factor in repair of radiation damage in endothelial cells. *Cancer Res.* **51**: 2552-2558.
Haimovitz-Friedman,A., Balaban,N., McLoughlin,M. *et al.* (1994a). Protein kinase C mediates basic fibroblast growth factor protection of endothelial cells against radiation-induced apoptosis. *Cancer Res.* **54**: 2591-2597.
Haimovitz-Friedman,A., Kan,C.C., Ehleiter,D. *et al.* (1994b) Ionising radiation acts on cellular membranes to generate ceramide and initiate apoptosis. *J. Exp.Med.* **180**: 525-535.
Haldar,S., Negrini,M., Monne,M. *et al.* (1994) Down-regulation of bcl-2 by p53 in breast cancer cells. *Cancer Res.* **54**: 2095-2097.
Halevy,O., Novitch,B.G., Spicer,D.B. *et al.* (1995) Correlation of terminal cell cycle arrest of skeletal muscle with induction of p21 by MyoD. *Science* **267**: 1018-1021.
Hall,E.J. (1988) In: Radiobiology for the radiologist (Hall, E.J., ed) pp.128-135, Lippincott, Philadelphia.
Hallahan,D.E., Spriggs,D.R., Beckett,M.A. *et al.* (1989) Increased tumor necrosis factor-α mRNA after cellular exposure to ionising radiation. *Proc.Natl.Acad.Sci. USA.* **86**: 10104-10107.
Hallahan,D.E., Virudachalam,S., Sherman,M.L. *et al.* (1991a) Tumor necrosis factor gene expression is mediated by protein kinase C following activation by ionising radiation. *Cancer Res.* **51**: 4565-4569.
Hallahan,D.E., Sukhatme,V.P., Sherman,M.L. *et al.* (1991b) Protein kinase C mediates x-ray inducibility of nuclear signal transducers EGR1 and Jun. *Proc.Natl.Acad.Sci. USA.* **88**: 2156-2160.
Hallahan,D.E., Gius,D., Kuchibhotla,J. *et al.* (1993) Radiation signaling mediated by jun activation following dissociation from a cell type-specific repressor. *J. Biol.Chem.* **268**: 4903-4907.
Hallahan,D.E., Dunphy,E., Virudachalam,S. *et al.* (1995) c-jun and Egr-1 participate in DNA synthesis and cell survival in response to ionising radiation exposure. *J.Biol.Chem.* **270**: 30303-30309.
Harper,J.W., Adami,G.R. Wei,N. *et al.* (1993) The p21 Cdk-interacting protein Cip1 is a potent inhibtor of G1 cyclin-dependent kinases. *Cell* **75**: 805-816.
Harper,J.W. and Elledge,S.J. (1996) Cdk inhibitors in development and cancer. *Curr. Opin. Genet. Dev.* **6**: 56-64.
Harrington,E.A., Bennett,M.R., Fanidi,A. *et al.* (1994) c -Myc-induced apoptosis in fibroblasts is inhibited by specific cytokines. *EMBO J.* **13**: 3286-3295.
Hartwell,L.H. and Kastan,M.B. (1994) Cell cycle control and cancer. *Science* **266**: 1821-1828.
Heidecker,G., Huleihel,M., Cleveland,J.L. *et al.* (1990) Mutational activation of c-raf-1 and definition of the minimal transforming sequence. *Mol.Cell.Biol.* **10**: 2503-2512.
Hockenbery,D., Nunez,G., Milliman,C. *et al.* (1990) Bcl-2 is an inner mitochondrial membrane protein that blocks programmed cell death. *Nature* **348**: 334-336.
Hockenberry,D., Oltivai,Z., Yin,X. *et al.* (1993) Bcl-2 functions in an antioxidant pathway to prevent apoptosis. *Cell* **75**: 241-251.
Hollander,M.C., Alamo,I., Jackman,J. *et al.* (1993) Analysis of the mammalian gadd45 gene and its response to DNA damage. *J.Biol.Chem.* **268**: 24385-24393.
Hollander,M.C., Zhan,Q., Bae,I. *et al.* (1997) Mammalian GADD34, an apoptosis- and DNA damage-inducible gene. *J.Biol.Chem.* **272**: 13731-13737.
Hong,J.H., Chiang,C.-S., Campbell,I.L. *et al.* (1995) Induction of acute phase gene expression by brain irradiation. *Int.J. Radiation Oncology Biol.Phys.* **33**: 619-626.
Horwitz,S.B. (1992) Mechanism of action of Taxol. *Trends Pharmacol.Sci.* **13**: 134-136.
Howard,A. and Pelc,S.R. (1953) Synthesis of deoxyribonucleic acid in normal and irradiated cells and its relation to chromosome breakage. *Heredity* **6**: 261-273.
Howe,L.R., Leevers,S.J., Gomez,N. *et al.* (1992) Activation of the MAP kinase pathway by the protein kinase Raf. *Cell* **71**: 335-342.
Huang,L., Clarkin,K.C. and Wahl,G.M. (1996) p53-dependent cell cycle arrests are preserved in DNA-activated protein kinase-deficient mouse fibroblasts. *Cancer Res.* **56**: 2940-2944.
Ikawa,S., Fukui,M., Ueyama,Y. *et al.* (1988) B-raf, a new member of the raf family, is activated by DNA rearrangement. *Mol.Cell.Biol.* **8**: 2651-2654.
Jeggo,P.A., Tesmer,J. and Chen,D.J. (1991) Genetic analysis of ionising radiation sensitive mutants of cultured mammalian cell lines. *Mut. Res.* **254**: 125-133.
Jelinek,T., Dent,T., Sturgill,T.W. *et al.* (1996) Ras-induced activation of Raf-1 is dependent on tyrosine phosphorylation. *Mol.Cell.Biol.* **16**: 1027-1034.
Kao,G.D., McKenna,W.G., Maity,A. *et al.* (1997) Cyclin B1 availability is a rate-limiting component of the radiation-induced G2 delay in HeLa cells. *Cancer Res.* **57**: 753-758.
Kasid,U., Pfeifer,A., Weichselbaum,R.R. *et al.* (1987) The raf oncogene is associated with a radiation-resistant human laryngeal cancer. *Science* **237**: 1039-1041.
Kasid,U., Pfeifer,A., Brennan,T. *et al.* (1989) Effect of antisense c-Raf-1 on tumorigenicity and radiation sensitivity of a human squamous carcinoma. *Science* **243**: 1354-1356.
Kasid,U., Pirollo,K., Dritschilo,A. *et al.* (1993) Oncogenic basis of radiation resistance. *Advances in Cancer Res.* **61**: 195-233.
Kasid,U., Suy,S., Dent,P. *et al.* (1996) Activation of Raf by ionising radiation. *Nature* **382:** 813-816.
Kasid,U., Wang, F-H. and Whiteside,T.L. (1997) Ionising radiation and TNFα stimulate gene expression of a Thr/Tyr-protein phosphatase HVH1 and inhibitory factor IκBα. *Mol.Cell. Biochem.* **173**: 193-197.
Kastan,M.B., Onyekwere,O., Sidransky,D. *et al.* (1991) Participation of p53 protein in the cellular response to DNA damage. *Cancer Res.* **51**: 6304-6311.

Kastan,M.B., Zhan,Q., El-Deiry,W.S. *et al.* (1992) A mammalian cell cycle checkpoint pathway utilizing p53 and GADD45 is defective in ataxia-telangiectasia. *Cell* **71**: 587-597.
Kato,J.Y., Matsuoka,M., Polyak,K. *et al.* (1994) Cyclic AMP-induced G1 phase arrest mediated by an inhibitor ($p27^{kip1}$) of cyclin dependent kinase 4 activation. *Cell* **79**: 487-496.
Kauffmann-Zeh,A., Rodriguez-Viciana,P., Ulrich,E. *et al.* (1997) Suppression of c-Myc-induced apoptosis by Ras signalling through PI(3) K and PKB. *Nature* **386**: 544-548.
Kennedy,S.G., Wagner,A.J., Conzen,S.D. *et al.* (1997) The PI 3-kinase/Akt signaling pathway delivers an anti-apoptotic signal. *Gen. Dev.* **11**: 701-713.
Kerkhoff,E. and Rapp,U.R. (1997) Induction of cell proliferation in quiescent NIH 3T3 cells by oncogenic c-Raf-1. *Mol.Cell.Biol.* **17**: 2576-2586.
Kharbanda,S., Saleem,A., Sharman,T. *et al.* (1994a) Activation of pp90rsk and mitogen-activated serine/threonine protein kinases by ionising radiation. *Proc.Natl.Acad.Sci. USA.* **91**: 5416-5420.
Kharbanda,S., Saleem,A., Datta,R. *et al.* (1994b) Ionising radiation induces rapid tyrosine phosphorylation of p34cdc2. *Cancer Res.* **54**: 1412-1414.
Kharbanda,S., Saleem,A., Shafman,T. *et al.* (1995a) Ionising radiation stimulates a Grb2-mediated association of the stress-activated protein kinase with phosphatidylinositol 3-kinase. *J. Biol.Chem.* **270**: 18871-18874.
Kharbanda,S., Ren,R., Pandley,P. *et al.* (1995b) Activation of the c-Abl tyrosine kinase in the stress response to DNA-damaging agents. *Nature* **376:** 785-788.
Kharbanda,S., Bharti,A., Pei,D. *et al.* (1996) The stress response to ionising radiation involves c-Abl-dependent phosphorylation of SHPTP1. *Proc.Natl.Acad.Sci. USA.* **93**: 6898-6901.
Kharbanda,S., Pandey,P., Schofield,L. *et al.* (1997) Role for Bcl-X_L as an inhibitor of cytosolic cytochrome C accumulation in DNA damage-induced apoptosis. *Proc.Natl.Acad.Sci. USA.* **94**: 6939-6942.
Khwaja,A., Rodriguez-Viciana,P., Wennstrom,S. *et al.* (1997) Matrix adhesion and Ras transformation both activate a phosphoinositide 3-OH kinase and protein kinase B/Akt cellular survival pathway. *EMBO J.* **16**: 2783-2793.
King,W.G., Mattaliano,M.D., Chan,T.O. *et al.* (1997) Phosphatidylinositol 3-kinase is required for integrin-stimulated AKT and Raf-1/mitogen-activated protein kinase pathway activation. *Mol.Cell.Biol.* **17**: 4406-4418.
Kinoshita,T., Yokota,T., Arai,K. *et al.* (1995) Suppression of apoptotic death in hematopoietic cells by signalling through the IL-3/GM-CSF receptors. *EMBO J.* **14**: 266-275.
Kirchgessner,C.U., Patil,C.K., Evans,J.W. *et al.* (1995) DNA-dependent kinase (p350) as a candidate gene for the murine SCID defect. *Science* **267**: 1178-1183.
Kitada,S., Takayama,S., DeRiel,K. *et al.* (1994) Reversal of chemoresistance of lymphoma cells by antisense-mediated reduction of Bcl-2 gene expression. *Antisense Res.Dev.* **4**: 71-79.
Kitada,S., Krajewski,S., Miyashita,T. *et al.* (1996) Gamma -radiation induces upregulation of bax protein and apoptosis in radiosensitive cells *in vivo*. *Oncogene* **12:** 187-192.
Kluck,R.M., Bossy-Wetzel,E., Green,D.R. *et al.* (1996). The release of cytochrome C from mitochondria: a primary site for Bcl-2 regulation of apoptosis. *Science* **275**: 1132-1136.
Kolch,W., Cleveland,J.L. and Rapp,U. (1990) Role of oncogenes in the abrogation of growth factor requirements of hemopoietic cells. *Crit.Rev.Cancer* **2**: 279-303.
Kolch,W., Heidecker,G., Lloyd,P. *et al.* (1991) Raf-1 protein kinase is required for growth of induced NIH 3T3 cells. *Nature* **349**: 426-428.
Kolch,W., Heidecker,G., Kochs,G. *et al.* (1993) Protein kinase C alpha activates Raf-1 by direct phosphorylation. *Nature* **364**: 249-252.
Kolesnick,R. and Golde,D.W. (1994) The sphingomyelin pathway in tumor necrosis factor and interleukin-1 signaling. *Cell* **77**: 325-328.
Koong,A.C., Chen,E.Y., Mivechi,N.F. *et al.* (1994) Hypoxic activation of nuclear factor-κB is mediated by a Ras and Raf signaling pathway and does not involve MAP kinase (ERK1 or ERK2). *Cancer Res.* **54**: 5273-5279.
Krajewski,S., Tanaka,S., Takayama,S. *et al.*(1993) Investigation of the subcellular distribution of the bcl-2 oncoprotein: residence in the nuclear envelope, endoplasmic reticulum, and outer mitochondrial membranes. *Cancer Res.* **53**: 4701-4714.
Kuerbitz,S.J., Plunkett,B.S., Walsh,W.V. *et al.* (1992) Wild type p53 is a cell cycle checkpoint determinant following irradiation. *Proc.Natl.Acad.Sci. USA.* **89**: 7491-7495.
Kuo,W.L., Abe,M., Rhee,J. *et al.* (1996) Raf, but not MEK or ERK is sufficient for differentiation of hippocampal neuronal cells. *Mol.Cell.Biol.* **16**: 1458-1470.
Kwak,S.P., Hakes,D.J., Martell,K.J. *et al.* (1994) Isolation and characterization of a human dual protein-tyrosine phosphatase gene. *J.Biol.Chem.* **269**: 3596-35604.
Kyriakis,J.M., App,H., Zhang,X.-F. *et al.* (1992) Raf-1 activates MAP kinase-kinase. *Nature* **358**: 417-421.
Lavoie,J.N., L'Allemain,G., Brunet,A. *et al.* (1996) Cyclin D1 expression is regulated positively by the p42/p44 MAPK and negatively by the p38/HOGMAPK pathway. *J.Biol.Chem.* **271**: 20608-20616.
Lee,L.-F., Haskill,J.S., Mukaida,N. *et al.* (1997) Identification of tumor-specific paclitaxel (Taxol)-responsive regulatory elements in the interleukin-8 promoter. *Mol.Cell.Biol.* **17**: 5097-5105.
Leevers,S.J., Paterson,H.F. and Marshall,C.J. (1994) Requirement for ras in raf activation is overcome by targeting raf to the plasma membrane. *Nature* **369**: 411-414.
Leigh,B.R., Hancock,S.L. and Knox,S.J. (1993) The effect of stem cell factor on irradiated human bone

marrow. *Cancer Res.* **53**: 3857-3859.
Lenormand,P., Sardet,C., Pages,G. *et al.* (1993) Growth factors induce nuclear translocation of MAP kinases (p42mapk/p44mapk) but not of their activator MAP kinase kinase (p45mapkk) in fibroblasts. *J.Cell Biol.* **122**: 1079-1088.
Li,P., Nijhawan,D., Budihardjo,I. *et al.* (1997) Cytochrome c and dATP-dependent formation of Apaf-1/caspase-9 complex initiates an apoptotic protease cascade. *Cell* **91**: 479-489.
Li,S.-F. and Sedivy,J.M. (1993) Raf-1 protein kinase activates the NF-κB transcription factor by dissociating the cytoplasmic NF-κB-IκB complex. *Proc.Natl.Acad.Sci.USA.* **90**: 9247-9251.
Littlewood,T.D., Hancock, D.C., Danielian, P.S. *et al.* (1995) A modified estrogen receptor ligand-binding domain as an improved switch for the regulation of heterologous proteins. *Nuc. Acids Res.* **23**: 1686-1690.
Liu,Y., Gorospe,M., Yang,C. *et al.* (1995) Role of mitogen-activated protein kinase phosphatase during the cellular response to genotoxic stress. Inhibition of c-Jun N-terminal kinase activity and AP-1-dependent gene activation. *J. Biol.Chem.* **270**: 8377-8380.
Liu,Z.-G., Baskaran,R., Lea-Chou,E.T. *et al.* (1996a) Three distinct signalling responses by murine fibroblasts to genotoxic stress. *Nature* **384**: 273-276.
Liu,Z-G., Hsu,H., Goeddel,D.V. *et al.* (1996b) Dissection of TNF receptor-1 effector functions:JNK activation is not linked to apoptosis while NF-kappaB activation prevents cell death. *Cell* **87**: 565-576.
Ludwig,S., Engel,K., Hoffmeyer,A. *et al.* (1996) 3pK, a novel mitogen-activated protein (MAP) kinase-activated protein kinase, is targeted by three MAP kinase pathways. *Mol.Cell.Biol.* **16:** 6687-6697.
Macleod,K.F., Sherry,N., Hannon,G. *et al.* (1995) p53-dependent and independent expression of p21 during cell growth, differentiation, and DNA damage. *Gen. Dev.* **9**: 935-944.
Malynn,B.A., Blackwell,T.K., Fulop,G.M. *et al.* (1988) The Scid defect affects the final step of the immunoglobulin VDJ recombinase mechanism. *Cell* **54**: 453-460.
Marais,R., Light,Y., Paterson,H.F. *et al.* (1995) Ras recruits raf-1 to the plasma membrane for activation by tyrosine phosphorylation. *EMBO J.* **14**: 3136-3145.
Marshall, C.J. (1995) Specificity of receptor tyrosine kinase signaling: transient versus sustained extracellular signal-regulated kinase activation. *Cell*, **80**, 179-185.
Marshall,C.J. (1996a) Cell signalling. Raf gets it together. *Nature* **383**: 127-128.
Marshall,C.J. (1996b) Ras effectors. *Curr.Opin.Cell Biol.* **8**: 197-204.
McKenna,W.G., Iliakis,G., Weiss,M.C. *et al.* (1991) Increase G2 delay in radiation-resistant cells obtained by transformation of primary rat embryo cells with the oncogenes H-Ras and v-Myc. *Radiation Res.***125**: 283-287.
McKenna,W.G., Bernhard,E.J., Markiewicz,D.A. *et al.* (1996) Regulation of radiation-induced apoptosis in oncogene-transfected fibroblasts: influence of H-ras on the G2 delay. *Oncogene* **12**: 237-245.
McLaughlin,M.M., Kumar,S., McDonnell,P.C. *et al.* (1996) Identification of mitogen-activated protein (MAP) kinase- activated protein kinase-3, a novel substrate of CSBP p38 MAP kinase. *J.Biol.Chem.* **271**: 8488-8492.
Metz,T., Harris,A.W. and Adams,J.M. (1995) Absence of p53 allows direct immortalization of hematopoietic cells by the myc and raf oncogenes. *Cell* **82**: 29-36.
Minn,A.J., Rudin,C.M., Boise,L.H. *et al.* (1995) Expression of Bcl-XL can confer a multiple resistance phenotype. *Blood* **86**: 1903-1910
Minn,A.J., Velez,P., Schendel,S.L. *et al.* (1997) Bcl-XL forms an ion channel in synthetic lipid membranes. *Nature* **385**: 353-357.
Miyashita,T., Krajewski,S., Krajewska,M. *et al.* (1994) Tumor suppressor p53 is a regulator of bcl-2 and bax gene expression *in vitro* and *in vivo*. *Oncogene* **9:** 1799-1805.
Moelling,K., Heimann,B., Beimling,P. *et al.* (1984) Serine and threonine-specific protein kinase activities of purified gag-mil and gag-raf proteins. *Nature* **312**: 558-561.
Monia,B.P., Johnston,J.F., Geiger,T. *et al.* (1996) Antitumor activity of a phosphorothioate antisense oligodeoxynucleotide targeted against c-raf kinase. *Nature Med.* **2**: 668-675.
Moodie,S.A., Willumsen,B.M., Weber,M.J. *et al.* (1993) Complexes of Ras·GTP with Raf-1 and mitogen-activated protein kinase kinase. *Science* **260**: 1658-1661.
Morgan,S.E. and Kastan,M.B. (1997a) p53 and ATM: Cell cycle, cell death, and cancer. *Adv.Cancer Res.* **71**: 1-25.
Morrison,D.K. (1990) The Raf-1 kinase as a transducer of mitogenic signals. *Cancer Cells* **2**: 377-382.
Murnane,J.P. (1995) Cell cycle regulation in response to DNA damage in mammalian cells: a historical perspective. *Cancer Metastasis Rev.* **14**: 17-29.
Muschel,R.J., Zhang,H.B., Iliakis,G. *et al.* (1991) Cyclin B expression in HeLa cells during the G2 block induced by ionising radiation. *Cancer Res.* **51**: 5113-5117.
Nakagawa,T., Mabry,M., DeBustros,A. *et al.* (1987) Introduction of v-Ha-Ras oncogene induces differentiation of cultured human medullary thyroid carcinoma cells. *Proc.Natl.Acad.Sci.USA.* **84:** 5923-5927.
Nakielny,S., Cohen,P., Wu,J. *et al.* (1992) MAP kinase activator from insulin-stimulated skeletal muscle is a protein threonine/tyrosine kinase. *EMBO J.* **11**: 2123-2129.
Naumovski,L. and Cleary,M.L. (1996). The p53-binding protein 53BP2 also interacts with Bcl-2 and impedes cell cycle progression at G2/M. *Mol.Cell.Biol.* **16**: 3884-3892.

Neta,R. and Oppenheim,J.J. (1991) Radioprotection with cytokines-learning from nature to cope with radiation damage. *Cancer Cells* **3**: 391-396.
Nguyen,M., Millar,D.G., Yong,V.W. *et al.* (1993) Targeting of Bcl-2 to the mitochondrial outer membrane by a COOH-terminal signal anchor sequence. *J.Biol.Chem.* **268**: 25265-25268.
Nunez,G., London,L., Hockenberry,D. *et al.* (1990) Deregulated Bcl-2 gene expression selectively prolongs survival of growth factor-deprived hemopoietic cell lines. *J. Immunol.* **144**: 3602-3610.
Obeid,L.M., Linardic,C.M., Karolak,L.A. *et al.* (1993) Programmed cell death induced by ceramide. *Science* **259**: 1769-1771.
Olivera,A and Spiegel,S. (1993) Sphingosine 1-phosphate as second messenger in cell proliferation induced by PDGF and FCS mitogens. *Nature* **365**: 557-560.
Oltvai,Z.N. and Korsmeyer,S.J. (1994) Checkpoints of dueling dimers foil death wishes. *Cell* **79**: 189-192.
Pandey,P., Raingeaud,J., Kaneke,M. *et al.* (1996) Activation of p38 mitogen-activated protein kinase by c-Abl-dependent and -independent mechanisms. *J.Biol.Chem.* **271**: 23775-23779.
Patel,B.K.R., Ray,S., Whiteside,T.L. *et al.* (1997) Correlation of Constitutive activation of Raf-1 with morphological transformation and abrogation of tyrosine phosphorylation of distinct sets of proteins in human squamous carcinoma cells. *Mol.Carcinogenesis* **18**: 1-6.
Patel,S., Wang,F.-W., Whiteside,T.L. *et al.* (1997) Constitutive modulation of Raf-1 protein kinase is associated with differential gene expression of several known and unknown genes. *Mol.Medicine* **3**: 647-685.
Patel,S., Wang,F.-W., Whiteside,T.L. and Kasid,U. (1998) Ionizing radiation and TNF-α stimulate expression of α1-antichymotrypsin gene in human squamous carcinoma cells. *Acta Oncologia* In Press.
Pfeifer,A.M.A., Mark,G.III, Malan-Shibley *et al.* (1989) Cooperation of c-raf-1 and c-myc protooncogenes in the neoplastic transformation of simian virus 40 large tumor antigen-immortalized human bronchial epithelial cells. *Proc.Natl.Acad.Sci. USA* **86**: 10075-10079.
Pines,J. (1995) Cyclins, CDKs and cancer. *Semin. Cancer Biol.* **6**: 63-72.
Pirollo,K.F., Garner,R., Yuan,S.Y. *et al.* (1989) Raf involvement in the simultaneous genetic transfer of the radioresistant and transforming phenotypes. *Int. J. Radiat. Biol.* **55**: 783-796.
Pirollo,K.F., Hao,Z., Rait,A. *et al.* (1997) Evidence supporting a signal transduction pathway leading to the radiation-resistant phenotype in human tumor cells. *Biochem. Biophys.Res.Com.* **230**: 196-201.
Polyak,K., Lee,M.H., Erdjument-Bromage,H. *et al.* (1994) Cloning of p27kip1, a cyclin-dependent kinase inhibitor and a potential mediator of extracellular antimitogenic signals. *Cell* **78**: 59-66.
Qureshi,S.A., Rim,M., Bruder, J. *et al.* (1991) An inhibitory mutant of c-raf-1 blocks v-src-induced activation of the egr-1 promoter. *J.Biol.Chem.* **266**: 20594-20597.
Radler-Pohl,A., Sachsenmaier,C., Gebel,S. *et al.* (1993) UV-induced activation of AP-1 involves obligatory extranuclear steps including Raf-1 kinase. *EMBO J.* **12**: 1005-1012.
Rajan,J.V., Wang,M., Marquis,S.T. *et al.* (1996) Brca2 is coordinately regulated with Brca1 during proliferation and differentiation in mammary epithelial cells. *Proc.Natl.Acad.Sci. USA.* **93**: 13078-13083.
Rapp,U.R., Goldsborough,M.D., Mark,G.E. *et al.* (1983) Structure and biological activity of v-raf, a unique oncogene transduced by a retrovirus. *Proc.Natl.Acad.Sci.USA.* **80**: 4218-4222.
Rapp,U.R. (1991) Role of Raf-1 serine/threonine protein kinase in growth factor signal transduction. *Oncogene* **6**: 495-500.
Ravi,R.K., Weber,E., McMahon,M. *et al.* (1998) Activated Raf-1 causes growth arrest in human small cell lung cancer cells. *J.Clin.Invest.* **101:** 153-159.
Reed, J. (1997) Double identity for proteins of the BCL-2 family. *Nature* **387**: 773-776.
Renshaw,M.W, Kipreos,E.T., Albrecht,M.R. *et al.* (1992) Oncogenic v-Abl tyrosine kinase can inhibit or stimulate growth, depending on cell context. *EMBO J.* **11**: 3941-3951.
Rudin,C.M. and Thompson,C.B. (1997) Apoptosis and disease: Regulation and clinical relevance of programmed cell death. *Annu.Rev.Med.* **48**: 267-281.
Sahijdak,W.M., Yang,C.-R., Zuckerman,J.S. *et al.* (1994) Alterations in transcription factor binding in radioresistant human melanoma cells after ionising radiation. *Radiation Res.* **138**: S47-S51.
Salvesen,G.S. and Dixit,V.M. (1997) Caspases: Intracellular signaling by proteolysis. *Cell* **91**: 443-446.
Samuels,M.L., Weber,M.J., Bishop,J.M. *et al.* (1993) Conditional transformation of cells and rapid activation of the mitogen-activated protein kinase cascade by an estradiol-dependent human Raf-1 protein kinase. *Mol.Cell.Biol.* **13**: 6241-6252.
Santana,P., Pena,L.A., Haimovitz-Friedman,A. *et al.* (1996) Acid sphingomyelinase-deficient human lymphoblasts and mice are defective in radiation-induced apoptosis. *Cell* **86**: 189-199.
Sato,N., Sakamaki,K., Terada,N. *et al.* (1993) Signal transduction by the high-affinity GM-CSF receptor: two distinct cytoplasmic regions of the common beta subunit responsible for different signalling. *EMBO J.* **12**: 4181-4189.
Schieven,G.L., Kirihara,J.M., Myers,D.E. *et al.* (1993) Reactive oxygen intermediates activate NF-κB in a tyrosine kinase-dependent mechanism and in combination with vanadate activate the p56lck and p59fyn tyrosine kinases in human lymphocytes. *Blood* **82**: 1212-1220.
Scully,R., Chen,J., Ochs,R.L. *et al.* (1997a) Dynamic changes of Brca1 subnuclear location and phosphorylation state are initiated by DNA damage. *Cell* **90**: 425-435.
Scully,R., Chen,J., Plug,A. *et al.* (1997b) Association of BRCA1 with Rad51 in mitotic and meiotic cells. *Cell* **88**: 265-275.

Seger,R. and Krebs,E.G. (1995) The MAPK signaling cascade. *FASEB J.* **9**: 726-735.
Selvakumaran,M., Lin,H., Miyashita,T. *et al.* (1994) Immediate early up-regulation of bax expression by p53 but not TGF beta 1: a paradigm for distinct apoptotic pathways. *Oncogene* **9**: 1791-1798.
Sentman,C.L., Shutter,J.R., Hockenbery,D. *et al.* (1991) Bcl-2 inhibits multiple forms of apoptosis but not negative selection in thymocytes. *Cell* **67**: 879-888.
Serrano,M., Lin,A.W., McCurragh,M.E. *et al.* (1997) Oncogenic Ras provokes premature cell senescene associated with accumulation of p53 and $p16^{INK4a}$. *Cell* **88**: 593-602.
Sewig,A., Wiseman,B., Lloyd,A.C. *et al.* (1997) High-intensity Raf signal causes cell cycle arrest mediated by $p21^{Cip1}$. *Mol.Cell.Biol.* **17**: 5588-5597.
Shafman,T. Khanna,K.K., Kedar,P. *et al.* (1997) Interaction between ATM protein and c-Abl in response to DNA damage. *Nature* **387**: 450-451.
Sharan,S.K., Morimatsu,M., Albrecht,U. *et al.* (1997a) Embryonic lethality and radiation hypersensivity mediated by Rad51 in mice lacking Brca2. *Nature* **386**: 804-810.
Sharan,S.K. and Bradley,A. (1997b) Murine Brca2: Sequence, map position and expression pattern. *Genomics* **40**: 234-241.
Sherman,M.L., Datta,R., Hallahan,D.E. *et al.* (1990) Ionising radiation regulates expression of the c-jun protooncogene. *Proc.Natl.Acad.Sci. USA.* **87**: 5663-5666.
Shiina,N., Goth,Y., Kubomura,N. *et al.* (1994) Microtubule severing by elongation factor 1 alpha. *Science* **266**: 282-285
Singleton,B.K., Priestley,A., Steingrimsdottir,H. *et al.* (1997) Molecular and biochemical characterization of xrs mutants defective in Ku80. *Mol.Cell. Biol.* **17**: 1264-1273.
Sipley,J.D., Menninger,J.C., Hartley,K.O. *et al.* (1995) Gene for the catalytic subunit of the human DNA-activated protein kinase maps to the site of the XRCC7 gene on chromosome 8. *Proc.Natl.Acad.Sci. USA.* **92**: 7515-7419.
Sklar,M.D., McQuiston,S., Terry,V.O. *et al.* (1986) *Radiat.Oncol.Biol.Phys.* **12**(Suppl. 1), 190-191.
Smith,D.M., Tran,H.M., Soo,V.W. *et al.* (1994) Enhanced synthesis of tumor necrosis factor-inducible proteins, plasminogen activator inhibitor-2, manganese superoxide dismutase, and protein 28/5.6, is selectively triggered by the 55kDa tumor necrosis factor receptor in human melanoma cells. *J.Biol.Chem.* **269**: 9898-9905.
Smith,M.L., Chen,I.-T., Zhan,Q. *et al.* (1994) Interaction of the p53-regulated protein Gadd4 with proliferating cell nuclear antigen. *Science* **266**: 1376-1380.
Smith,M.R., Heidecker,G., Rapp,U.R. *et al.* (1990) Induction of transformation and DNA synthesis after microinjection of Raf proteins. *Mol.Cell.Biol.* **10**: 3828-3833.
Smith,S.A., Easton,D.G., Evans,D.G.R. *et al.* (1992) Allele losses in the region 17q12-21 in familial breast and ovarian cancer involve the wild type chromosome. *Nature Genet.* **2**: 128-131.
Soldatenkov,V.A., Dritschilo,A., Wang,F.-W *et al.* (1997) Inhibition of Raf-1 protein kinase by antisense phosphorothioate oligodeoxyribonucleotide is associated with sensitization of human laryngeal squamous carcinoma cells to gamma radiation. *Cancer J.Sci.Am.* **3**: 13-20.
Solomon,M.J., Glotzer,M., Lee,T.H. *et al.* (1990) Cyclin activation of $p34^{cdc2}$. *Cell* **63**: 1013-1024.
Songyang,Z., Baltimore,D., Cantley,L. *et al.* (1997). Interleukin 3-dependent survival by the Akt protein kinase. *Proc. Natl. Acad. Sci. USA* **94**: 11345-11350.
Stanton,V.P.Jr., Nichols,D.W., Laudano,A.P. *et al.* (1989) Definition of the human raf amino-terminal regulatory region by deletion mutagenesis. *Mol.Cell.Biol.* **9**: 639-647.
Stevenson,M.A., Pollock,S.S., Coleman,C.N. *et al.* (1994) X-irradiation, phorbol esters, and H2O2 stimulate mitogen-activated protein kinase activity in NIH-3T3 cells through the formation of reactive oxygen intermediates. *Cancer Res.* **54**: 12-15.
Stewart,C.E. and Rotwein,P. (1996) Growth, differentiation, and survival:multiple physiological functions for insulin-like growth factors. *Physiol.Rev.* **76**: 1005-1026.
Storm,S.M., Cleveland,J.L. and Rapp,U.R. (1990) Expression of raf family proto-oncogenes in normal mouse tissues. *Oncogene* **5**: 345-351.
Strasser,A., Harris,A.W., Jacks,T. *et al.* (1994) DNA damage can induce apoptosis in proliferating lymphoid cells via p53-independent mechanisms inhibitable by Bcl-2. *Cell* **79**: 329-339.
Suy,S., Anderson,W.B., Dent,P. *et al.* (1997) Association of Grb2 with Sos and Ras with Raf-1 upon gamma irradiation of breast cancer cells. *Oncogene* **15**: 53-61.
Taccioli,G.E., Gottlieb,T.M., Blunt,T. *et al.* (1994) Ku80: product of the XRCC5 gene and its role in DNA repair and V(D)J recombination. *Science* **265**: 1442-1445.
Tatsuka,M., Owada,M.K. and Mitsui,M. (1992) Expression of platelet-derived growth factor-independent phenotypes in BALB/c3T3 cell variant with high susceptibility to chemically or physically induced neoplastic cell transformation: dissociation from activation of protein kinase C. *Cancer Res.* **52**: 4232-4241.
Thompson,L.H. and Jeggo,P.A. (1995) Nonmenclature of human genes involved in ionising radiation sensitivity. *Mutat.Res.* **337**: 131-134.
Toyoshima,H. and Hunter,T. (1994) p27, a novel inhibitor of G1 cyclin-cdk protein kinase activity, is related to p21. *Cell* **78**: 67-74.
Treisman,R. (1996) Regulation of transcription by MAP kinase cascades. *Curr. Opin.Cell Biol.* **8**: 205-215.
Troppmair,J. Cleveland,J.L., Askew,D.S. *et al.* (1992) v-Raf/v-Myc synergism in abrogation of IL-3

dependence: v-Raf suppresses apoptosis. *Curr. Top. Microbiol. Immunol.* **182**: 453-460.
Uckun,F.M., Tuel-Ahlgren,L., Song,C.W. *et al.*(1992) Ionising radiation stimulates unidentified tyrosine-specific protein kinases in human B-lymphocyte precursors, triggering apoptosis and clonogenic cell death. *Proc.Natl.Acad.Sci. USA.* **89**: 9005-9009.
Uckun,F.M., Schieven,G.L., Tuel-Ahlgren,L.M. *et al.* (1993) Tyrosine phosphorylation is a mandatory proximal step in radiation-induced activation of the protein kinase C signaling pathway in human B-lymphocyte precursors. *Proc.Natl.Acad.Sci. USA.* **90**: 252-256.
Uckun,F.M., Waddick,K.G., Mahajan,S. *et al.* (1996) BTK as a mediator of radiation-induced apoptosis in DT-40 lymphoma B Cells. *Science* **273**: 1096-100.
Van Antwerp,D.J., Martin,S.J., Kafri,T. *et al.* (1996) Suppression of TNF-alpha-induced apoptosis by NF-kappaB. *Science* **274**: 787-789.
Vaux,D.L., Cory,S., Addams,J.M. (1988) Bcl-2 gene promotes haemopoietic cell survival and cooperates with c-myc to immortalise pre-B cells. *Nature* **335**: 440-442.
Verheij,M., Bose,R., Lin,X.H. *et al.* (1996) Requirement for ceramide-initiated SAPK/JNK signalling in stress-induced apoptosis. *Nature* **380**: 75-79.
Waddick,K.G. and Uckun,F.M. (1993) Effects of recombinant interleukin-3 and recombinant interleukin-6 on radiation survival of normal human bone marrow progenitor cells. *Radiat.Oncol. Invest.* **1**: 34-40.
Wakabayashi-Ito,N. and Nagata,S. (1994) Characterization of the regulatory elements in the promoter of the human elongation factor-1 alpha gene. *J.Biol.Chem.* **269**: 29831-29837.
Wang,C.-Y., Mayo,M.W. and Baldwin,A.S. Jr. (1996) TNF-and cancer therapy-induced apoptosis: Potentiation by inhibition of NF-κB. *Science* **274**: 784-787.
Wang,H.G., Rapp,U.R., Reed,J.C. *et al.* (1996a). Bcl-2 targets the protein kinase Raf-1 to mitochondria. *Cell* **87**: 629-638.
Wang,H.G., Takayama,S., Rapp,U.R. *et al.* (1996b). Bcl-2 interacting protein, BAG-1, binds to and activates the kinase Raf-1. *Proc. Natl. Acad. Sci. USA* **93**: 7063-7068.
Wang,X.W., Yeh,H., Schaeffer,L. *et al.* (1995) p53 modulation of TFIIH-associated nucleotide excision repair activity. *Nature Genet.* **10**: 188-195.
Wang,Y.A., Elson,A. and Leder,P. (1997) Loss of p21 increases sensitivity to ionising radiation and delays the onset of lymphoma in atm-deficient mice. *Proc.Natl.Acad.Sci.USA.* **94**: 14590-14595.
Wasylyk,C., Wasylyk,B., Heidecker,G. *et al.* (1989) Expression of raf oncogenes activates the PEA1 transcription factor motif. *Mol.Cell.Biol.* **9**: 2247-2250.
Weichselbaum,R.R., Beckett,M.A., Schwartz,J.L. *et al.*(1988) Radioresistant tumor cells are present in head and neck carcinomas that recur after radiotherapy. *Int. J. Radiat. Oncol.Biol.Phys.* **15**: 575-579.
Weissinger,E.M., Eissner,G., Grammar,C. *et al* .(1997) Inhibition of the Raf-1 kinase by cyclic AMP agonists causes apoptosis of v-abl-transformed cells. *Mol.Cell.Biol.* **17**: 3229-3241.
Williams,G.T., Smith,C.A., Spooncer,E. *et al.* (1990) Haemopoietic colony stimulating factors promote cell survival by suppressing apoptosis. *Nature* **343**: 76-79.
Winston,J.T., Coats,S.R., Wang,Y.Z. *et al.* (1996) Regulation of the cell cycle machinery by oncogenic ras. *Oncogene* **12**: 127-134.
Woloschak,G.E., Chang-Liu,C.-M and Shearin-Jones,P. (1990) Regulation of protein kinase C by ionzing radiation. *Cancer Res.* **50**: 3963-3967.
Wong,K.-K., Zou,X., Merrell,K.T. *et al.* (1995) v-Abl activates c-myc transcription through the E2F site. *Mol.Cell.Biol.* **15**: 6535-6544.
Wood,K.W., Qi,H., D'Arcangelo,G. *et al.* (1993) The cytoplasmic raf oncogene induces a neuronal phenotype in PC12 cells: a potential role for cellular raf kinases in neuronal growth factor signal transduction. *Proc.Natl.Acad.Sci. USA.* **90**: 5016-5020.
Woods,D., Parry,D., Cherwinski,H. *et al.* (1997) Raf-induced proliferation or cell cycle arrest is determined by the level of Raf activity with arrest mediated by $p21^{Cip1}$. *Mol.Cell.Biol.* **17**: 5598-5611.
Wooster,R., Neuhausen,S.L., Mangion,J. *et al.* (1994) Localization of a breast cancer susceptibility gene, BRCA 2, to chromosome 13q12-13. *Nature* **265**: 2088-2090.
Wu,J., Dent,P., Jelinek,T. *et al.* (1993) Inhibition of the EGF-activated MAP kinase signalling pathway by adenosine 3',5'-monophosphate. *Science* **262**: 1065-1069.
Wu,J., Spiegel,S. and Sturgill,T.W. (1995) Sphingosine 1-phosphate rapidly activates the mitogen-activated protein kinase pathway by a G protein-dependent mechanism. *J.Biol.Chem.* **270**: 11484-11488.
Xia,Z., Dickens,M., Raingeaud,J. *et al.* (1995) Opposing effects of ERK and JNK-p38 MAP kinases on apoptosis. *Science* **270**: 1326-1331.
Yang,E., Zha,J., Jockel,J. *et al.* (1995) Bad, a heterodimeric partner for Bcl-XL and Bcl-2, displaces Bax and promotes cell death. *Cell* **80**: 285-291.
Yang,J., Liu,X., Bhalla,K. *et al.* (1997) Prevention of apoptosis by Bcl-2 release of cytochrome C from mitochondria blocked. *Science* **275**: 1129-1132.
Yuan,Z.-M., Huang,Y., Whang,Y. *et al.* (1997) Role for c-Abl tyrosine kinase in growth arrest response to DNA damage. *Nature* **382**: 272-274.
Zampetti-Bosseler,F. and Scott,D. (1981) Cell death, chromosome damage and mitotic delay in normal human, ataxia telangectasia and retinoblastoma fibroblasts after X-irradiation. *Int.J.Radiat.Biol.* **39**: 547-558.
Zamzami,N. Susin,A., Marchetti,P. *et al.* (1996). Mitochondrial control of nuclear apoptosis. *J. Exp.*

Med. **183**: 1533-1544.
Zha,J., Harada,H., Yang,E. *et al.* (1996) Serine phosphorylation of death agonist BAD in response to survival factor results in binding to 14-3-3 not Bcl-XL. *Cell* **87**: 619-628.
Zhan,Q., Fan,S., Bae,I. *et al.* (1994) Induction of bax by genotoxic stress in human cells correlates with normal p53 status and apoptosis. *Oncogene* **9**: 3743-3751.
Zhan,Q., Alamo,I., Yu,K. *et al.* (1996) The apoptosis-associated γ-ray response of Bcl-X_L depends on normal p53 function. *Oncogene* **13**: 2287-2293.
Zhou,G. and Kuo,M.T. (1997) NF-κB-mediated induction of *mdr1b* expression by insulin in rat hepatoma cells. *J.Biol.Chem.* **272**: 15174-15183

Chapter 5

Control of Apoptosis Through Gene Regulation.

Yue Eugene Chin and Xin-Yuan Fu.
Department of Pathology, Yale University School of Medicine,310 Cedar Street, New Haven, CT 06520-8023.

1.0 INTRODUCTION

Apoptosis plays an important role in both physiological and pathological processes. Apoptosis occurs during both normal embryonal development and homeostasis in multicellular organisms, and possibly in the prevention of neoplastic transformation. In addition, derangement or failure of apoptosis may be involved in the pathogenesis of a number of human diseases including cancer. Recent research has begun to elucidate the molecular mechanism underlying the initiation and manifestation of the apoptotic process. It is now widely accepted that apoptotic cells are killed via a highly ordered and controlled program. In brief, extracellular and intracellular signals activate a signal transduction cascade involving proteases, ultimately resulting in the activation of DNA endonucleases that cleave chromosomal DNA into fragments.

Tumour necrosis factor-α (TNF-α) and Fas have been most extensively studied for their capability to induce apoptosis by triggering a cascade of protease activation (Nagata and Goldstein, 1995; Fraser and Evan, 1996). Although events of *de novo* gene regulation and protein synthesis are considered not essential in TNF/Fas-induced apoptosis, over-expression of the effector(s) involved in their apoptotic signalling pathways usually results in significant cell death (Hsu *et al.* 1996). In contrast to TNF/Fas, a multitude of cytokines, growth factors and soluble factors act as cell survival factors (Thompson, 1995). However, many of these cytokines and growth factors have been also found to trigger apoptosis under certain physiological conditions. Interleukins (IL) such as IL-2, IL-4, IL-6 and IL-10, growth factors such as epidermal growth factor (EGF), nerve growth factor (NGF) and platelet-derived growth factor (PDGF), and interferons (IFN) all have been reported to induce apoptosis in a variety of cell types (Lenardo, 1991; Manabe *et al.* 1994; Fluckiger *et al.* 1994; Kim *et al.* 1995; Casaccia-Bonnefil *et al.* 1996; Chin *et al.* 1997). Apoptosis induced by cytokines like IFN or EGF may require transactivation of one or more cell death related genes (Chin *et al.* 1997). Thus, protein modification and gene regulation (up and down) are both involved in cytokine-induced cell death. The programme of cell death by apoptosis can be positively and negatively regulated by the products of many different genes.

Recently, a number of gene products involved in the regulation of apoptosis have been identified, although little is known about their precise mechanism of action in apoptosis. The interleukin-1β converting enzyme (ICE), or caspase (cysteine-specific proteases that cleave after aspartic acid), family has been characterised as an executor of cell death (Yuan *et al.* 1993; Alnemri *et al.*1996). Over-expression of caspase family members in cells causes apoptosis (Miura *et al.* 1996), and some of them have been shown to mediate apoptosis induced by different cytokines or ligands (Chin *et al.* 1997). The caspases normally exist as inactive proenzymes which are cleaved immediately prior to apoptotic cell death to heterodimers that constitute the active proteases. Apoptosis is also regulated by the Bcl-2 family proteins. Some of this family's members (Bax, Bcl-xS, Bad) accelerate death while others (Bcl-2 and Bcl-xL) inhibit death (Reed 1995; Gajewski and Thompson 1996; Chittenden - this volume).

In order to induce apoptosis-related gene expression, a cytokine or ligand needs to trigger a specific signalling pathway which activates specific transcription factor(s). Regulation of the apoptotic death genes constitutes one possible scenario how the activation of specific signal transduction pathways can lead to cell death. The combination and extent of gene activation or repression might reflect the determining apoptosis-specific feature. Despite identification of many of the important players in the apoptosis process, as yet very little information has been obtained concerning how these genes are regulated.

Recently, an increasing number of transcription activators have been identified in the initiation and manifestation of the programmed cell death process (Table 1). Some types of cells are particularly susceptible to the apoptotic process whereas others are resistant to the induction of apoptosis. These differences may depend on expression levels of apoptosis related gene products and precursors which are controlled by *de novo* RNA and protein synthesis. For example, apoptosis in rodent thymocytes induced by ionising radiation requires new gene expression products, and for activation-induced apoptosis in T cell hybridomas, the transcription factor nur-77 is critical (Weih *et al.* 1996). Glucocorticoid receptor (GR), a transcription factor, was found to induce apoptosis in thymocytes. This GR-induced apoptosis requires new mRNA and protein synthesis (Wyllie 1980; Zhou and Thompson 1996). Transfection of different types of cells with cDNA expressing TNF-α receptor 1, c-Myc, or caspase-1 triggers significant apoptosis (Hsu *et al.* 1996; Miura *et al.* 1993; Evan *et al.* 1992). This indicates that apoptosis can be induced by increasing the expression level of a specific apoptosis gene product. Therefore, transcriptional control of these apoptosis genes can play a key role in the regulation of apoptosis.

Table I. Transcription Factors and Their Target Genes Involved in Apoptosis

Transcription Factor	Target Genes	References
I. Growth Inhibition		
IRF-1	*iNOS, ICE, AT2-R,* $p21^{WAF1/CIP1}$	Tamura *et al.* 1995
		Spink and Evans 1997
		Kamijo *et al.* 1994
p53	*bax, PIGs, Fas,* $p21^{WAF1/CIP1}$	Polyak *et al.* 1997
		Miyashita and Reed 1995
STAT1	*IRF-1, iNOS, ICE,* $p21^{WAF1/CIP1}$	Chin *et al.* 1997
		Kumar *et al.* 1997
	Fas/FasL	Chin *et al.* 1996
		Chon *et al.* 1996
		Xu *et al.* 1998
STAT3		Minami *et al.* 1996
		Naka *et al.* 1997
II. Growth Stimulation		
AP-1	*cyclin D1*	Herber *et al.* 1994
	c-*Jun*	Bossy-Wetzel *et al.* 1997
		Ham *et al.* 1995
	c-*Fos*	Smeyne *et al.* 1993
		Preston *et al.* 1997
E2F		Field *et al.* 1996
GR	c-*Jun*	Zhou and Thompson 1996
Myc	*ODC, p53, cdc25A, FasL*	Evan *et al.* 1992
		Reisman *et al.* 1993a
		Packham and Cleveland 1994b
		Galaktionov *et al.* 1996
		Shi *et al.* 1992
nur77		Weih *et al.* 1996
EGR-1	p53	Nair *et al.* 1997

This review, discusses the roles of the transcription factors in apoptotic gene regulation. In particular, it emphasises those transcription factors that are also involved in cell cycle control. Interestingly, transcription factors involved in both the stimulation and inhibition of cell growth have been shown to induce apoptosis. Different cell types may have different expression patterns of apoptotic genes and therefore may be regulated by different transcription factors. Here, we focus on cell growth inhibitory transcription factors (i.e. IRF-1, STAT, and p53) and cell growth stimulatory transcription factors (i.e. c-Myc and AP-1), and recent discoveries of their effects in apoptosis-related target gene regulation.

2.0 CYTOKINES HAVE DUAL EFFECTS ON CELLS: SURVIVAL AND APOPTOSIS

Growing evidence indicates that many cytokines, growth factors or ligands have dual effects on cell growth and survival: maintaining survival and promoting growth in some types of cells, while inducing growth arrest and/or apoptosis in others. For instance, EGF is a growth factor promoting cell survival and growth in the classical sense, but EGF has been extensively reported to produce cell growth arrest and induce apoptosis in a number of cell lines such as the rat fibroblastic line Rat-1 and NRK keratinocytes, the human epidermal cell line A431, and the human breast cancer cell line MDA-MB-468 (Kim *et al.* 1995; Chin *et al.* 1997; Armstrong *et al.* 1994). Aberrant expression of the EGF receptor or its closely related receptor erB2/neu has been found frequently in breast and ovary cancer cells and is correlated to cellular transformation. Like EGF, PDGF is a potent mitogen *in vitro*. PDGF also signals a diversity of other important cellular responses, including chemotaxis, survival, and transformation. However, PDGF induces apoptosis in NRK cells (Kim *et al.* 1995). NGF promotes growth in neurons but induces the death of mature oligodendrocytes (Casaccia-Bonnefil *et al.* 1996). IL-6 is produced by a variety of cells including fibroblasts, endothelial cells, keratinocytes, T/B lymphocytes and monocytes/ macrophages. It is thought to be a major molecular mediator of inflammatory conditions, acute phase protein synthesis in the liver, and various immunological reactions. IL-6 can stimulate proliferation and prevent apoptosis from occurring in hepatocytes; But IL-6 inhibits cell growth and induces differentiation or apoptosis in some myeloma cell lines (Altmeyer *et al.* 1997). IL-2 is a well known growth factor for lymphocytes but it can evoke apoptosis in some thymocytes and peripheral T cells (Lenardo, 1991; Migliorati *et al.* 1993, 1994). Like IL-2, IL-4 and IL-10 also have been reported to have dual effects on B cells (survival and apoptosis) (Manabe *et al.* 1994; Fluckiger *et al.* 1994; Li *et al.* 1997). In fact, TNF-α, is also a dual effect cytokine and has been found to initiate at least two opposing signalling pathways: Induction of cell death through the activation of a protease signalling pathway, and activation of transcription factor NF-κB which suppresses apoptosis and leads to cell growth (Beg and Baltimore, 1996; Liu *et al.* 1996; Wang *et al.* 1996; Van Antwerp *et al.* 1996).

An essential question is whether a single ligand-bound receptor activates two different signal transduction pathways leading to death gene and survival gene expression, respectively. The possibility exists that the signalling pathways required for apoptosis gene induction differ from those involved in cell proliferation and/or survival. Recent work found that all these cytokines can simultaneously trigger more than one signalling pathway and turn on multiple genes after binding to their respective receptor in the cell membrane (Van der Greer *et al.* 1994; Ihle, 1995). Among the signalling pathways, the Ras-MAP kinase family has received particular attention (See Kasid and Suy - this volume). In the Ras-MAP kinase cascade, ligand-bound receptors can dimerise and activate intrinsic receptor or associated tyrosine kinases resulting in a phosphotyrosine dependent recruitment of signalling molecules containing SH2 domains, i.e., Grb2 and SHC. The subsequent interaction between Grb2-SHC and SOS mediates Ras activation. The pathway ultimately involves Raf and MAP kinase. Activated MAP kinases translocate into the nucleus and phosphorylate transcription factors (Karin and Hunter 1995; Hill and Treisman, 1995) which are believed to be critical for both stimulation of cell growth and suppression of cellular suicide programs (Thompson, 1995). Recently, it has been found that signalling through the lipid products of PI3K is also important for cell survival and proliferation. Activation of PI3K has been shown to prevent apoptosis in several cell types (Franke *et al.* 1997).

Is there any signalling pathway for cell growth arrest and/or apoptosis? Recently, the activation of STAT has been linked to the regulation of cell growth arrest and apoptosis (Chin *et al.* 1997). The STAT (signal transduction and activator of transcription) pathway is a recently discovered direct signalling route, through which the occupancy of a cell surface receptor by its ligand can trigger activation of STAT proteins which translocate to the nucleus to regulate gene expression (Fu, 1992; Schindler *et al.* 1992; Darnell *et al.* 1994). Like the Ras-MAP kinase pathway, latent STAT can be activated by receptor associated tyrosine kinase (i.e., JAK kinases) or receptor intrinsic tyrosine kinases (e.g. EGFR) (Muller *et al.* 1993; Velazquez *et al.* 1992; Fu and Zhang, 1993; Quelle *et al.* 1995). STATs are regulated by phosphorylation on a tyrosine residue and have a domain specialised for interacting with other proteins containing phosphorylated tyrosines, the Src homology (SH2) domain. The phosphorylated STAT proteins dimerise and translocate to the nucleus where they bind to DNA in a sequence specific manner and activate gene transcription. The STAT pathway was initially found to respond to the interferon family. JAK kinases associated with IFN receptors become activated after receptor-ligand interaction to catalyse phosphorylation of a specific tyrosine on the STAT proteins (Muller *et al.* 1993; Shuai *et al.* 1993). Recently, other tyrosine kinases such as Src kinase, EGF receptor, and PDGF receptor all have been found to activate STAT proteins directly (Leaman *et al.* 1996).

It was shown that STAT activation is crucial for interferon and EGF signalling, especially in cell growth arrest and apoptosis. Selection of cell lines resistant to interferon or EGF action aided in the proof that STAT is required for the action of IFN or EGF. Although initial studies of STAT1 knock-out (-/-) mice indicated that these mice have no gross developmental abnormalities except that these STAT null mice are extremely sensitive to infections and can die from liver necrosis (Durbin *et al.* 1996; Meraz *et al.* 1996). However, recent studies showed that STAT1 null mice have significant defects in systems other than interferons. In particular, these mice are defective in the induction of apoptosis e.g. the apoptosis of underlying keratinocytes that follows the ablation of corneal epithelia cells (Kumar *et al.* 1997).

Following the cloning of the genes encoding p91 and p113 (Schindler *et al.* 1992; Fu *et al.* 1992), components of ISGF3 (interferon stimulated gene factor), it became evident that these subunits are members of the larger STAT family of transcription regulatory proteins. To date, seven members of the STAT family have been identified and characterised and they are widely expressed in different cell types and tissues. In addition to interferons, a large number of cytokines activate the PTK-STAT intracellular signalling pathway. Many interleukins have been found to selectively activate more than one kind of STAT protein. For instance, STAT1 is activated by most of the interleukins including IL-2, IL-3, IL-5, IL-6, IL-7, IL-13, and IL-15; STAT3 is activated by the IL-6 family of cytokines, STAT5 is activated by prolectin, IL-2 and IL-5, and STAT6 is activated by IL-4 and IL-13. The genes regulated by the STAT family of transcription factors are diverse and include those involved in immune function, inflammatory response, and cell adhesion (Darnell, 1997).

3.0 TRANSCRIPTION FACTORS AND THE GENES THEY REGULATE IN APOPTOSIS

3.1 STAT1

3.1.1 Caspase Family

Caspase-1 (ICE) gene expression can be influenced by a number of factors. Up-regulation of caspase-1 gene expression has been observed in, and correlated with, DNA damage-induced and IRF-1-mediated apoptosis (Tamura *et al.* 1995), granzyme B-mediated apoptosis (Shi *et al.* 1996), and lipopolysaccharide (LPS)-induced, or basement membrane extracellular matrix(ECM) degradation-induced, apoptosis in mammary epithelial cells (Boudreau *et al.* 1995). Cytokines such as EGF, interferons, and TNF-α, all have been found to increase *caspase-1* gene expression (Chin *et al.* 1997; Chin and Fu unpublished observation). Tetrapeptide inhibitors of caspase-1 are able to prevent the apoptosis of interdigital cells during the formation of fingers, the apoptosis of motoneurons both *in vitro* as a result of trophic factor deprivation and *in vivo* around chick embryonic day 8, and also apoptosis induced by growth factor withdrawal (Livingston, 1997; Kondo *et al.* 1996). Although *caspase-1-/-* thymocytes undergo apoptosis normally when treated with dexamethasone and γ-irradiation, they are partially resistant to Fas-induced apoptosis (Kuida *et al.* 1995).

In A431 and MDA-MB-468 cells, both EGF and IFN–γ can cause apoptosis (Chin *et al.* 1997; Armstrong *et al.* 1994). STAT1 activation and *caspase-1* gene expression were observed in response to both EGF and IFN-γ in these cells (Chin *et al.* 1997). However, in an EGF resistant A431 cell line, both EGF and IFN-γ failed to induce apoptosis presumably due to their failure to activate STAT1 and *caspase-1* gene expression in this cell line (Chin *et al.* 1997). In order to determine whether STAT1 is required for IFN-γ-induced apoptosis and ICE expression, the effects of STAT1 deficiency were examined in U3A cells which is a STAT1 null cell line. In this cell line, there is no detectable *caspase-1* mRNA, and IFN-γ failed to induce *caspase-1* mRNA expression and apoptosis (Chin *et al.* 1997; Kumar *et al.* 1997). However, both *caspase-1* expression and apoptosis can be readily induced by IFN-γ in the parental cell line 2fTGH and in STAT1 re-introduced U3A cells (Chin *et al.* 1997). Therefore, STAT1 is essential for *caspase-1* gene expression and apoptosis induction by IFN-γ. Since the STAT1 binding site SIE has not been identified yet in the promoter region (1kb) of this gene, involvement of other transcription factor(s) in the regulation of this gene is not completely excluded. Recently, Stark and co-workers showed *caspase-1* cannot be induced in U3A cells that were transfected with the IRF-1 gene and that constitutive expression of caspases requires STAT1 (Kumar *et al.* 1997).

Like *caspase-1*, *caspase-2* (*ICH-1/NEDD-2*) gene expression can be up-regulated by IFN-γ in several cell types (Ossina *et al.* 1997). In STAT1 deficient cells, *caspase-2* gene expression has been shut down and cannot be induced by IFN-γ (Kumar *et al.* 1997; Chin and Fu unpublished observation). Like caspase-1, caspase-3 (CPP32) and caspase-8 (FLICE), caspase-2 is also considered to play an important role in TNF-α induced apoptosis. Therefore, reduced expression of these caspase family members will influence the apoptosis triggered by TNF-α or other cytokines (Kumar *et al.* 1997; Chin and Fu, unpublished observation).

Although caspase-3 is considered to play a pivotal role in apoptosis, compared with *caspase-1,* the *caspase-3* gene expression pattern is much stable and less effected by extracellular environmental changes. In HT29 cells, treatment with IFN-γ weakly (1.9 fold) increased *caspase-3* expression as measured by its mRNA level (Ossina *et al.* 1997). In STAT1 deficient U3A cells, *caspase-3* is still expressed or only weakly affected (Chin *et al.* 1997; Kumar *et al.* 1997). However, it is possible that caspase-3 could be enzymatically activated by auto-cleaved caspase-1 which could be induced by the STAT signalling pathway (Chin *et al.* 1997).

3.1.3 Fas/TNF-α Receptor 1

Fas is constitutively expressed in many types of cells but is most abundant in hepatocytes and epithelial cells. IFN-γ is able to stimulate tumour and immune cells to express Fas (Maciejewski *et al.* 1995; Xu *et al.* 1998). The ability of IFN-γ to induce Fas expression suggests that IFN-γ may inhibit cell growth by Fas-mediated apoptosis. Some breast cancer cell lines become more sensitive to Fas-mediated apoptosis upon treatment with IFN-γ. IFN-γ not only upregulates the expression of Fas but also FasL in U937 cells subsequently resulting in apoptosis (Xu *et al.* 1998). Upregulation of Fas and FasL requires the expression and activation of STAT1 protein. Haematopoietic progenitor cells are hypersensitive to IFN-γ-mediated growth arrest or apoptosis in mice with a targeted disruption of the *Fanconi anemia group C* (*FA-C*) gene (Rathbun *et al.* 1997). IFN-γ induced progenitor cells undergo programmed cell death at least in part through its capacity to induce Fas expression in these cells (Rathbun *et al.* 1997). In addition, IFN-γ can also upregulate *TNF receptor 1* mRNA level in some types of cells, presumably through STAT1 activation (Chin and Fu, unpublished observation). TNF-α-induced apoptosis can be greatly enhanced by pre-treatment of cells with IFN-γ which may partly account for the increased levels of TNF-α R1 and caspases in these cells (Ossina *et al.* 1997).

3.1.4 iNOS

Nitric oxide synthesised by nitric oxide synthase (NOS), can cause apoptosis in some types of cells including tumour cells (Kitajima *et al.* 1994). Mouse macrophages, and many other cell types express an inducible isoform of NOS (iNOS) that accumulates after cell activation by stimuli such as IFN-γ plus bacterial LPS. The mouse *iNOS* promoter has been extensively characterised and is known to contain several binding sites for transcription factors such as NF-κB and IRF-1 (Spink *et al.* 1997; Xie *et al.* 1993). The STAT1 binding site GAS has also been identified in the promoter region and is indeed involved in the induction of *iNOS* expression by IFN-γ, or IFN-γ plus LPS (Gao *et al.* 1997). STAT1 has been shown to bind to the GAS site of the *iNOS* gene promoter directly in IFN-γ-stimulated RAW264.7 mouse macrophages (Gao *et al.* 1997). These data provide evidence for the direct involvement of STAT1 in the transcriptional induction of the mouse *iNOS* gene.

3.1.5 Other genes

STAT1 has been implicated in the up-regulation of the expression of cell cycle inhibitor $p21^{WAF1/CIP1}$, indoleamine 2,3-dioxygenase (IDO), and 2-5A synthetase (Chin *et al.* 1996; Chon *et al.* 1996). All of these gene products have been shown to be related to cell growth arrest or to cell death, either directly or indirectly. $p21^{WAF1/CIP1}$-mediated cell growth arrest may sensitise the cells undergoing apoptosis.

Retinoic acids (RA), a group of vitamin A related compounds, have been found to induce apoptosis in a number of tumour cells. Upon entry into cells, RA activates retinoic acid receptors (RAR), which then interact with RA response elements to induce target gene expression. A combination of IFNs and RA often results in significant additive or synergistic apoptotic activity. Recent data showed that RA can induce *STAT1* gene expression; hence enhanced transcriptional induction of interferon-stimulated genes is due to an increase in the STAT1 levels in RA-treated cells (Kolla *et al.* 1996).

3.2 Other STATs

STAT3 can be activated by the IL-6 family cytokines which share a common receptor β unit, gp130. Terminal differentiation of B cells by IL-6 or LIF is coupled to cell death. In melanocytes and early stage melanoma cells, IL-6 inhibits growth and induces apoptosis (Altmeyer *et al.* 1997). Down-regulated expression of *bcl-2* and *bax* genes in M1 myeloid leukaemic cells due to treatment with IL-6 is associated with increased susceptibility to induction of apoptosis (Lotem and Sachs, 1995). The expression of the three related genes *bcl-2*, *bax*, and *bcl-xL* can be differentially regulated suggesting that the response to apoptosis-inducing treatments is controlled by the balance between the expression of these and possibly other apoptosis-inducing and apoptosis-suppressing genes.

There is some evidence to show that STAT3 plays an important role in IL-6 or LIF- induced apoptosis. Over-expression of wild-type STAT3 in M1 cells results in apoptosis without terminal differentiation (Minami *et al.* 1996). Furthermore, IL-6 and LIF-mediated growth arrest and apoptosis of M1 cells are blocked by constitutive expression of the dominant negative tyrosine mutant STAT3/Y705F ((Minami *et al.* 1996). More recently, an inhibitor of STAT activation, SSI-1 (STAT-induced STAT inhibitor) was identified and characterised (Naka *et al.* 1997). SSI-1 has a STAT-like SH2 domain and can competitively bind to JAK kinase, preventing STAT activation by this enzyme. Over-expression of *SSI-1* cDNA by transfection blocked IL-6 and LIF-induced apoptosis and macrophage differentiation of M1 cells, as well as IL-6 induced tyrosine phosphorylation of a receptor glycoprotein component, gp130, and of STAT3 (Naka *et al.* 1997).

Since activated STAT3 and STAT1 can form heterodimers and bind to the same DNA sequence as STAT1 homodimers binds (i.e. SIE), it is therefore believed that STAT3 may join STAT1 in the regulation of apoptotic gene expression. STAT1-inducible genes such as *iNOS* and *IRF-1* were induced by IL-6 in a STAT3 activity dependent manner (Yamanaka *et al.* 1996). STAT3-dependent up-regulation of apoptosis-related oncogenes such as Jun B and EGR-1 were also observed (Yamanaka *et al.* 1996). STAT5 may play a role in apoptosis through its association with other transcription factors, such as steroid hormone receptors. Formation of specific protein-protein interactions between steroid hormone receptors and other transcription factors have been commonly observed. GR is believed to be a regulator of apoptosis in T cells, the involuting mammary gland cells, and other epithelial cells (Wyllie, 1980; Chapman *et al.* 1996). The proteolytic degradation of ECM occurring during mammary gland involution leads to the loss of the differentiated state. Induction of caspase-1 expression and activity, and ultimately apoptotic cell death were observed in post-lactational involution of the mammary gland (Boudreau *et al.* 1995). Activation of GR by glucocorticoids and STAT5 by prolactin, is required for the efficient induction of the *β-casein* gene and other genes in the mammary epithelium (Lechner *et al.* 1997). The discovery of the interaction between STAT5 and GR in transcriptional regulation (Stocklin *et al.* 1996) suggests a possible role for STAT5 in GR-mediated apoptosis. STAT5-mediated $p21^{WAF1/CIP1}$ induction and growth arrest by thrombopoinetin (TPO) in the leukaemic cell line CMK (Matsumura *et al.* 1997) and synergistic inhibition of prolactin and IFN-γ on U937 cell growth were observed (Sedo *et al.* 1996). However, direct evidence that STAT5 activation by these cytokines leads to apoptosis is still lacking.

3.3 IRF-1

IRF-1 belongs to the interferon regulatory factor (IRF) family which were originally identified as DNA binding factors to regulatory elements in interferon promoters (Taniguchi *et al.* 1995). IRF-1 plays a role in the IFN response, since the expression of the *IRF-1* gene itself is strongly induced by IFN-α/β and IFN-γ (Harada *et al.* 1993). This property is due to the presence of a GAS (gamma-interferon activated site) sequence within the *IRF-1* gene promoter region and this GAS element can be bound by activated STAT (Pine *et al.* 1994). The *IRF-1* gene promoter also contains NF-κB sites which are involved in the basal or TNF-α–dependent transcriptional activation of this gene (Pine, 1997; Ohmori *et al.* 1997).

IRF-1 has been described as a positive transcription factor and may act as a tumour-suppressor gene (Taniguchi *et al.* 1995; Tanaka *et al.* 1994). Interestingly, IRF-1 has been shown experimentally to play a role in the induction of apoptosis (Tamura *et al.* 1995; Taniguchi *et al.* 1995). IRF-1 DNA binding sites are present in the promoters of genes such as *iNOS* and *caspase-1*.

Enhanced expression of oncogenes can promote cells to undergo apoptosis, particularly under conditions of low serum concentration or high cell density. Tanaka *et al.* reported that IRF-1 may be a critical determinant of oncogene-induced cell transformation or apoptosis in mouse embryonic fibroblasts (EFs) (Tanaka *et al.* 1994). They demonstrated that Ras signalling under conditions of low serum or at high cell density, or following treatment by anti-cancer drugs or ionising radiation, induces the death of EFs derived from wild type mouse but not the death of EFs from *IRF1-/-* mice. These data demonstrated that the IRF-1 may be involved in Ras-mediated apoptosis.

Activation of the angiotensin II (AT2) receptor has been reported to induce apoptosis in certain cell types, such as a rat pheochromocytoma cell line (PC12W) and a confluent mouse fibroblast cell line (R3T3) (Yamada *et al.* 1996). Horiuchi *et al* (1997) reported that in R3T3 cells, serum starvation can upregulate the AT2 receptor and induce apoptosis. IRF-1 has been shown to induce *AT2 receptor* gene expression presumably due to its ability to bind to the ISRE motif that has been identified in the *mouse AT2 receptor* gene promoter region. Transfected antisense oligonucleotides for *IRF-1* attenuated apoptosis and abolished the upregulation of the AT2 receptor. The increased IRF-1 expression level after serum starvation stimulated AT2 receptor expression and enhanced AT2 receptor-mediated apoptosis.

Furthermore, Giandomenico *et al.* showed that the induction of IRF-1 by both IFN-γ and RA correlates with the subsequent induction of apoptosis in both cervical carcinoma cells and squamous carcinoma cell line SiHa cells (Giandomenico *et al.* 1997). Tamura *et al* (1995) showed that IRF-1 is involved in apoptosis following radiation- or chemically-induced DNA damage in activated lymphocytes. Splenocytes from IRF-1-deficient mice were markedly less sensitive to apoptosis than were wild type. This event has been correlated with activation of caspase-1 during mitogenesis. Caspase-1 activation was completely abrogated in splenocytes from the IRF-1-deficient mice strongly suggesting that this gene is indeed IFN-γ inducible (Tamura *et al.* 1995). This suggestion is further supported by the presence of an IRF-1 binding site in this promoter which can be bound by IRF-1 and is able to confer IRF-1-dependent transactivity when inserted into the promoter region of a luciferase reporter plasmid (Tamura *et al.* 1995). Additionally, ISRE (IFN-stimulated regulatory element) sequences have been found in the *caspase-1* promoter, but their function has not yet been put to direct test (Chin and Fu, unpublished observation). There was evidence that p53 and IRF-1 can co-operate in response to DNA damage-induced cell cycle arrest, as indicated by $p21^{WAF1/CIP1}$ induction (Tanaka *et al.* 1996). IFN-γ has no effect on *p53* gene expression but is able to augment the *IRF-1* expression (Tanaka *et al.* 1996). Over-expression of IRF-1 results in the activation of the endogenous genes for caspase-1 and p21WAF1/CIP1 and enhances the sensitivity of cells to radiation-induced apoptosis. Thus, the anti-oncogenic transcription factors p53 and IRF-1 are required for distinct apoptotic pathways in T lymphocytes. Although NF-κB is considered in general for cell survival, the fact that *IRF-1* can be upregulated by TNF-α through NF-κB activation suggests a potential contribution of NF-κB to apoptosis (Pine, 1997; Ohmori *et al.* 1997).

Activity of iNOS has been linked to tissue damage in arthritis, nephritis, and septic shock. Cloning of the *iNOS* promoter from human and mouse elucidated the presence of potential IRF-1 binding motifs. Deletion and mutational analyses showed that the binding of IRF-1 to these sequences constitutes the critical molecular determinants of the action of IFN-γ on the *iNOS* promoter (Spink and Evans 1997; Kamijo *et al.* 1994). Moreover, macrophages from mice with a targeted disruption of the *IRF-1* gene produced little or no nitric oxide and synthesised barely detectable *iNOS* messenger RNA in response to stimulation. Additionally, IRF-1 plays an important role in inflammation and autoimmunity *in vivo* by controlling Th1 cell differentiation (Lohoff *et al.* 1997). Therefore, IRF-1 may have potential as a therapeutic agent due to its role in regulating nitric oxide production and nitric oxide-induced toxicity.

3.4 p53

p53 protein is best known for its anti-oncogenic role. p53 induces both stable cell growth arrest and programmed cell death, probably due to its activation of different genes (Polyak *et al.* 1997; Levine 1997). The inhibitory effect of p53 on cell growth is attributed to, at least in part, the transcriptional induction of $p21^{WAF1/CIP1}$, a cell cycle inhibitor which blocks cyclin kinases, thus inhibiting the function of cyclins required for cell proliferation (Levine 1997). The role of p53 in apoptosis induction is apparent because thymocytes or hematopoietic cells with either *p53* mutations or knockouts became resistant to radiation-induced apoptosis. Using single cell micro-injection of loss-of-function and chimeric gain-of-function mutants demonstrated that transcriptional activation events are critical for both G1 arrest and apoptosis processes (Attardi *et al.* 1996). Although there is evidence that cytokine (e.g., IFN-α)-induced cell cycle arrest and apoptosis result from two distinct transcriptional events, cell cycle arrest may sensitise the cell to undergo apoptosis (Chin *et al.* 1997).

Reed's group obtained evidence that restoration of p53 in a murine leukaemia cell, M1, was associated with increases in *bax* mRNA and protein (Miyashita and Reed 1995). Moreover, they showed that p53 is a direct transcriptional activator of the human *bax* gene. The *bax* gene promoter contains a 39-bp region with a p53 binding site consensus sequence. To further confirm the role of Bax in p53-dependent apoptosis, McCurrach *et al.* (1997) examined the effects of Bax deficiency in primary fibroblasts expressing the *E1A* oncogene, a setting where apoptosis is dependent on endogenous p53. They demonstrated that Bax can function as an effector of p53 in chemotherapy-induced apoptosis and contributes to a p53 pathway to suppress oncogenic transformation. Furthermore, they showed that additional p53-controlled factors act synergistically with Bax to promote a full apoptotic response, and their action can be blocked by Bcl-2.

By using SAGE (serial analysis of gene expression), a technique that allows the quantitative and simultaneous evaluation of cellular mRNA populations, Vogelstein's group identified a number of genes which are markedly increased in p53-expressing cells compared with control cells (Polyak *et al.* 1997). Strikingly, many of these genes were predicted to encode proteins that could generate or respond to oxidative stress. There is evidence indicating that reactive oxygen species may be involved in programmed cell death. Bcl-2 is known to protect cells from the lethal effects of γ-irradiation, H_2O_2, and t-butyl hydroperoxide. According to these findings, a three-step process is proposed: the transcriptional induction of redox-related genes, the generation of reactive oxygen species, and finally oxidative degradation of mitochondrial components culminating in cell death.

p53 is also involved in c-Myc- and Fas-FasL-induced apoptosis (Owen-Schaub *et al.* 1995). Wild type human p53 and a temperature sensitive mutant induce Fas expression in the absence of *de novo* protein synthesis, suggesting that Fas is a direct transcriptional target for p53 (Owen-Schaub *et al.* 1995; Reinke and Lozano *et al.* 1997). p53 has been found to co-operate with another tumour-suppressor, transcription factor IRF-1, in response to DNA damage-induced gene (e.g., $p21^{WAF1/CIP1}$) expression (Tanaka *et al.* 1996).

These findings demonstrate that p53-mediated apoptosis involves the activation of a number of target genes and co-ordinated activities of multiple effectors. However, Karin's group provided evidence that *de novo* protein synthesis is not required in p53-dependent induction of apoptosis by DNA damage in cells. Shifting to the permissive temperature triggered apoptosis following DNA damage, but this is independent of new RNA or protein synthesis, as actinomycin D (transcriptional inhibitor) and cycloheximide (protein synthesis inhibitor) did not block the apoptosis induced by p53 (Caelles *et al.* 1994). Therefore, it seems that p53 either represses genes necessary for cell survival or is a component of a large complex controlling the enzymatic machinery for apoptotic cleavage or repair of DNA.

3.5 c-Myc

The c-Myc protein is a basic helix-loop-helix leucine zipper transcription factor. It has been implicated in the positive control of cell proliferation. c-Myc is sufficient to promote cell cycle progression, inhibit differentiation and cause transformation of some fibroblast cell lines. However, recent work showed clearly that c-Myc promotes apoptosis in some cells transfected with *c-myc* cDNA gene (Evan *et al.* 1992). Thus c-Myc is able to induce apoptosis as well as cell cycle progression. Several genes which co-operate with *c-myc* in tumourigenesis, e.g. *bcl-2*, *pim1*, and *v-raf*, have been shown to suppress apoptosis, suggesting that c-Myc can induce apoptosis *in vivo* and that suppression of this activity contributes to tumourigenesis. Although the apoptotic function of c-Myc has been well described, its downstream effectors are still largely unknown. Among the putative c-Myc target genes are *ornithine decarboxylase* (*ODC*), *p53*, *prothymosin a*, *cdc25A*, *ECA39*, *MrDb*, and *rcl* (Lewis *et al.* 1997; Resiman *et al.* 1993a).

ODC is considered a mediator of c-Myc function. ODC catalyses the decarboxylation of ornithine to putrescine and is a key regulator of polyamine biosynthesis. c-Myc has been found to be an essential transcription activator for *ODC* gene induction, because forced expression of c-Myc gene in cells results in the constitutive growth factor-independent expression of *ODC* mRNA (Packham and Cleveland 1994b). Furthermore, c-Myc is a potent activator of *ODC* gene reporter fusion constructs and site-directed mutagenesis demonstrated that this effect is mediated by two converted CACGTG sequences motifs located in *ODC* intron 1 (Packham and Cleveland 1994b). c-Myc has been shown to directly regulate the endogenous ODC without requiring *de novo* protein synthesis. Like c-Myc, ODC is sufficient to compromise cell survival; *ODC*-transfected clones showed accelerated loss of viability following IL-3 withdrawal (Packham and Cleveland 1994a). Therefore, enforced expression of *ODC*, like c-Myc, is sufficient to induce accelerated apoptosis in 32D.3 cells.

Activation of c-Myc is able to upregulate *p53* gene expression and stabilise p53 protein in mouse fibroblasts (Hermeking and Eick 1994). The *p53* promoter contains a conserved c-Myc consensus recognition sequence which is required for full promoter activity (Reisman and Rotter 1993b). In co-transfection assays, c-Myc transactivates the *p53* promoter and the thymidine kinase promoter containing multiple copies of the c-Myc binding site derived from the *p53* gene promoter (Reisman *et al.* 1993a). The *p53* promoter lacking the Myc binding site no longer responds to c-Myc, suggesting that c-Myc directly regulates *p53* gene expression (Reisman *et al.* 1993a). These findings may explain why c-Myc elevates p53 expression level in some tumour cells (Roy *et al.* 1994). To determine whether p53 is required for c-Myc-dependent apoptosis, the effects of p53 deficiency were examined in fibroblasts transfected with c-Myc-ER, a setting where c-Myc activation-dependent apoptosis can be controlled by the addition of β-oestradiol. Hermeking and Eick (1994) showed clearly that c-Myc activation in *p53*-null cells fails to induce apoptosis. These findings demonstrate that c-Myc-induced apoptosis is indeed mediated by p53.

c-Myc has also been found to transactivate *cdc25A*, which encodes a G1-specific phosphatase required for the initiation of the cell cycle and its progression through G1 (Galaktionov *et al.* 1996). Suppression of cdc25A expression using antisense RNA substantially diminished the ability of c-Myc to promote apoptosis. Moreover, cdc25A expression induces apoptosis in serum-deprived fibroblasts, exactly as does c-Myc. Although c-Myc is an oncogenic transcription factor, site-directed mutagenesis studies indicate that c-Myc promotes apoptosis by transactivation of its downstream genes

(Evan *et al.* 1992). Recently, it has been found that the Fas-FasL interaction is involved in c-Myc-induced apoptosis (Hueber *et al.* 1997). This effect is not due to the transactivation of *Fas* expression, although c-Myc may act upstream of Fas and FasL by inducing expression of their cognate genes. ODC and p53, transactivated by c-Myc, may function in the induction of, or sensitisation to, Fas-FasL interactions. The possibility that c-Myc may also operate through enhancement of FasL expression cannot be excluded.

3.6 AP1

The AP-1 transcriptional complex is composed of c-Fos and c-Jun which have been implicated in a number of biological processes, including control of cell proliferation. c-Fos and c-Jun are also involved in the apoptosis process in certain types of cells (Smeyne *et al.* 1993; Bossy-Wetzel *et al.* 1997).

c-fos and *c-jun* mRNA are induced rapidly during apoptosis triggered by deprivation of growth factors, hormone depletion, or dexamethasone addition (Estus *et al.* 1994; Grassilli *et al.* 1992). Antisense oligonucleotides against *c-fos* and *c-jun* genes are capable of increasing the viability of lymphocytes deprived of growth factors (Colotta *et al.* 1992). Using a *c-fos* promoter transgenic mouse it was shown that continuous expression of a *β-galactosidase* reporter gene begins hours or days before morphological signs of apoptosis are apparent i.e. precedes programmed cell death *in vivo* (Smeyne *et al.* 1993). Stress-induced apoptosis may require the activity of Jnk3, a member of the Jun kinase (JNK) family selectively expressed in the nervous system. Yang *et al.* (1997) showed that disruption of the *jnk3* gene in mice caused the mice to be resistant to apoptosis triggered by the glutamate-receptor agonist kainic acid. The phosphorylation of c-Jun and the transcriptional activity of the AP-1 transcription factor complex were markedly reduced and hippocampal neuron apoptosis was prevented in these Jnk3 deficient mice in response to kainic acid, despite the same level of noxious stress.

Enhanced expression of *c-fos* and *c-jun* genes and AP-1 transactivity are associated with UV-induced cell death and certain forms of neuronal cell death (Smeyne *et al.* 1993; Ham *et al.* 1995). But what might be the down stream factor(s) by which AP-1 triggers apoptosis? It seems conceivable that AP-1 may trigger apoptosis, either by inducing the expression of specific death genes such as members of caspase family of proteases, or by repressing the expression of genes that code for proteins that exhibit survival-promoting activities such as Bcl-2 protein. Results from *in vitro* and *in vivo* studies are consistent with this notion. Microinjection of a *c-jun* dominant negative mutant (D169) into sympathetic neurons prevented them from undergoing NGF withdrawal-induced apoptosis (Ham *et al.* 1995). c-Jun D169, lacking the trans-activation domain, binds to the AP-1 site and prevents gene regulation. Significant apoptosis will occur by forced overexpression of wild type c-Jun but not c-Jun D169 (Bossy-Wetzel *et al.* 1997). By screening suspect genes for initiating the execution of apoptosis, some genes were found to be induced upon c-Jun activation by β-oestradiol e.g. the *bad* gene and *tissue transglutaminase* genes (Bossy-Wetzel *et al.* 1997). Hence c-Jun increases the gene products that are capable of stimulating apoptotic cell death.

Cyclin D1 gene is a potential target of c-Jun transactivation, since the *cyclin D1* promoter contains AP-1 binding sites and can be activated by overexpression of c-Jun in luciferase assay experiments (Herber *et al.* 1994). In neurons, c-Jun protein levels start to increase 4-8 hr after NGF withdrawal, before cyclin D1 levels peak (15-20hr) (Ham *et al.* 1995; Freeman *et al.* 1994). Over-expression of cyclin D1 triggers apoptosis in neuronal or rat embryo cells (Kranenburg *et al.* 1996; Pardo *et al.* 1996). These findings may be related to the fact that the activities of c-Jun and cyclin D1 are essential during neuronal apoptosis.

It seems that c-Fos-induced apoptosis is p53-dependent, since activation of c-Fos induces apoptosis in *p53+/+* tumour cell line but not in a *p53*-null tumour cell line (Preston *et al.* 1997). Apoptosis induced by overexpressing c-Fos or c-Jun can be blocked by the co-expression of bcl-2 or tripeptide inhibitor of caspase proteases (Bossy-Wetzel *et al.* 1997; Preston *et al.* 1997). Like the case of p53 or c-Myc, c-Fos-mediated apoptosis cannot be blocked by cycloheximide, as showed by Barrett and coworkers, suggesting that c-Fos-induced apoptosis does not require *de novo* protein synthesis (Preston *et al* 1997). Furthermore, Jacobson *et al* also showed that cells can undergo apoptosis even in the absence of a nucleus (Jacobson *et al.* 1994). However, one cannot exclude the possibility that in some situations the cell death machinery can be more easily activated by elevating the level of cell death effectors through transactivation and new protein synthesis.

4.0 CONCLUSION

The onset and regulation of apoptotic cell death processes are just beginning to be understood. It is becoming clear now that many cytokines and growth factors can trigger different pathways for growth and growth arrest. In the case of cell death, both overactive survival pathways (e.g. overexpression of Ras, or c-Myc) and overactive inhibitory pathways (e.g. overexpression of p53) lead to apoptosis.

In a cell, creating harmony for survival and growth requires a balance of opposing forces. To proceed through a normal cell cycle, a cell needs both positive and negative effectors. The relationship between these opposing forces is epitomised by transactivation of different genes. Apparently, disruption of this balance can lead to either cell death or cancer. Figure 1 provides an overview of the potential involvement of various transcription factors in these processes, especially those closely related to the cell cycle. Transcription factors which inhibit cells from entering S phase, such as p53, STAT1, and IRF-1, can trigger the cell apoptosis machinery by upregulating some apoptosis genes. Whereas, transcription factors which promote cell growth and stimulate cells to enter S phase will also trigger cell death without progressing further to M phase. p53 can induce apoptosis in both G1/S and G2/M phases while c-Myc induces apoptosis throughout the cell cycles. Apparently, either "too slow" or "too fast" cell cycle will trigger the genetic program for death.

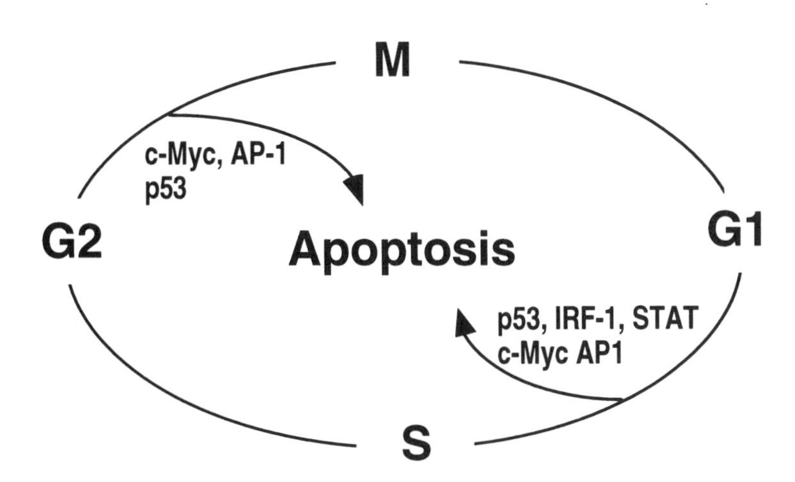

Figure 1. Cell cycle stimulatory and inhibitory transcription factors can induce apoptosis in different phases of the cell cycle,

The basic mechanisms for proliferation and transformation are linked to apoptosis. This concept of an active suppression of death via specific survival signals arose from the observation that c-Myc and AP-1 drive apoptosis in cells unless survival factors are present. Likewise the anti-apoptotic effects of Bcl-2 are mediated by the diversion of c-Myc or AP-1 signals toward cell proliferation rather than cell death. Although the biochemical roles of transcription factors in apoptosis have been well described, the identities of downstream effectors are just beginning to emerge. However, apoptosis and the molecular mechanisms involved in its induction are most probably not reflected by linear signal transduction pathways. Cross-talk between different components normally involved in the regulation of such vital processes as proliferation and differentiation may be combined and interconnected in a new and sometimes completely unexpected mode. This leaves a number of question still unanswered and often provides researchers with new insights into the basic molecular mechanisms regulating the life and death of cells and organisms.

Acknowledgments

The authors are grateful to Dr. Susan Glantz for her critical reading of this manuscript. X.-Y. Fu is supported by the National Institutes of Health grants RO1 AI34522 and RO1 GM55590. Y.E. Chin was supported by an National Institutes of Health postdoctoral fellowship award.

5.0 REFERENCES

Alnemri,E.S., Livingston,D.J., Nicholson,D.W. *et al.* (1996) Human ICE/CED-3 protease nomenclature. *Cell* **87**: 171.

Altmeyer,A., Simmons,R.C., Krajewski,S., *et al.* (1997) Reversal of EBV immortalization precedes apoptosis in IL-6-induced human B cell terminal differentiation. *Immunity* **7**: 667-677.

Armstrong,D.K., Kaufmann,S.H., Ottaviano,Y.L. *et al.* (1994) Epidermal growth factor-mediated apoptosis of MDA-MB-468 human brest cancer cells. *Cancer Res.* **54**: 5280-5283.

Attardi,L.D., Lowe,S.W., Brugarolas,J. and Jacks,T. (1996) Transcriptional activation by p53, but not induction of the p21 gene, is essential for oncogene-mediated apoptosis. *EMBO J.* **15**: 3693-3701.

Beg,A.A. and Baltimore,D. (1996) An essential role for NF-κB in preventing TNF-induced cell death. *Science* **274**: 782-784.

Bossy-Wetzel,E., Bakiri,L. and Yaniv,M. (1997) Induction of apoptosis by the transcription factor c-Jun. *EMBO J.* **16**: 1695-1709.

Boudreau,N., Sympson,C.J., Werb,Z. and Bissell,M.J. (1995) Suppression of ICE and apoptosis in mammary epithelial cells by extracellular matrix *Science* **267**: 891-893.

Caelles,C., Helmberg,A. and Karin,M. (1994) p53-dependent apoptosis in the absence of transcriptional activation of p53-target genes. *Nature* **370**: 220-223.

Casaccia-Bonnefil,P., Carter,B.D., Dobrowsky,R.T. and Chao,M.V. (1996) Death of oligodendrocytes mediated by the interaction of nerve growth factor with its receptor p75. *Nature* **383**: 716-719.

Chapman,M.S., Askew,D.J., Kuscuoglu,U., and Miesfeld,R.L. (1996) Transcriptional control of steroid-regulated apoptosis in murine thymoma cells. *Mol. Endocrinol.* **10**: 967-978.

Chin,Y.E., Kitagawa,M., Kuida,K. *et al.* (1997) Activation of the STAT signaling pathway can cause expression of caspase 1 and apoptosis. *Mol. Cell. Biol.* **17**: 5328-5337.

Chin,Y.E., Kitagawa,M., Su,W-C.S., *et al.* (1996) Cell growth arrest and induction of cyclin-dependent kinase inhibitor p21WAF1/CIP1 mediated by STAT1. *Science* **272**: 719-722.

Chon,S.Y., Hassanain,H.H. and Gupta,S.L. (1996) Co-operative role of interferon regulatory factor 1 and p91 (STAT1) response elements in interferon-gamma-inducible expression of human indoleamine 2,3-dioxygenase gene. *J.Biol.Chem.* **271**: 17247-17252.

Colotta,F., Polentarutti,N., Sironi,M. and Mantovani,A. (1992) Expression and involvement of c-fos and c-jun protoomcogenes in programmed cell death induced by growth factor deprivation in lymphoid cell lines. *J. Biol. Chem.* **267**: 18278-18283.

Darnell,J.E.,Jr., Kerr,I.M. and Stark,G.R. (1994) Jak-STAT pathways and transcriptional activation in response to IFNs and other extracellular signaling proteins *Science* **264**: 1415-1421.

Darnell,J.E.J. (1997) STATs and gene regulation. *Science* **277**: 1630-1635.

Durbin,J.E., Hackenmiller,R., Simon,M.C. and Levy,D.E. (1996) Targeted disruption of the mouse Stat1 gene results in compromised innate immunity to viral disease. *Cell* **84**: 443-450.

Estus,S., Zaks,W.J., Freeman,R.S. *et al.* (1994) Altered gene expression in neurons during programmed cell death: identification of c-jun as necessary for neuronal apoptosis. *J. Cell. Biol.* **127**: 1717-1727.

Evan,G.I., Wyllie,A.H., Gilbert,C.S. *et al.* (1992) Induction of apoptosis in fibroblasts by c-myc protein. *Cell* **69**: 119-128.

Field,S.J., Tsai,F.Y., Kuo,F. *et al.* (1996) E2F-1 functions in mice to promote apoptosis and suppress proliferation. *Cell* **85**: 549-561.

Fluckiger,A.C., Durand,I. and Banchereau,J. (1994) Interleukin 10 induces apoptotic cell death of B-chronic lymphocytic leukemia cells. *J. Exp .Med.* **179**: 91-99.

Franke,T.E., Kaplan,D.R. and Cantley,L.C. (1997) PI3K: downstream AKTion blocks apoptosis. *Cell* **88**: 435-437.

Fraser,A. and Evan,G. (1996) A license to kill. *Cell* **85**: 781-784.

Freeman,R.S., Estus,S. and Johnson,E.M.Jr. (1994) Analysis of cell cycle-related gene expression in postmitotic neurons: selective induction of Cyclin D1 during programmed cell death. *Neuron* **12**: 343-355.

Fu,X-Y. (1992) A transcription factor with SH2 and SH3 domains is directly activated by an interferon α-induced cytoplasmic protein tyrosine kinase(s) *Cell* **70**: 323-335.

Fu,X-Y. and Zhang,J-J. (1993) Transcription factor p91 interacts with the epidermal growth factor receptor and mediates activation of the c-fos gene promoter. *Cell* **74**: 1135-1145.

Fu,X-Y., Schindler,C., Improta,T. *et al.* (1992) The proteins of ISGF-3, the interferon α-induced transcriptional activator, define a gene family involved in signal transduction. *Proc. Natl. Acad. Sci.USA* **89**: 7840-7843

Gajewski,T.F. and Thompson,C.B. (1996) Apoptosis meets signal transduction: elimination of a BAD influence. *Cell* **87**: 589-592.

Galaktionov,K., Chen,X. and Beach,D. (1996) Cdc25 cell-cycle phosphatase as a target of c-myc. *Nature* **382**: 511-517.

Gao,J., Morrison,D.C., Parmely,T.J. *et al.* Interferon-γ–activated site (GAS) is necessary for full expression of the mouse iNOS gene in response to interferon-γ and lipopolysaccharide. *J. Biol. Chem.* **272**: 1226-1230.

Giandomenico,V., Lancillotti,F., Fiorucci,G. *et al.* (1997) Retinoic acid and IFN inhibition of cell proliferation is asscoiated with apoptosis in squamous carcinoma cell lines: role of IRF-1 and TGas II-dependent pathways. *Cell Growth Diff.* **8**: 91-100.

Grassilli,E., Carcereri de Prati,A., Monti,D *et al.* (1992) Studies of the relationship between cell proliferation and cell death. II. Early gene expression during concanavalin A-induced proliferation or dexamethasone-induced apoptosis of rat thymocytes. *Biochem. Biophys. Res. Comm.* **188**: 1261-1266.

Ham,J., Babij,C., Whitfield,J *et al.* (1995) A c-Jun dominant negative mutant protects sympathetic neurons against programmed cell death. *Neuron* **14**: 927-939.

Harada,H., Kitagawa,M., Tanaka,N. *et al* (1993) Anti-oncogenic and oncogenic potentials of interferon regulatory factors-1 and -2. *Science* **259**: 971-974.

Herber,B., Truss,M., Beato,M. and Muller,R. (1994) Inducible regulatory elements in the human cyclin D1 promoter. *Oncogene* **9**: 1295-1304.

Hermeking,H. and Eick,D. (1994) Mediation of c-Myc-induced apoptosis by p53. *Science* **265**: 2091-2093.

Hill,C. and Treisman,R. (1995) Transcriptional regulation by extracellular signals: Mechanisms and specificity. *Cell* **80**: 199-211..

Horiuchi,M., Yamada,T., Hayashida,W. and Dzau,V.J. (1997) Interferon regulatory factor-1 up-regulates angiotensin II type receptor and induces apoptosis. *J. Biol. Chem.* **272**: 11952-11958.

Hsu,H., Shu,H.-B., Pan,M.-G. and Goeddel,D.V. (1996) TRADD-TRAF2 and TRADD-FADD interaction define two distinct TNF receptor 1 signal transduction pathways. *Cell* **84**: 299-308.

Hueber,A.-O., Zornig,M., Lyon,D., *et al.* (1997) Requirement for the CD95 receptor-ligand pathway in c-Myc-induced apoptosis. *Science* **278**: 1305-1309.

Ihle,J.N. (1995) Cytokine receptor signalling. *Nature* **377**: 591-594.

Jacobson,M.D., Burne,J.F. and Raff,M.C. (1994) Programmed cell death and Bcl-2 protection in the absence of a nucleus *EMBO J.* **13**: 1899-1910.

Kamijo,R., Harada,H., Matsuyama,T. *et al* (1994) Requirement for transcription factor IRF-1 in NO synthase induction in macrophages. *Science* **263**: 1612-1615.

Karin,M. and Hunter,T. (1995) Transcriptional control by protein phosphorylation: signal transmission from the cell surface to the nucleus. *Curr. Biol.* **5**: 747-757.

Kim,H.R., Upadhyay,S., Li,G. *et al.* (1995) Platelet-derived growth factor induces apoptosis in growth-arrested murine fibrolast. *Proc. Natl. Acad. Sci. USA* **92**: 9500-9504.

Kitajima,I., Kawahara,K., Nakajima,T. *et al.* (1994) Nitric oxide-mediated apoptosis in murine mastocytoma. *Biochem. Biophy. Res. Comm.* **204**: 244-251.

Kolla,V., Lindner,D.J., Weihua,X. *et al.* (1996) Modulation of interferon (IFN)-inducible gene expression by retinoic acid - Upregulation of STAT1 protein in IFN-unresponsive cells. *J. Biol. Chem.* **271**: 10508-10514.

Kondo,S., Kondo,Y., Yin,D *et al.* (1996) Involvement of interleukin-1 beta-converting enzyme in apoptosis of bFGF-deprived murine aortic endothelial cells. *FASEB J.* **10**: 1192-1197.

Kranenburg,O., van der Eb,A.J. and Zantema,A. (1996) Cyclin D1 is an essential mediator of apoptotic neuronal cell death. *EMBO J.* **15**: 46-54.

Kuida,K., Lippke,J.A., Ku,G. *et al* (1995) Altered cytokine export and apoptosis in mice deficient in interleukin-1 beta converting enzyme. *Science* **267**: 2000-2003.

Kumar,A., Commane,M., Flickinger,T.W. *et al.* (1997) Defective TNFα-induced apoptosis in STAT1-null cells due to low constitutive levels of caspases, *Science* **278**: 1630-1632.

Leaman,D.W., Leung,S., Li,X. and Stark,G.R. (1996) Regulation of STAT-dependent pathways by growth factors and cytokines. *FASEB J.* **10**: 1578-1588.

Lechner,J., Welte,T., Tomasi,J K. *et al.* (1997) Promoter-dependent synergy between glucocorticoid receptor and Stat5 in the activation of beta-casein gene transcription. *J. Biol. Chem.* **272**: 20954-20960.

Lenardo,M.J. (1991) Interleukin-2 programs mouse alpha beta T lymphocytes for apoptosis. *Nature* **353**: 858-861.
Levine,A.J. (1997) p53, the cellular gatekeeper for growth and division. *Cell* **88**: 323-331.
Lewis,B.L., Shim,H., Li,Q. *et al.* (1997) Identification of putative c-Myc-responsive genes: characterization of rcl, a novel growth-related gene. *Mol. Cell. Biol.* **17**: 4967-4978.
Li,L., Krajewski,S., Reed,J.C. and Choi,Y.S. (1997) The apoptosis and proliferation of SAC-activated B cells by IL-10 are associated with changes in Bcl-2, Bcl-xL, and Mcl-1 expression. *Cell Immunol.* **178**: 33-41.
Liu,Z.-G., Hsu,H., Goeddel,D.V. and Karin,M. (1996) Dissection of TNF receptor 1 effector functions: JNK activation is not linked to apoptosis, while NF-κB activation prevents cell death. *Cell* **87**: 565-575.
Livingston,D.J. (1997) *In vitro* and *in vivo* studies of ICE inhibitors. *J. Cell. Biochem.* **64**: 19-26.
Lohoff,M., Ferrick,D., Mittrucker,H.W. *et al.* (1997) Interferon regulatory factor-1 is required for a T helper 1 immune response *in vivo*. *Immunity* **6**: 681-689.
Lotem,J. and Sachs,L. (1995) Regulation of bcl-2, bcl-XL and bax in the control of apoptosis by hematopoietic cytokines and dexamethasone. *Cell Growth Different.* **6**: 647-653.
Maciejewski,J., Selleri,C., Anderson,S. and Young,N.S. (1995) Fas antigen expression on CD34+ human marrow cells is induced by interferon gamma and tumor necrosis factor alpha and potentiates cytokine-mediated hematopoietic suppression *in vitro*. *Blood* **85**: 3183-3190.
Manabe,A., Coustan-Smith,E., Kumagai,M *et al.* (1994) Interleukin-4 induces programmed cell death (apoptosis) in cases of high-risk acute lymphoblastic leukemia. *Blood* **83**: 1731-1737.
Matsumura,I., Ishikawa,J., Nakajima,K., *et al* (1997) Thrombopoietin-induced differentiation of a human megakaryoblastic leukemia cell line, CMK, involves transcriptional activation of p21WAF1/Cip1 by STAT5. *Mol. Cell. Biol.* **17**: 2933-2943.
McCurrach,M.E., Connor,T.M., Knudson,C.M. *et al.* (1997) bax-deficiency promotes drug resistance and oncogenic transformation by attenuating p53-dependent apoptosis. *Proc. Natl. Acad. Sci. USA* **94**: 2345-2349.
Meraz,M.A., White,J.M., Sheehan,K. *et al.* (1996) Targeted disruption of the Stat1 gene in mice reveals unexpected physiological specificity in the JAK-STAT signaling pathway. *Cell* **84**: 431-442.
Migliorati,G., Nicoletti,I., Pagliacci,M.C. *et al.* (1993) Interleukin-2 induces apoptosis in mouse thymocytes. *Cell. Immunol.* **146**: 52-61.
Migliorati,G., Nicoletti,I., Nocentini,G. (1994) Dexamethasone and interleukins modulate apoptosis of murine thymocytes and peripheral T-lymphocytes. *Pharmacol. Res.* **30**: 43-52.
Minami,M., Inoue,M., Wei,S. *et al* (1996) STAT3 activation is a critical step in gp130-mediated terminal differentiation and growth arrest of a myeloid cell line. *Proc. Natl. Acad. Sci. USA.* **93**: 3963-3966.
Miura,M., Zhu,H., Rotello,R. *et al.* (1993) Induction of apoptosis in fibroblasts by IL-β-converting enzyme, a mammalian homolog of the *C. elegans* cell death gene ced-3. *Cell* **75**: 653-660.
Miyashita,T. and Reed,J.C. (1995) Tumor suppressor p53 is a direct transcriptional activator of the human bax gene. *Cell* **80**: 293-299.
Muller,M., Briscoe,J., Laxton,C. *et al.* (1993) The protein tyrosine kinase JAK1 complements defects in interferon-α/β and -γ signal transduction. *Nature* **366**: 129-135.
Nagata,S. and Golstein,P. (1995) The Fas death factor. *Science* **267**: 1449-1456.
Nair,P., Muthukkumar,S., Sells,S.F., *et al.* (1997) Early growth response-1-dependent apoptosis is mediated by p53. *J. Biol. Chem.* **272**: 20131-20138.
Naka,T., Narazaki,M., Hirata,M. *et al* (1997) Structure and function of a new STAT-induced STAT inhibitor. *Nature* **387**: 924-929.
Ohmori,Y., Schreiber,R.D. and Hamilton,T.A. (1997) Synergy between interferon-gamma and tumor necrosis factor-alpha in transcriptional activation is mediated by co-operation between signal transducer and activator of transcription 1 and nuclear factor kappaB. *J. Biol. Chem.* **272**: 14899-14907.
Ossina,N.K., Cannas,A., Powers,V.C. *et al.* (1997) Interferon-gamma modulates a p53-independent apoptotic pathway and apoptosis-related gene expression. *J. Biol. Chem.* **272**: 16351-16357.
Owen-Schaub,L.B., Zhang,W., Cusack,J.C. *et al.* (1995) Wild-type human p53 and a temperature-sensitive mutant induce Fas/APO-1expression. *Mol. Cell. Biol.* **15**: 3032-3040.
Packham,G. and Cleveland,J.L. (1994a) The role of ornithine decarboxylase in c-myc-induced apoptosis. *Curr. Top. Microbiol. Immunol.* **194**: 283-290.
Packham,G. and Cleveland,J.L. (1994b) Ornithine decarboxylase is a mediator of c-Myc-induced apoptosis. *Mol. Cell. Biol.* **14**: 5741-5747.
Pardo,F.S., Su,M. and Borek,C. (1996) Cyclin D1 induced apoptosis maintains the integrity of the G1/S checkpoint following ionizing radiation irradiation. *Soma. Cell Mol. Genet.* **22**: 135-44.
Pine,R. (1997) Convergence of TNFalpha and IFNgamma signalling pathways through synergistic induction of IRF-1/ISGF-2 is mediated by a composite GAS/kappaB promoter element. *Nuc. Acids Res.* **25**: 4346-4354.
Pine,R., Canova,A. and Schindler,C. (1994) Tyrosine phosphorylated p91 binds to a single element in the ISGF2/IRF-1 promoter to mediate induction by IFN alpha and IFN gamma, and is likely to autoregulate the p91 gene. *EMBO J.* **13**: 158-67.
Polyak,K., Xia,Y., Zweier,J.L. *et al* (1997) A model for p53-induced apoptosis *Nature* **389**: 300-305.
Preston,G.A., Lyon,T.T., Yin,Y. *et al.* (1997) Induction of apoptosis by c-Fos protein. *Mol. Cell. Biol.* **16**: 211-218.

Quelle,F.W., Thierfelder,W., Witthuhn,B.A *et al.* (1995) Phosphorylation and activation of the DNA binding activity of purified Stat1 by the Janus protein-tyrosine kinases and the epidermal growth factor receptor. *J. Biol. Chem.* **270**: 20775-207780.
Rathbun,R.K., Faulkner,G.R., Ostroski,M.H. *et al.* (1997) Inactivation of the Fanconi anemia group C gene augments interferon gamma-induced apoptotic responses in hematopoietic cells. *Blood* **90**: 974-85.
Reed,J.C. (1995) Regulation of apoptosis by bcl-2 family proteins and its role in cancer and chemoresistance. *Curr. Opin. Oncol.* **7**: 541-546.
Reinke,V. and Lozano,G. (1997) The p53 targets mdm2 and Fas are not required as mediators of apoptosis *in vivo*. *Oncogene* **15**: 1527-1534.
Reisman,D., and Rotter,V. (1993b) The helix-loop-helix containing transcription factor USF binds to and transactivates the promoter of the p53 tumor suppressor gene. *Nuc. Acids. Res.* **21**: 345-350.
Reisman,D., Elkind,N.B., Roy,B. *et al* (1993a) c-Myc trans-activates the p53 promoter through a required downstream CACGTG motif. *Cell Growth Diff.* **4**: 57-65.
Roy,B., Beamon,J., Balint,E. and Reisman,D. (1994) Transactivation of the human p53 tumor suppressor gene by c-Myc/Max contributes to elevated mutant p53 expression in some tumors. *Mol. Cell. Biol.* **14**: 7805-7815.
Schindler,C., Fu,X.-Y., Improta,T. *et al.* (1992) Proteins of transcription factor ISGF-3: one gene encodes the 91- and 84-kDa ISGF-3 proteins that are activated by interferon α. *Proc. Natl. Acad. Sci. USA* **89**: 7836-7839.
Schindler,C., Shuai,K., Prezioso,V.R. and Darnell,J.E.,Jr. (1992) Interferon-dependent tyrosine phosphorylation of a latent cytoplasmic transcription factor. *Science* **257**: 809-813.
Sedo,A., Weyenbergh,J.V., Rouillard,D. and Bauvois,B. (1996) Synergistic effect of prolectin on IFN-γ–mediated growth arrest in human monoblastic cells: correlation with the up-regulation of IFN-γ receptor gene expression. *Immunol. Let.* **53**: 125-130.
Shi,L., Chen,G., MacDonald,G. *et al* (1996) Activation of an interleukin 1 converting enzyme-dependent apoptosis pathway by granzyme B. *Proc. Natl. Acad. Sci. USA* **93**: 11002-11007.
Shi,Y., Glynn,J.M., Guilbert,L.J. *et al* (1992) Role for c-myc in activation-induced apoptotic cell death in T cell hybridomas. *Science* **257**: 212-214.
Shuai,K., Stark,G.R., Kerr,I.M. and Darnell,J.E.Jr. (1993) A single phosphortyrosine residue of Stat91 required for gene activation by interferon-γ. *Science* **261**: 1744-1746.
Smeyne,R.J., Vendrell,M., Hayward,M. *et al.* (1993) Continuous c-fos expression precedes programmed cell death *in vivo*. *Nature* **363**: 166-169.
Smeyne,R.Y., Vendrell,M., Hayward,M. *et al.* (1993) Continous c-fos expression precedes programmed cell death *in vivo*. *Nature* **363**: 166-169.
Spink,J. and Evans,T. (1997) Binding of the transcription factor interferon regulatory factor-1 to the inducible nitric-oxide synthase promoter. *J. Biol. Chem.* **272**: 24417-24425.
Stocklin,E., Wissler,M., Gouilleux,F. and Groner,B. (1996) Functional interactions between Stat5 and the glucocorticoid receptor. *Nature* **383**: 726-728.
Tamura,T., Ishihara,M., Lamphier, M.S. *et al.* (1995) An IRF-1-dependent pathway of DNA damage-induced apoptosis in mitogen-activated T lymphocytes. *Nature* **376**: 596-599.
Tanaka,N., Ishihara,M., Lamphier,M.S. *et al* (1996) Co-operation of the tumour suppressors IRF-1 and p53 in response to DNA damage. *Nature* **382**: 816-818.
Tanaka,N., Ishihara,M., Kitagawa,M. *et al* (1994) Cellular commitment to oncogene-induced transformation or apoptosis is dependent on the transcription factor IRF-1. *Cell* **77**: 829-39.
Taniguchi,T., Harada,H. and Lamphier,M. (1995) Regulation of the interferon system and cell growth by the IRF transcription factors. *J. Cancer Res. Clin. Oncol.* **121**: 516-20.
Thompson,C.B., (1995) Apoptosis in the pathogenesis and treatment of disease *Science* **267**: 1456-1462.
Van Antwerp,D.J., Martin,S.J., Kafri,T., *et al* (1996) Suppression of TNF-α-induced apoptosis by NF-κB. *Science* **274**: 787-789.
Van der Geer,P., Hunter,T. and Lindberg,R. (1994) Receptor protein tyrosine kinases and their signal transduction pathways. *Ann. Rev. Cell. Biol.* **10**: 251-337.
Velazquez,L., Fellous,M., Stark,G.R. and Pellegrini,S. (1992) A protein tyrosine kinase in the interferon α/β signaling pathway. *Cell* **70**: 313-322.
Wang,C.Y., Mayo,M.W. and Baldwin,A.S.J. (1996) TNF- and cancer therapy-induced apoptosis: potentiation by inhibition of NF-kappaB. *Science* **274**: 784-787.
Weih,F., Ryseck,R-P, Che,L. and Bravo,R. (1996) Apoptosis of nur-77/N-10-transgenic thymocytes involves the Fas/Fas ligand pathway. *Proc. Natl. Acad. Sci. USA.* **93**: 5533-5538.
Wyllie,A.H. (1980) Glucocorticoid-induced thymoctye apoptosis is associated with endogenous endonuclease activity. *Nature* **284**: 555-556.
Xie,Q., Kashiwabara,Y. and Nathan,C. (1993) Role of transcription factor NF-κB/Rel in induction of nitric oxide/synthase. *J. Biol. Chem.* **269**: 4705-4708.
Xu,F., Fu,X.Y., Plate,J. and Chong,A.S. (1998) IFN-γ induces cell growth inhibition by Fas-mediated apoptosis: requirement of STAT1 protein for up-regulation of Fas and FasL expression. *Cancer Res.* **58**: 2832-2837.
Yamada,T., Horiuchi,M. and Dzau,V.J. (1996) Angiotensin II type 2 receptor mediates programmed cell death. *Proc. Natl. Acad. Sci. USA.* **93**: 156-160.

Yamanaka,Y., Nakajima,K., Fukada,T. *et al.* (1996) Differentiation and growth arrest signals are generated through the cytoplasmic region of gp130 that is essential for Stat3 activation. *EMBO J.* **15**: 1557-1565.

Yang,D.D., Kuan,C.Y., Whitmarsh,A.J. *et al.* (1997) Absence of excitotoxicity-induced apoptosis in the hippocampus of mice lacking the Jnk3 gene. *Nature* **389**: 865-870.

Yuan,J., Shaham,S., Ledoux,S. *et al.* (1993) The *C. elegans* cell death gene ced-3 encodes a protein similar to mammalian interleukin-1 beta-converting enzyme. *Cell* **75**: 641-52.

Zhou,F. and Thompson,E.B. (1996) Role of c-jun induction in the glucocorticoid-evoked apoptotic pathway in human leukemic lymphoblasts, *Mol. Endocrinol.* **10**: 306-316.

Chapter 6

Adhesion and Apoptosis

Andrew P. Gilmore and Charles H. Streuli
School of Biological Sciences, University of Manchester, 3.239 Stopford Building, Oxford Road, M13 9PT, UK

1.0 INTRODUCTION.

Tissue architecture in multicellular organisms is maintained through adhesive interactions between cells and their neighbours, and between cells and the underlying extracellular matrix (ECM). These interactions impart obvious mechanical properties to tissues and organs, such as the impermeability of epithelial cell layers. They are also important in the dynamic regulation of tissue organisation and the control of cell proliferation (Zhu *et al.*, 1996), differentiation (Streuli *et al.*, 1991) and apoptosis (Meredith *et al.*, 1993). The ultimate goal of this regulation is to promote cell growth and differentiation only when the cell is in the correct location, and to delete cells which have become displaced from their proper environment. This removal of cells is mediated by apoptosis. Apoptosis is a fundamental part of cell regulation and its deregulation has implications for degenerative disease (where undesired apoptosis occurs), or neoplasia (where there is failure to delete transformed cells). The particular type of apoptosis associated with loss of cell adhesion has been termed anoikis, from the Greek word for homelessness (Frisch and Francis, 1994). As well as eliminating cells which are detached from their ECM, this mechanism of apoptosis also plays an important role in development and tissue remodelling.

Very few cells in the body are able to grow in the absence of adhesion, the exception being cells which circulate in the bloodstream. Most other cells are dependent upon adhesion to the ECM in order to survive, and the development of anchorage independence forms an important aspect of tumour progression. Cells which lose their regulation of proliferation will form a tumour, but it is the spread of that tumour to secondary sites which signifies malignancy, and this spreading requires cells to survive away from their normal environment.

Cells adhere to their underlying ECM primarily through members of a family of heterodimeric transmembrane glycoproteins, the integrins (Hynes, 1992). The adhesion of cells to ECM via integrins was first observed to suppress apoptosis in endothelial cells (Meredith *et al.*, 1993) and subsequently in MDCK epithelial cells (Frisch and Francis, 1994). This is now accepted to be a general phenomenon. The role of integrins in survival is inextricably linked to their function as a cytoskeletal component, linking the underlying ECM to actin stress fibres. This cytoskeletal role is required for the multitude of signalling pathways activated in response to cell/ECM attachment. The many different responses to adhesion include reorganisation of the actin cytoskeleton and cell spreading, activation of tyrosine kinases such as pp125FAK and c-Src and the tyrosine phosphorylation of associated proteins, activation of serine/threonine kinases such as protein kinase C and

MAP kinase, elevation of intracellular H^+ and Ca^{2+} concentrations, and the stimulation of phosphoinositide synthesis and hydrolysis (Schwartz *et al.*, 1995). Many of these pathways have been implicated in the regulation of both cell growth and survival.

The purpose of this chapter is not so much to list multiple examples of how cell adhesion plays a role in survival, but to address the far more important question of how cells perceive their environment and interpret those signals in the context of life and death decisions. What becomes apparent is that there is an extraordinary variety of messages being provided by such a superficially simple phenomena as cell adhesion.

2.0 ADHESION MEDIATED APOPTOSIS IN DEVELOPMENT AND DISEASE.

The majority of experimental studies which have illuminated the signal transduction pathways associated with cell adhesion have involved the use of cell culture. Before that, this section will look at *in vivo* examples of apoptosis regulation and their role in aspects of development.

2.1 ECM degradation as a regulator of epithelial apoptosis

Both soluble and insoluble factors are required to provide the survival signals necessary to suppress the inherent apoptosis programme. Changes in soluble factors (such as growth factors) can be brought about through altered levels of expression. Survival signals from insoluble factors (such as ECM) may work in other ways. For example, a cell moving from its correct location becomes detached from the ECM, and this will result in apoptosis. In developmental situations, cells may become separated from their ECM through the production of proteases, with the subsequent degradation of the underlying substratum inducing apoptosis through reduced adhesion.

The developing mammary gland has provided an excellent model system for examining changes in cell growth, differentiation and involution as these changes occur during the reproductive cycle. The mammary gland consists of three types of epithelia. The milk collecting ducts are lined with luminal epithelial cells, and these contact a layer of myoepithelial cells separating them from the stroma. Milk is produced in alveolar structures which develop during pregnancy and then are removed following weaning by apoptosis. Alveoli are lined with a layer of milk-producing epithelial cells which contact both myoepithelial cells and the basement membrane that separates them from the stroma. The proliferation-differentiation-apoptosis cycle is ultimately under hormonal control. However, there is a major requirement for cell/ECM interactions for each stage. Mammary epithelial cells are dependent upon specific integrin/basement membrane interactions for milk protein expression (Streuli *et al.*, 1991) and survival (Boudreau *et al.*, 1995; Pullan *et al.*, 1996). During weaning, modulation of these interactions has been implicated in the regulation of apoptosis, which may be brought about by the expression of ECM degrading metalloproteinases (MMPs) (Talhouk *et al.*, 1992; Lund *et al.*, 1996).

During involution, expression of MMPs increases for up to 4 days post weaning causing matrix degradation (Talhouk *et al.*, 1992; Martinez-Hernadez *et al.*, 1976).) at which stage apoptosis is at a catastrophic level (Strange *et al.*, 1992). The result is that the mammary gland is remodelled, returning it to a 'resting' state in readiness for the next round of pregnancy. The case for involvement of MMPs in involution has been strengthened by a number of *in vivo* experiments. The overexpression of the preactivated MMP stromelysin-1 in a transgenic mouse model induced premature apoptosis during late pregnancy, and resulted in a lack of mature alveoli (Sympson *et al.*, 1994; Alexander *et al.*, 1996). Furthermore, the crossing of stromelysin-1 transgenic mice with those overexpressing tissue inhibitor of metalloproteinase-1 (TIMP-1) resulted in the rescue of the stromelysin-1 phenotype, with normal alveolar volume compared with wild-type mammary glands. The effects of stromelysin-1 were specifically on the rate of apoptosis in late pregnancy, and did not affect the rate of proliferation of mammary epithelial cells. In wild-type mammary glands, stromelysin-1 is expressed strongly in myoepithelial cells (Dickson and Warburton, 1992), where levels of expression increase during involution co-incident with loss of tissue function (Talhouk *et al.*, 1992; Strange *et al.*, 1992). The implantation of slow release pellets of TIMP-1 delayed the onset of involution, suggesting that stromelysin-1 may also play a role in apoptosis in wild-type animals (Talhouk *et al.*, 1992).

2.2 Melanoma cell adhesion - a mechanism for tumour progression.

The specificity of a cell for its preferred ECM is primarily determined by the repertoire of integrins it expresses. The integrins are a large family of receptors for ECM proteins, showing overlapping ligand specificities, although a number of integrins with different ECM preferences may ultimately feed into the same signalling pathways (Hynes, 1992; Schwartz *et al.*, 1995). Examples of this are the fibronectin receptor, $\alpha5\beta1$, and the vitronectin receptor, $\alpha v\beta3$, (the latter being quite promiscuous with regard to ECM ligands). The expression of integrin $\alpha v\beta3$ has been observed as a marker for progression of melanoma from radial to vertical growth phase (Albelda *et al.*, 1991). This results in a change in cell behaviour allowing migration and survival in an environment which would otherwise promote apoptosis. In particular, melanoma cells must migrate through the underlying layer of dermal collagen. In nude mice, αv deficient melanoma cells are less tumourigenic than those expressing αv (Felding-Habermann *et al.*, 1992). In culture αv deficient melanoma cells die in collagen gels by apoptosis, but survival can be restored by transfection of the αv subunit (Montgomery *et al.*, 1994). Survival could also be abrogated in αv positive cells with antibodies to αvβ3, but not α2β1, the receptor for type 1 collagen which is also expressed on these cells but does not appear to give a survival signal (Montgomery *et al.*, 1994).

However, survival and progression is more complex than simply a change in αvβ3 expression. Melanoma cells must also alter the dermal ECM in order to survive within it, and this most likely involves MMPs. αvβ3 allows melanoma cells to survive and adhere to dermal collagen, but does not recognise native collagen type 1 (Montgomery *et al.*, 1994). Instead, the cells utilise the collagen receptor α2β1 which binds to native collagen through the peptide motif Asp-Glu-Gly-Ala. This initial adhesion and spreading within collagen gels can be blocked with antibodies to this integrin, but not with those to αvβ3 (Montgomery *et al.*, 1994). Over time however, adhesion becomes α2β1 independent and αvβ3 dependent, indicating that an alteration of collagen structure occurs which alters its integrin specificity. Interestingly, α2β1 does not recognise proteolysed or heat denatured collagen, unlike which αvβ3 recognises an otherwise cryptic site within denatured collagen. Thus, melanoma cells which initially adhere to collagen via α2β1 alter its conformation to expose this cryptic site. Tumour conditioned media was found to bring about just this change, allowing αvβ3 mediated adhesion (Montgomery *et al.*, 1994). This could be inhibited by tissue inhibitor of metalloproteinase-1 (TIMP-1), restoring dependence on α2β1. Melanoma cells therefore appear to use the expression of alternative integrins coupled with the proteolytic modification of dermal collagen in order to invade the underlying tissue.

This is a superb example of the complex nature of changes required during tumour progression in order for cells to suppress apoptosis in inappropriate environments. However, circumventing the apoptotic machinery at any point between the adhesive interaction and the ultimate commitment to death is able to bring about survival. Other levels of transformation may involve a constitutive activation of the survival signals downstream of integrin ligation. An understanding of the ways in which integrins actually transmit information from the ECM to the interior of the cell is therefore vital to an understanding of adhesion mediated apoptosis.

This itself is inextricably linked with the mechanical aspects of integrins and the cytoskeleton.

3.0 INTEGRINS AND APOPTOSIS.

Integrins were found to be mediators of anchorage dependant survival as soon as ECM was recognised to deliver a strong anti-apoptotic signal (Meredith *et al.*, 1993). Endothelial cells are exquisitely sensitive to detachment from ECM, with almost all cells becoming apoptotic after 24 hours in suspension. The involvement of integrins was shown by the absence of apoptosis when cells were plated onto an antibody against β1 integrin, whereas antibodies to other cell surface molecules (HLA and VCAM-1) failed to inhibit cell death. Similar effects were observed on MDCK cells. RGD containing peptides, which mimic the recognition sequence for integrins within fibronectin and vitronectin and therefore block adhesion of cells on these substrata, induced apoptosis in MDCK cells plated onto ECM, thus implicating integrins in survival (Frisch and Francis, 1994). Like other systems in which apoptosis has been studied, cells had to be removed from ECM for a defined period of time (3-5 hours) before they had committed to the apoptotic programme (Frisch and Francis, 1994). Before this, cells were still viable when replated onto ECM, but after this time, replating onto ECM could not rescue them from an apoptotic fate. To understand the influence of cell adhesion on the ability of cells to survive we must look within the myriad of signalling pathways emanating from integrins at the cell membrane.

3.1 Cell type and ECM specificity

Most cells are dependent upon adhesion for survival, although different degrees of susceptibility are observed. Although epithelial and endothelial cells undergo apoptosis readily when in suspension, fibroblasts and mesothelial cells are less susceptible and will survive in suspension if they receive survival signals from growth factors (Meredith *et al.*, 1993; Re *et al.*, 1994; Zhang *et al.*, 1995; Hungerford *et al.* 1996). Fibroblasts will also survive (although not proliferate) in the absence of growth factors if they receive survival signals from the ECM. Fibroblasts, although resistant to apoptosis in suspension, withdraw from the cell cycle (Folkman and Moscona, 1978).

Different cell types are therefore influenced to differing degrees by the number of external signals which they receive, which in turn is related to the particular signalling pathways activated by these signals. Transforming oncogenes demonstrate this by altering the requirement of ECM for survival. For example, transformation of epithelial cells and endothelial cells by oncogenes such as v-Src and v-Ha-ras allows survival in the absence of adhesion (Frisch, 1994; Re *et al.*, 1994). Conversely, transformed cells, which are anchorage independent, can be made adhesion dependent by reversing transformation through transfection of the adenovirus protein E1a (Frisch, 1994; Frisch, 1997).

The differences in dependence upon adhesion for survival may relate to morphology. The more migratory behaviour of fibroblasts is consistent with an ability to survive within a wide variety of matrices. Epithelial cells, on the other hand, are required to remain within a defined tissue layer. Scatter factor (HGF) is able to convert epithelial cells to a fibroblastic morphology, promoting loss of cell/cell junctions and an increase in migration (Frisch and Francis, 1994). HGF also induces epithelial cells to become more resistant to detachment induced apoptosis. Transfection of E1a, which can revert transformed cells to an anchorage dependent state, suppresses expression of a number of fibroblast genes whilst at the same time inducing expression of epithelial markers. Similarly, the formation of cell/cell junctions increases the sensitivity of epithelial cells to loss of ECM contact. Furthermore, MDCK cells grown in sparse conditions are less inclined to undergo apoptosis when trypsinised and replated onto a non-adhesive substrata than cells which have been grown as a confluent monolayer (Frisch and Francis, 1994). This suggests that as epithelial cells organise themselves into defined tissue layers they become primed for apoptosis through the formation of cell contacts. Epithelial cells grown in sparse culture, with an absence of cell/cell contact, express Bcl-2, an apoptosis regulatory protein which promotes survival (Frisch *et al.*, 1996a). As cell-cell contacts form when cell density increase, Bcl-2 expression becomes downregulated.

Individual cell types can be very specific for different ECM constituents, which in turn directs cell survival. For example, myotubes will survive on a basement membrane of merosin (laminin-2 and -4), but undergo apoptosis on laminin-1 (Vachon *et al.*, 1996). However, parental myoblasts will survive and proliferate on laminin-1, the process of differentiation switching this specificity. Mammary epithelial cells require a specific basement membrane for survival. In a cell culture system, mammary cells will differentiate when grown on Englebreth-Holm-Swarm (EHS) matrix, but not on collagen, fibronectin or plastic (Streuli *et al.*, 1991). This matrix also suppresses of apoptosis, whereas cells grown on other types of ECM enter apoptosis after several days in culture (Boudreau *et al.*, 1995; Pullan *et al.*, 1996; Boudreau *et al.*, 1996). This suppression of apoptosis by EHS matrix involves an interaction with a $\beta1$ integrin, as anti-$\beta1$ antibodies or synthetic integrin ligands block the survival effect (Pullan *et al.*, 1996; Boudreau *et al.*, 1995).

Adhesion to a particular ECM substratum *per se* does not automatically endow cells the capacity to survive upon it. Different integrins have differing abilities to suppress apoptosis, even though they may recognise the same substratum. The expression of α5β1, an integrin which recognises the RGD tripeptide motif, in CHO (Chinese hamster ovary) cells, suppressed apoptosis when cells were plated on fibronectin (Zhang *et al.*, 1995). However, αvβ1 expressed on CHO cells, which recognises fibronectin via the same RGD site (Zhang *et al.*, 1993), was unable to promote cell survival, and apoptosis resulted. The critical determinant was found to be the α5 cytoplasmic domain, since expression of an α5β1 mutant in which the α5 cytoplasmic domain had been deleted abrogated the survival effect (Zhang *et al.*, 1995). Such subunit-specific effects may relate to how particular integrins are able to link with the downstream signalling pathways within the cell in response to adhesion, the so called "outside-in" signalling (Schwartz *et al.*, 1995).

In contrast, some integrins are quite promiscuous with regard to the types of ECM on which they can promote survival. αvβ3 allows growth within a variety of substrata and is also associated with invasive cell types. For example, it is upregulated on vascular endothelial cells fourfold during FGF or TNFα induced angiogenesis (Brooks *et al.*, 1994a). αvβ3 and could therefore allow the survival of vascular cells through the variety of ECM types that might be encountered during angiogenesis. Blocking αvβ3 adhesion (with inhibitory antibodies or cyclic peptides that mimic the αvβ3 binding site on ECM molecules) induced apoptosis, and inhibited angiogenesis (Brooks *et al.*, 1994b; Brooks *et al.*, 1995). αvβ3 also contributes to melanoma cell invasion by suppressing apoptosis in dermal collagen. The important question is therefore how are integrins are able to transmit signals from the underlying substratum to the cell's survival machinery?

3.2 Integrins and the cytoskeleton - a requirement for adhesion mediated signal transduction.

Integrins provide both a structural link with the actin cytoskeleton and a mechanism for interpreting the nature of the ECM into intracellular signals (Burridge and Chrzanowska-Wodnicka, 1996; Schwartz *et al.*, 1995). However, these two roles are interdependent. Mere ligand occupancy is not sufficient for survival : Endothelial cells plated onto non-adhesive substrata undergo apoptosis even in the presence of soluble fibronectin or vitronectin, even though these soluble ligands bind integrins on the cell surface (Re *et al.*, 1994). Furthermore, cells plated onto surfaces coated with varying densities of either fibronectin or vitronectin showed levels of survival proportional to the density of ECM protein. At low densities of either substratum (up to 2.5ng/cm^2), cells adhered but did not spread, and underwent apoptosis. At densities at which cells began to spread (10ng/cm^2), a reduction in the rate of apoptosis was seen, and at higher concentrations when cells were fully spread, survival was virtually 100%. It is unlikely that the number of integrins engaged is the critical determinant with increasing ECM density, as the experiments with soluble ECM ligands used saturating concentrations of fibronectin or vitronectin. However, the engagement of the cytoskeleton and its associated signalling proteins through cell spreading may regulate the critical survival signals.

These observations have been extended by the use of microfabricated adhesive surfaces of precise dimensions where the surface area onto which a cell adhered, and the shape assumed following spreading, could be controlled (Chen *et al.*, 1997). Endothelial cells attached to small islands of fibronectin, 5μm in diameter, spaced on a grid pattern 10μm apart, and spread across several neighbouring islands. However, cells plated on larger islands of 20μm diameter, and spaced further apart such that cells could only contact a single region, did not spread and remained rounded though attached. Although in both cases the actual surface area of attachment, and therefore the potential number of integrins bound to fibronectin was identical, those cells that spread across several smaller islands survived and proliferated, whilst those on the single, larger island underwent apoptosis.

Cell spreading on substrata of varying densities has previously been correlated with the formation of stress fibres and focal adhesions (Massia and Hubbell, 1991), regions at which integrins have become clustered on the cell surface in response to both integrin ligation on the extracellular side and attachment to the cytoskeleton on the cytoplasmic face (Burridge and Chrzanowska-Wodnicka, 1996). The formation of focal adhesions and stress fibres correlates strongly with the activation of a number of signal transduction pathways linked to integrin mediated cell adhesion, including a dramatic increase in tyrosine phosphorylation of a number of focal adhesion proteins (Kornberg *et al.*, 1991; Lipfert *et al.*, 1992; Burridge *et al.*, 1992).

4.0 SIGNALLING PATHWAYS FROM INTEGRINS TO APOPTOSIS

One of the earliest signalling events to occur when a wide variety of cell types are plated onto suitable ECM proteins is an increase in tyrosine phosphorylation of a number of proteins. This coincides with cell spreading, reorganisation of the actin cytoskeleton, and clustering of integrins into focal adhesions. These are regions at which cells are most closely associated with the underlying substratum, and provide a direct link between the actin cytoskeleton and the ECM (Burridge *et al.*, 1988). Tyrosine phosphorylation in adherent cells is localised to focal adhesions, and in whole cell lysates a large proportion of tyrosine phosphorylation has been identified to be on focal adhesion proteins (Burridge *et al.*, 1992; Petch *et al.*, 1995). Cells rapidly lose this tyrosine phosphorylation when they are detached from the underlying ECM, a process that is associated with an increase in tyrosine phosphatase activity (Maher, 1993). Cells taken into suspension could be protected from apoptosis by treating them with the tyrosine phosphatase inhibitor orthovanadate, (thereby preventing the dephosphorylation of a number of the tyrosine phosphorylated proteins seen in adherent cells) (Meredith *et al.*, 1993). Thus, tyrosine phosphorylation signalling cascades may be critically important for survival.

Current evidence suggests that the association of integrins with the ECM and the subsequent cell spreading results in a reorganisation of the actin cytoskeleton and clustering of integrins (Burridge and Chrzanowska-Wodnicka, 1996). This in turn brings together a plethora of signal transduction pathways in a manner analogous to the dimerisation of growth factor receptors following growth factor binding (Miyamoto *et al.*, 1995a,b)(Figure 1). A number of studies have demonstrated that the reorganisation of the cytoskeleton is essential for these subsequent signalling events, as disruption of the actin cytoskeleton with cytochalasin D prevents the association and activation of many of these pathways, including tyrosine phosphorylation and cell cycle progression (Miyamoto *et al.*, 1995b; Zhu and Assoian, 1995; Zhu *et al.*, 1996). The linkage between integrins and the cytoskeleton is mediated by their cytoplasmic domains. A number a proteins have been implicated in mediating this linkage. *In vitro* studies have demonstrated physical interactions, and *in vivo* work has revealed co-localisation with integrins within focal adhesions.

4.1 Pp125FAK and tyrosine phosphorylation

One of the first signalling events to occur following adhesion to ECM is the tyrosine phosphorylation of pp125FAK (Focal Adhesion Kinase), a non-receptor tyrosine kinase which localises to focal adhesions in adherent cells (Schaller *et al.*, 1992; Hanks *et al.*, 1992). The N-terminus of pp125FAK can associate with peptides mimicking the cytoplasmic domain of β1 integrin *in vitro* (Schaller *et al.*, 1995), although this has not been demonstrated *in vivo*. However, the C-terminus of the protein appears also to contain all the determinants required to localise to focal adhesions (Hildebrand *et al.*, 1993).

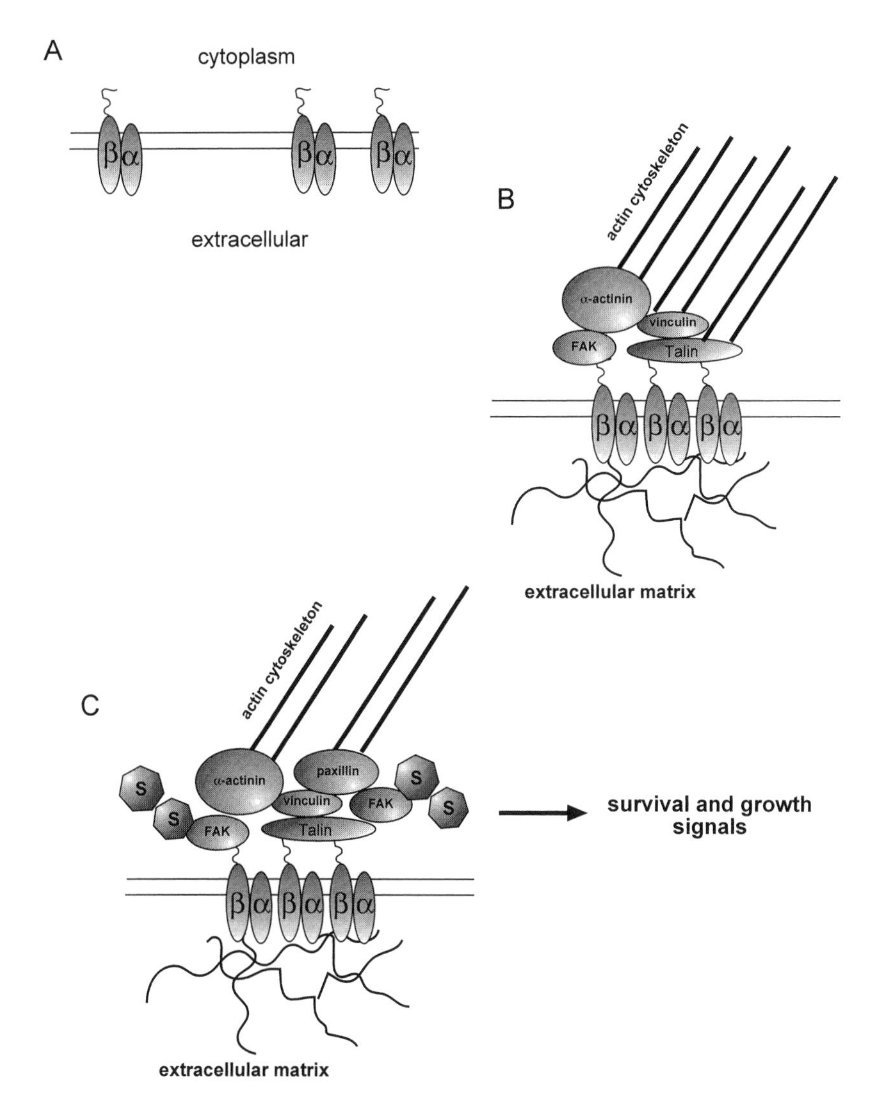

Figure 1. Integrin-mediated adhesion to the extracellular matrix induces focal adhesion formation and clustering of multiple signalling pathways.

A. In a nonadherent cell, non-ligated integrins are diffuse on the cell surface and are not associated with the cytoskeleton or signalling complexes.

B. Upon adhesion extracellular matrix molecules, integrins become associated with the actin cytoskeleton to suitable, which clusters them into focal adhesions (Burridge and Chrzanowska-Wodnicka 1996). Some signalling molecules are closely associated with integrins and are recruited to these complexes early, e.g. pp125FAK (Miyamoto *et al.*, 1995a). pp125FAK recruitment and clustering results in its autophosphorylation, which both increases pp125FAK activity and generates binding sites for a number of signalling proteins containing phosphotyrosine binding SH2 domains (Hanks and Polte, 1997).

C. More peripheral signalling molecules are recruited to focal adhesions, propagating signals through a multitude of pathways to regulate survival and growth (Miyamoto *et al.*, 1995b).

This same region contains binding sites for two focal adhesion proteins, talin and paxillin (Chen *et al.*, 1995; Hildebrand *et al.*, 1995). Whatever the mechanism for pp125FAK targeting to focal adhesions, it appears that this targeting and clustering within integrin/cytoskeleton complexes at the membrane results in it becoming autophosphorylated and activated in a manner analogous to that of growth factor receptors (Hanks and Polte, 1997; Parsons, 1996). Autophosphorylation of pp125FAK results in the recruitment of other signalling proteins, including PI3-kinase (Chen and Guan, 1994), and members of the src tyrosine kinase family (Schlaepfer *et al.*, 1994; Schlaepfer and Hunter, 1996; Xing *et al.*, 1994; Cobb *et al.*, 1994) (Figure 2). Proteins like paxillin and p130CAS, which are substrates for pp125FAK, bind to it and act as adapter proteins for yet more signalling components (Hildebrand *et al.*, 1995; Polte and Hanks, 1995).

pp125FAK is also likely to play a role linking integrins and the suppression of apoptosis. Epithelial cells transfected with a constitutively activated CD2/pp125FAK chimera showed the constitutive phosphorylation of paxillin and c-src, whereas the endogenous pp125FAK was only phosphorylated when cells were adherent (Frisch *et al.*, 1996b). The effect on cell survival was dramatic. CD2/pp125FAK transfectants in suspension showed 10 fold less apoptosis than the untransfected controls (Frisch *et al.*, 1996b). This survival potential was dependent upon the chimeric protein's kinase activity, as well as its ability to associate with other signalling molecules. CD2/pp125FAK expressing cells were able to grow in soft agar and formed tumours in nude mice. Consistent with this, pp125FAK was found to be overexpressed in a number of primary tumours (Owens *et al.*, 1995). Moreover, down regulation using antisense oligonucleotides to the FAK message resulted in apoptosis of the tumour cells in culture (Xu *et al.*, 1996). Inhibition of endogenous pp125FAK in primary fibroblasts was found to result in apoptosis even though the cell were attached to a fibronectin substratum (Hungerford *et al.*, 1996). Endothelial cells and 3T3 fibroblasts injected with a dominant negative pp125FAK resulted in blockage of cell cycle progression and the death of a significant number of cells although that study did not confirm that death was by apoptosis (Gilmore and Romer, 1996). The expression of CD2/pp125FAK chimeras did not remove the requirement for serum growth factors for the survival of epithelial cells, again consistent with the varying sensitivities of different cell types to anoikis (Frisch *et al.*, 1996b). It has been observed that during apoptosis induced by c-myc or FAS, pp125FAK undergoes cleavage by caspases (Crouch *et al.*, 1996; Wen *et al.*, 1997). It may be that separation of the kinase domain from the sequences targeting FAK to the cytoskeleton is a mechanism for inhibiting its activity during apoptosis.

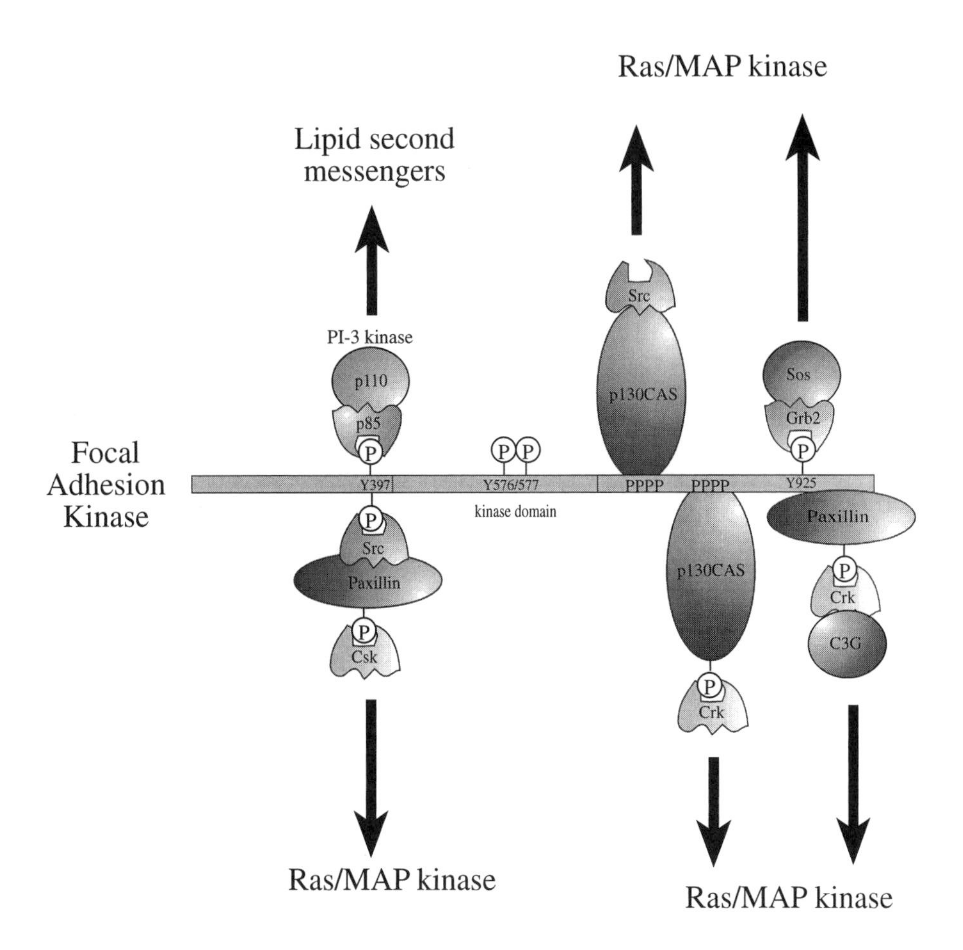

Figure 2. *The tyrosine kinase pp125FAK has a central role in regulating a number of signalling pathways.*

Blocking pp125FAK signalling inhibits cell migration, cell growth and induces apoptosis (Gilmore and Romer, 1996; Hungerford *et al.*, 1996). A number of proteins containing SH2 domains, which bind to phosphorylated tyrosine residues, associate with pp125FAK (Hanks and Polte, 1997). These include PI-3 Kinase (via its p85 regulatory subunit), members of the Src family of protein kinases, and the adapter protein Grb2, which links with guanine nucleotide exchange factor Sos which in turn activates Ras. Other proteins, paxillin and p130CAS, bind to other sites in pp125FAK, but are also substrates for it and Src kinases. Phosphorylated tyrosine residues in paxillin and p130CAS can in turn recruit other signalling molecules, such as Crk, an adapter protein similar to Grb2, which associates another nucleotide exchange factor. All these interactions have been observed to occur *in vivo*.

4.2 Phosphatidylinositol-3-kinase - linking membrane signalling to the apoptotic machinery

Much evidence has recently appeared linking PI3-kinase with apoptosis. PI3-kinase inhibitors induce apoptosis in PC12 cells and MDCK cells, whereas transfection of MDCK with a constitutively activated PI3 kinase allowed survival in suspension. The p85 regulatory subunit of PI3-kinase contains an SH2 domain which interacts with the major autophosphorylation site in pp125FAK in response to cell adhesion, recruiting PI3-kinase to the membrane (Chen and Guan, 1994; Chen *et al.*, 1996). PI3-kinase activation results in increasing cellular concentrations of PtdIns-3,4-P_2 and PtdIns-3,4,5-P_3 (Toker and Cantley, 1997), which act as second messengers in a variety of pathways. With particular regard to apoptosis, a protein serine/threonine kinase, Akt (also called protein kinase B, PKB) is activated by growth factor stimulation of PI3 kinase (Hemmings, 1997; Dudek *et al.*, 1997; Franke *et al.*, 1997) (Figure 3). Indeed, PI3 kinase activation directly due to integrin mediated adhesion causes Akt activation (Khwaja *et al.*, 1997). Akt contains a pleckstrin homology (PH) domain, which binds phosphatidylinositols (Klippel *et al.*, 1997; Franke *et al.*, 1997). PtdIns-3,4-P_2 can target Akt to the cell membrane through this PH domain whereas Akt lacking the PH domain cannot be activated by PI3 kinase. However, if the PH domain is substituted with a myristoylation site, constitutively targeting Akt to the membrane, kinase activity is high independent of PI3 kinase (Franke *et al.*, 1997). Akt may be activated at the membrane in two ways: Direct binding of PtdIns-3,4-P_2 *in vitro* induces a conformational change which can increase activity, though not to the extent seen in *vivo* (Klippel *et al.*, 1997). Primary responsibility for activation appears to be due to the fact that Akt is also serine/threonine phosphorylated following membrane targeting (Kohn *et al.*, 1996). As Akt does not autophosphorylate itself, the kinase responsible is yet to be identified.

A known Akt substrate is glycogen synthase kinase 3 (GSK3). GSK3 is implicated in regulating some transcription factors (e.g. AP1), and the tumour suppressor APC (adenomatous polyposis coli). However, it is not yet known if these are relevant for the protection against apoptosis. More relevant is the recent finding that some members of the Bcl-2 family are regulated by phosphorylation (Zha *et al.*, 1996; Wang *et al.*, 1996). Proteins of this family (discussed in detail elsewhere in this volume), can either promote cell survival (Bcl-2, Bcl-X_L, Bcl-w) or death (Bax, Bad, Bak). The homo- or hetero-dimerisation of survival and death promoting proteins determines whether or not a cell will enter the apoptotic programme.

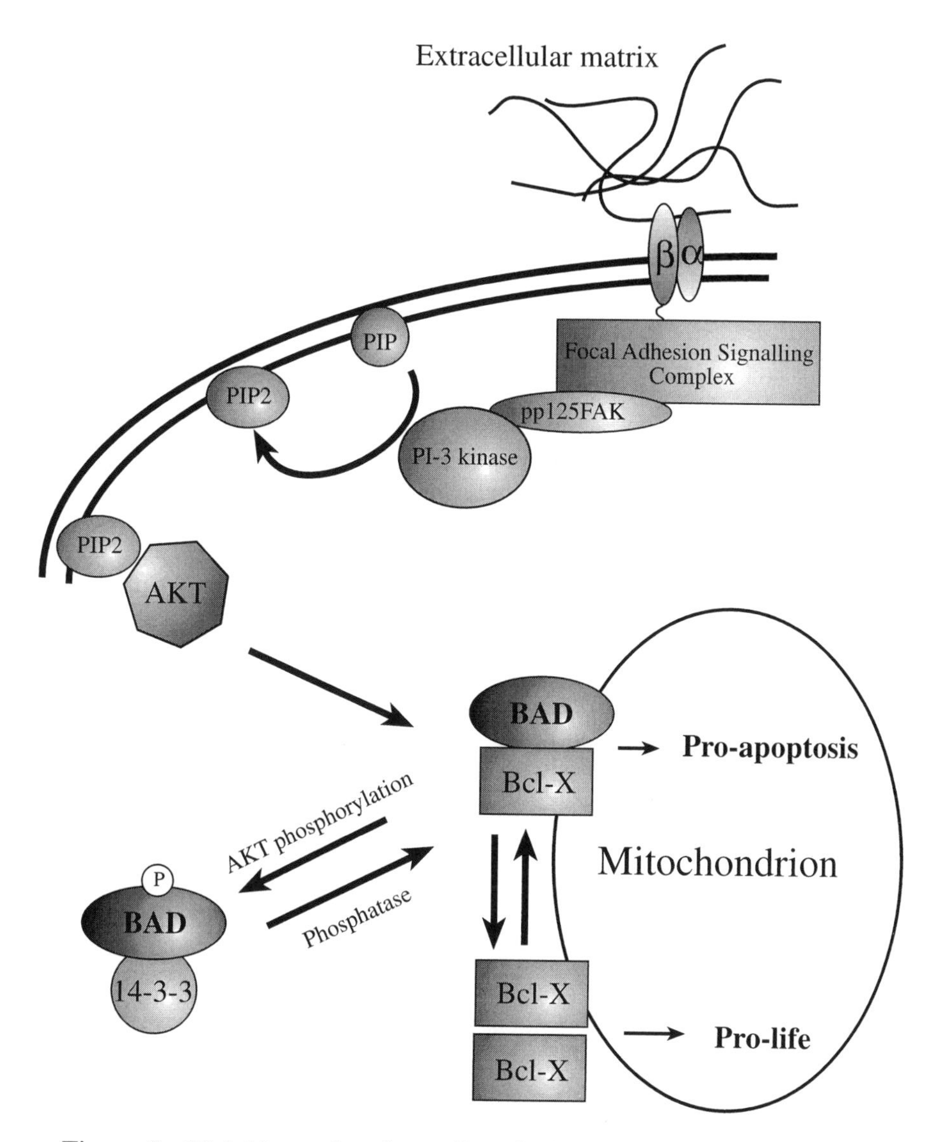

Figure 3. *PI-3 kinase has been directly linked with components of the apoptotic machinery.*

PI-3 kinase associates with pp125FAK during adhesion, and is activated by it. PI-3 kinase phosphorylates phosphatidyinositol-4-phosphate (PIP) to generate phosphatidyinositol-3,4-bisphosphate (PIP2). PIP2 recruits AKT to the membrane, where it is activated by phosphorylation. AKT in turn phosphorylates the pro-apoptotic protein BAD on two serine residues (del Peso *et al.*, 1997). Non-phosphorylated BAD interacts with the apoptosis suppresser protein Bcl-X, possibly on mitochondrial membranes, which inhibits the latters survival effect. AKT phosphorylation of BAD inhibits its interaction with Bcl-X and instead promotes its interaction with 14-3-3, sequestering BAD in the cytoplasm, allowing Bcl-X to suppress apoptosis. Upon loss of adhesion, removal of AKT kinase activity must allow an as yet unknown phosphatase to dephosphorylate BAD, thus inducing apoptosis.

Recent work has suggested that the association of some of these molecules can be regulated by phosphorylation, and that the kinase Akt is responsible (Franke and Cantley, 1997).

The death promoting protein Bad can heterodimerise with the survival proteins Bcl-X_L and Bcl-2, which neutralises their survival effects and promotes apoptosis. Both Bcl-X_L and Bcl-2 associate with internal cell membranes, but Bad is usually cytoplasmic. The survival factor IL-3 was found to induce Bad phosphorylation at two serine residues, Ser-112 and -136 (Zha *et al.*, 1996). Phosphorylation on one or both residues inhibited Bad from interacting with Bcl-X_L, and resulted in it becoming sequestered in the cytoplasm through an interaction with 14-3-3, a highly conserved protein with multiple roles. Substitution of these serine residues for alanines resulted in the inability of IL-3 to induce Bad phosphorylation, and and increased Bad's death promoting activity. The survival effect of IL-3 has been shown to work through PI3-kinase (Parrizas *et al.*, 1997). Exposure to IL-3 results in the rapid activation of Akt, and the expression of oncogenic forms of Akt protected cells against IL-3 withdrawl (Songyang *et al.*, 1997). The final connection between IL-3 and Bad was made when it was shown that Akt could phosphorylate Bad on Ser-112 and Ser-136 (del Peso *et al.*, 1997). Again, activated Akt was unable to phosphorylate Bad in which Ser-112 and -136 had been exchanged for alanines. The key survival role of PI-3 kinase and Akt in adhesion mediated signalling raises the exciting possibility that similar mechanisms for regulating Bad and possibly other Bcl-2 family proteins occur here (Figure 3).

4.3 MAP kinase pathways and integrin suppression of apoptosis.

MAP kinases have been shown to have differing effects on apoptosis (Xia *et al.*, 1995). NGF stimulates the activation of Erk-1 and Erk-2, and upon withdrawl these are rapidly deactivated. At this time other MAP kinases, c-Jun-N-terminal kinase (JNK) and p38 MAP kinase are turned on, and apoptosis ensues. Expression of dominant negative JNK prevented apoptosis in response to NGF withdrawl. Conversely, activating JNK in cells in the presence of NGF induced apoptosis. Positive signals from Erk-1 and -2 , and negative signals from JNK and p38 MAP kinase, could contribute to a cells decision on whether or not to undergo apoptosis. In a similar way constitutive activation of Erk could keep cells alive in the absence of NGF. Maundrell *et al.* (1997) have suggested a mechanism by which JNK activation could influence apoptosis. They found the JNK, but not other MAP kinases, could phosphorylate recombinant Bcl-2 *in vitro*. This phosphorylation was found to occur on two threonine and two serine residues. When Bcl-2 was expressed in COS-7 cells along with a constitutively activated Rac mutant, the same residues were phosphorylated, suggesting that this may be a regulatory mechanism *in vivo*.

The Erk kinases can be activated by integrin mediated adhesion (Chen *et al.*, 1994). pp125FAK has been implicated in the activation of the Ras/MAP kinase pathway (Schlaepfer *et al.*, 1994; Schlaepfer and Hunter, 1996; Schlaepfer and Hunter, 1997). Adhesion caused the association of c-Src and the adapter protein Grb-2 with pp125FAK, resulting in the activation of MAP kinase. Grb-2 was found to bind directly to tyrosine phosphorylated pp125FAK. However, it was recently suggested that the integrin dependent activation of MAP kinase may not act through pp125FAK (Wary *et al.*, 1996; Lin *et al.*, 1997). Integrin mutants which still activated MAP kinase when ligated on the surface of cells were not able to activate pp125FAK. Integrin activation of MAP kinase was demonstrated to be occurring through Shc, an adapter protein which itself binds Grb-2 and therefore activates Ras. The activation of Shc was shown to be dependent on the integrin α-subunit, but not by its cytoplasmic domain. Instead the interaction between integrin and Shc appeared to be indirect, mediated through caveolin. In this study, only integrins which activated Shc could support proliferation and suppress apoptosis (Wary *et al.*, 1996).

A role for JNK in suspension induced apoptosis has recently been suggested (Frisch *et al.*, 1996a; Cardone *et al.*, 1997). Although MDCK cells do not appear to activate Erk kinases in response to integrin mediated adhesion, adhesion does apparently suppress the activation of JNK (Frisch *et al.*, 1996a). JNK becomes rapidly activated in response to detachment from the substratum. Expression of a dominant negative form of JNK was able to suppress this suspension induced apoptosis, even after several hours. However, JNK activation was dependent upon the activation of caspases, the proteases which execute the apoptotic programme. Caspase inhibitors and expression of Bcl-2 could both block JNK activation, suggesting that the activation of JNK is actually downstream of the decision to undergo apoptosis (Frisch *et al.*, 1996a). An activator of JNK, MEKK, was also cleaved by caspase's in response to detachment from ECM (Cardone *et al.*, 1997). This truncation in MEKK activates it, thereby activating JNK, but the same conceptual problem remains. If the activation of proteases within the apoptotic programme is a requirement for JNK activation, then JNK is unlikely to be involved in the decision to apoptose. Indeed, Khwaja and Downward (1997) have shown a lack of correlation between factors resulting in JNK activation and those that protect from suspension induced apoptosis. For example, activated PI3-kinase or Akt could protect MDCK cells from apoptosis in suspension, although neither was able to suppress JNK activation. Likewise, a variety of activated proteins in the Ras pathway suppressed JNK without protecting from apoptosis following detachment from ECM. The role of integrin regulation of MAP kinases in suppression of apoptosis is, therefore, still unclear.

4.4 Abl and Integrin Linked Kinase

The tyrosine kinase Abl has been implicated by a number of sources in the attainment of anchorage independence. Transformation by bcr/abl results in anchorage independent growth and survival (Cortez *et al.*, 1996). Transformation by Abl results in activation of the Ras pathway, and antagonists of Ras signalling induce apoptosis in Abl transformed cells. Recent evidence has implicated a role for integrin regulation of c-Abl (Lewis *et al.*, 1996). A mouse fibroblast line was found to have reduced c-Abl activity in both the nucleus and the cytoplasm following detachment from the substratum. Replating the cells on fibronectin restored Abl activity, and resulted in a transient co-localisation of c-Abl with integrin containing focal contacts, followed by translocation to the nucleus. Both the focal adhesion proteins paxillin and p130CAS, adapter proteins for pp125FAK signalling, are possible substrates for c-Abl, and may link these two signalling pathways.

Integrin linked kinase (ILK) is a 59kDa serine/threonine kinase whose activity is suppressed when cells are plated on fibronectin, though not on plastic (Hannigan *et al.*, 1996). ILK was isolated in a yeast two hybrid screen using the integrin β1 cytoplasmic domain as bait, and was found to associate with integrins *in vivo*. Its sequence contains four repeats, homologous to those originally identified in the erythrocyte cytoskeletal proteins ankyrin, and may be involved in mediating protein/protein interactions. Overexpression of ILK in epithelial cells produced a phenotype which was anchorage independent for both survival and proliferation (Hannigan *et al.*, 1996; Radeva *et al.*, 1997). However, its effects were not just to prevent apoptosis when cells were deprived of ECM contact. Overexpression of ILK also disrupted epithelial monolayers and produced a more fibroblastic morphology. ILK therefore seems to affect cell adhesion as well as regulating adhesion dependent behaviour.

4.5 Crosstalk between growth factors and integrins.

As already discussed, many cell types require growth factor signals in addition to an integrin mediated signal. Evidence suggests that these two signals may be quite closely associated. As can be seen above, many of the signalling proteins activated in response to integrin ligation, such as src and PI3-kinase, are also associated with growth factor signalling. Indeed, the roles of many of them in the regulation of apoptosis has been characterised more extensively in response to growth factors response than integrins.

There may be close associations between growth factor receptors and integrin complexes in cells (Plopper *et al.*, 1995; Miyamoto *et al.*, 1996). The integrin αvβ3 has been shown to associate with elements of the PDGF (Bartfeld *et al.*, 1993) and insulin (Vuori and Ruoslahti, 1994) receptor pathways. αvβ3 was found to co-immuneprecipitate with Insulin Receptor Substrate-1 (IRS-1), a known transducer of signals from both the insulin and insulin -like growth factor receptors. The mitogenic effects of signalling though insulin and IGF was found to be enhanced following αvβ3 mediated adhesion (Vuori and Ruoslahti, 1994).

Primary mammary epithelial cells require insulin in addition to ECM/integrin interactions to suppress apoptosis (Farrelly *et al.*, MS in preparation). These cells require specific adhesion to basement membrane to survive, and adhesion to collagen I results in apoptosis. There appears to be crosstalk between the specific integrins utilised on basement membrane and the insulin receptors. Mammary cells adherent on collagen I did not show phosphorylation of IRS-1 in response to insulin, whereas IRS-1 was phosphorylated on cells growing on basement membrane. The survival effects of both adhesion and insulin could be abrogated by inhibiting both PI3-kinase or MAP kinase. It was recently suggested that a direct association exists between activated insulin and PDGF receptors with αvβ3 integrin following growth factor stimulation (Schneller *et al.*, 1997). If such interactions only occur between receptors and certain integrins, then it might account for the crosstalk seen between growth factors and specific ECM types.

5.0 SUMMARY AND PERSPECTIVES

The majority of cells exist in communities. They form adhesive interactions with the ECM and with other cells. Such associations are essential for maintaining normal tissue homeostasis, and where this breaks down cells are deleted by apoptosis. Indeed, the development of an adhesion–based survival mechanism may have been the most important driving force in the evolution of multi–cellular organisms. So, although it is well established that cells require extracellular signals in order to survive, it is no longer sufficient to consider that apoptosis is solely regulated by soluble factors such as growth factors and cytokines. Cellular adhesion plays an essential, if not dominant, role.

The mechanism of adhesion–dependent suppression of apoptosis is only just beginning to be worked out. However, unravelling how ECM regulates survival is inexorably linked to a knowledge of how integrins interact with intracellular signalling pathways, and there is still much work to be done in learning how integrins function. Moreover, there are virtually no studies that investigate survival signalling induced by cell–cell adhesions, a mode of communication that is likely to be equally important in life and death decisions. Thus, although much progress has recently been made in identifying apoptosis proteins, there are undoubtedly exciting times ahead in unravelling how adhesion–based systems communicate with the apoptosis machine.

6.0 REFERENCES.

Albelda,S.M., Mette,S.A., Elder,D.E., *et al.*.(1991) Integrin distribution in malignant-melanoma - association of the beta-3-subunit with tumour progression. *Cancer Res.* **50**: 6757-6764.

Alexander,C.M., Howard,E.W., Bissell,M.J. and Werb,Z. (1996) Rescue of mammary epithelial cell apoptosis and entactin degradation by a tissue inhibitor of metalloproteinases-1 transgene. *J. Cell Biol.***135**: 1669-1677.

Bartfeld,N.S., Pasquale,E.B., Geltosky,J.E. and Languino,L.R (1993) The alpha-v-beta-3 integrin associates with a 190-kda protein that is phosphorylated on tyrosine in response to platelet-derived growth-factor. *J Biol. Chem.* **268**: 17270-17276.

Boudreau,N., Sympson,C.J., Werb,Z. and Bissell,M.J. (1995) Suppression of ice and apoptosis in mammary epithelial-cells by extracellular-matrix. *Science* **267**: 891-893.

Boudreau,N., Werb,Z. and Bissell,M.J. (1996) Suppression of apoptosis by basement-membrane requires 3-Dimensional tissue organization and withdrawal from the cell-cycle. *Proc. Natl. Acad. Sci. USA* **93**: 3509-3513.

Brooks,P.C., Clark,R.A.F. and Cheresh,D.A. (1994a) Requirement of vascular integrin $\alpha v\beta 3$ for angiogenesis. *Science* **264**: 569-571.

Brooks,P.C., Montgomery,A.M.P., Rosenfeld,M. *et al.* (1994b) Integrin Alpha(V)Beta(3) antagonists promote tumor-regression by inducing apoptosis of angiogenic blood-vessels. *Cell* **79**: 1157-1164.

Brooks,P.C., Stromblad,S., Klemke,R., *et al.* (1995) Antiintegrin Alpha-V-Beta-3 blocks human breast-cancer growth and angiogenesis in human skin. *J. Clin. Invest.* **96**: 1815-1822.

Burridge,K. and Chrzanowska-Wodnicka,M (1996) Focal adhesions, contractility, and signalling. *Ann. Rev. Cell and Dev. Biol.* **12**: 463-518.

Burridge,K., Fath,K., Kelly,T., *et al.* (1988) Focal adhesions: transmembrane junctions between the extracellular matrix and the cytoskeleton. *Ann. Rev. Cell. Biol.*_ **4**: 487-525.

Burridge,K., Turner,C.E. and Romer,L.H. (1992) Tyrosine phosphorylation of Paxillin and pp125(Fak) accompanies cell- adhesion to extracellular-matrix - a role in cytoskeletal assembly. *J. Cell Biol.* **119**: 893-903.

Cardone,M.H., Salvesen,G.S., Widmann,C. *et al.* (1997) The regulation of anoikis: MEKK-1 activation requires cleavage by caspases. *Cell* **90**: 315-323.

Chen,C.S., Mrksich,M., Huang,S. *et al.* (1997) Geometric control of cell life and death. *Science* **276**: 1425-1428.

Chen,H.C., Appeddu,P.A., Isoda,H. and Guan,J.L. (1996) Phosphorylation of tyrosine 397 in focal adhesion kinase is required for binding phosphatidylinositol 3-kinase. *J. Biol. Chem.* **271**: 26329-26334.

Chen, H.C., Appeddu,P.A., Parsons,J.T. *et al.* (1995) Interaction of focal adhesion kinase with cytoskeletal protein talin. *J. Biol. Chem.* **270**: 16995-16999.

Chen,H.C. and Guan,J.L. (1994) Association of focal adhesion kinase with its potential substrate phosphatidylinositol 3-kinase. *Proc. Natl. Acad. Sci. USA* **91**: 10148-10152.

Chen,Q.M., Kinch,M.S., Lin,T.H. *et al.* (1994) Integrin-mediated cell-adhesion activates mitogen-activated protein-kinases. *J. Biol. Chem.* **269**: 26602-26605.

Cobb,B.S., Schaller,M.D., Leu,T.H. and Parsons,J.T. (1994) Stable association of pp60(Src) and pp59(Fyn) with the focal adhesion-associated protein-tyrosine kinase, pp125(Fak). *Mol. Cell. Biol.* **14**: 147-155.

Cortez,D., Stoica,G., Pierce,J.H. and Pendergast,A.M.. (1996) The BCR-ABL tyrosine kinase inhibits apoptosis by activating a Ras-dependent signalling pathway. *Oncogene* **13**: 2589-2594.

Crouch,D.H., Fincham,V.J. and Frame,M.C.. (1996) Targeted proteolysis of the focal adhesion kinase pp125FAK during c-Myc-induced apoptosis is suppressed by integrin signalling. *Oncogene.* **12**: 2689-2696.

delPeso,L., Gonzalez-Garcia,M., Page,C. *et al.* (1997). Interleukin-3-induced phosphorylation of BAD through the protein kinase Akt. *Science* **278**: 687-689.

Dickson,S.R. and Warburton,M.J. (1992) Enhanced synthesis of gelatinase and stromelysin by myoepithelial cells during involution of the rat mammary-gland. *J. Histo. Cyto.* **40**: 697-703.

Dudek,H., Datta,S.R., Franke,T.F. *et al.*(1997) Regulation of neuronal survival by the serine-threonine protein kinase Akt. *Science* **275**: 661-664.
Felding-Habermann,B., Mueller,B.M., Romerdahl,C.A. and Cheresh,D.A.. (1992) Involvement of the integrin αv gene expression in human melanoma tumorigenicity. *J. Clin. Invest.* **89**: 2018-2022.
Folkman,J. and Moscona,A. (1978) Role of cell shape in growth control. *Nature* **273**: 345-349.
Franke,T.F. and Cantley,L.C. (1997) A bad kinase makes good. *Nature* **390**: 116-117.
Franke,T.F., Kaplan,D.R., Cantley,L.C. and Toker,A. (1997) Direct regulation of the Akt proto-oncogene product by phosphatidylinositol-3,4-bisphosphate. *Science* **275**: 665-668.
Frisch,S.M. (1994) E1a induces the expression of epithelial characteristics. *J. Cell Biol.* **127**: 1085-1096.
Frisch,S.M. (1997) The epithelial cell default-phenotype hypothesis and its implications for cancer. *Bioessays* **19**: 705-709.
Frisch,S.M. and Francis,H. (1994) Disruption of epithelial cell-matrix interactions induces apoptosis. *J. Cell Biol.* **124**: 619-626.
Frisch,S.M., Vuori,K., Kelaita,D. and Sicks,S. (1996a) A role for Jun-N-terminal kinase in anoikis - suppression by Bcl-2 and Crma. *J. Cell Biol.* **135**: 1377-1382.
Frisch,S.M., Vuori,K., Ruoslahti,E. and Chanhui,P.Y. (1996b) Control of adhesion-dependent cell-survival by focal adhesion kinase. *J. Cell. Biol.* **134**: 793-799.
Gilmore,A.P. and Romer,L.H. (1996) Inhibition of focal adhesion kinase (FAK) signaling in focal adhesions decreases cell motility and proliferation. *Mol. Biol. Cell* **7**: 1209-1224.
Hanks,S.K., Calalb,M.B., Harper,M.C. and Patel,S.K. (1992) Focal adhesion protein-tyrosine kinase phosphorylated in response to cell attachment to fibronectin. *Proc. Natl. Acad. Sci. USA* **89**: 8487-8491.
Hanks,S.K. and Polte,T.R. (1997) Signaling through focal adhesion kinase. *Bioessays* **19**: 137-145.
Hannigan,G.E., Leung-Hagesteijn,C., Fitz-Gibbon,L. *et al.* (1996) Regulation of cell adhesion anchorage-dependent growth by a new β1-integrin-linked protein kinase. *Nature* **379**: 91-96.
Hemmings,B.A. (1997) Akt signalling: linking membrane events to life and death decisions. *Science* **275**: 628-630.
Hildebrand,J.D., Schaller,M.D. and Parsons,J.T. (1993) Identification of sequences required for the efficient localization of the focal adhesion kinase, Pp125(Fak), to cellular focal adhesions. *J. Cell. Biol.* **123**: 993-1005.
Hildebrand,J.D., Schaller,M.D, and Parsons,J.T. (1995) Paxillin, a tyrosine-phosphorylated focal adhesion-associated protein binds to the carboxyl-terminal domain of focal adhesion kinase. *Mol. Biol. Cell* **6**: 637-647.
Hungerford,J.E., Crompton,M.T., Matter,M.L. *et al.* (1996) Inhibition of pp125FAK in cultured fibroblasts results in apoptosis. *J. Cell Biol.* **135**: 1383-1390.
Hynes,R.O. (1992) Integrins - versatility, modulation, and signaling in cell-adhesion. *Cell* **69**: 11-25.
Khawaja,A. and Downward,J. (1997) Lack of correlation between activation of Jun-NH2-terminal kinase and induction of apoptosis after detachment of epithelial cells. *J. Cell Biol.* **139**: 1017-1023.
Khwaja,A., Rodriguez Viciana,P., Wennstrom,S., *et al.* (1997) Matrix adhesion and Ras transformation both activate a phosphoinositide 3-OH kinase and protein kinase B/Akt cellular survival pathway. *EMBO J.* **16**: 2783-2793.
Klippel,A., Kavanaugh,W.M., Pot,D. and Williams,L.T. (1997) A specific product of phosphatidylinositol 3-kinase directly activates the protein kinase Akt through its pleckstrin homolgy domain. *Mol. Cell Biol.* **17**: 338-344.
Kohn,A.D., Summers,S.A. and Roth,R.A.. (1996) Akt, a pleckstrin homolgy domain -containing kinase, is activated primarily by phosphorylation. *J. Biol. Chem.* **271**: 31372-31378.
Kornberg,L.J., Earp,H.S., Turner,C.E. *et al.* (1991) Signal transduction by integrins - increased protein tyrosine phosphorylation caused by clustering of beta-1 integrins. *Proc. Natl. Acad. Sci. USA* **88**: 8392-8396.
Lewis,J.M., Baskaran,R., Taagepera,S. *et al.* (1996) Integrin regulation of c-Abl tyrosine kinase activity and cytoplasmic-nuclear transport. *Proc. Natl. Acad. Sci. USA* **93**: 15174-15179.
Lin,T.H., Aplin,A.E., Shen,Y. *et al.* (1997) Integrin-mediated activation of MAP kinase is independent of FAK: Evidence for dual integrin signaling pathways in fibroblasts. *J. Cell Biol.* **136**: 1385-1395.
Lipfert,L., Haimovich,B., Schaller,M.D. *et al.* (1992) Integrin-dependent phosphorylation and activation of the protein tyrosine kinase pp125(Fak) in platelets. *J. Cell Biol.* **119**: 905-912.
Lund,L.R., Romer,J., Thomasset,N. *et al.* (1996) 2 Distinct phases of apoptosis in mammary-gland involution - proteinase-independent and proteinase-dependent pathways. *Development* **122**: 181-193.
Maher,P.A. (1993). Activation of phosphotyrosine phosphatase activity by reduction of cell-substrate adhesion. *Proc. Natl. Acad. Sci. USA* **90**: 11177-11181.
Massia,S.P. and Hubbell,J.A. (1991) An RGD spacing of 440n is sufficient for integrin avb3-mediated fibroblast spreading and 140nm for focal contact and stress fiber formation. *J. Cell Biol.* **114:** 1089-1100.
Martinez-Hernandez,A., Fink,L.M., and Pierce,G.B. (1976) Removal of basement membrane in the involuting breast. *Lab. Invest.* **34**: 455-461.
Maundrell,K., Antonsson,B., Magnenat,E., *et al.* (1997) Bcl-2 undergoes phosphorylation by c-Jun N-terminal kinase/stress activated protein kinase in the presence of the constitutively active GTP-binding protein Rac1. *J. Biol. Chem.* **272**: 25238-25242.
Meredith,J.E., Fazeli,B. and Schwartz,M.A. (1993) The extracellular-matrix as a cell-survival factor. *Mol. Biol. Cell* **4**: 953-961.

Miyamoto,S., Akiyama,S.K., and Yamada,K.M. (1995a) Synergistic roles for receptor occupancy and aggregation in integrin transmembrane function. *Science* **267**: 883-885.
Miyamoto,S., Teramoto,H., Coso,O.A. *et al.* (1995b) Integrin function - molecular hierarchies of cytoskeletal and signaling molecules. *J. Cell Biol.* **131**: 791-805.
Miyamoto,S.H., Teramoto,H., Gutkind,J.S. and Yamada,K.M. (1996) Integrins can collaborate with growth factors for phosphorylation of receptor tyrosine kinases and MAP kinase activation: role of integrin aggregation and occupancy of receptors. *J. Cell Biol.* **135:** 1633-1642.
Montgomery,A.M.P., Reisfeld,R.A. and Cheresh,D.A. (1994) Integrin-alpha(V)beta(3) rescues melanoma-cells from apoptosis in 3- Dimensional dermal collagen. *Proc. Natl. Acad. Sci. USA* **91**: 8856-8860.
Owens,L.V., Xu,L.H., Craven,R.J *et al.* (1995). Overexpression of the focal adhesion kinase (p125FAK) in invasive human tumors. *Cancer Res.* **55**: 2752-2755.
Parrizas,M., Saltiel,A.R., and LeRoith,D. (1997). Insulin-like growth factor 1 inhibits apoptosis using the phosphatidylinositol 3'-kinase and mitogen-activated protein kinase pathways. *J. Biol. Chem.* **272**: 154-161.
Parsons,J.T. (1996). Integrin-mediated signalling: regulation by protein tyrosine kinases and small GTP-binding proteins. *Curr. Op. Cell Biol.* **8**: 146-152.
Petch,L.A., Bockholt,S.M., Bouton,A., *et al.* (1995) Adhesion-induced tyrosine phosphorylation of the p130 Src substrate. *J. Cell Sci.* **108**: 1371-1379.
Plopper,G.E., McNamee,H.P., Dike,L.E. *et al.* (1995) Convergence of integrin and growth-factor receptor signaling pathways within the focal adhesion complex. *Mol. Biol. Cell* **6**: 1349-1365.
Polte,T.R. and Hanks,S.K. (1995). Interaction between focal adhesion kinase and Crk-associated tyrosine kinase substrate p130Cas. *Proc. Natl. Acad. Sci. USA* **92**: 10678-82.
Pullan,S., Wilson,J. Metcalfe,A. *et al.* (1996) Requirement of basement-membrane for the suppression of programmed cell-death in mammary epithelium. *J. Cell Sci.* **109**: 631-642.
Radeva,G., Petrocelli,T., Behrend,E. (1997) Overexpression of the integrin-linked kinase promotes anchorage-independent cell cycle progression. *J. Biol. Chem.* **272**: 13937-13944.
Re,F., Zanetti,A., Sironi,M. *et al.* (1994) Inhibition of anchorage-dependent cell spreading triggers apoptosis in cultured human endothelial-cells. *J. Cell Biol.* **127**: 537-546.
Schaller,M.D., Borgman,C.A., Cobb,B.S. (1992) pp125fak, a structurally distinctive protein-tyrosine kinase associated with focal adhesions. *Proc. Natl. Acad. Sci. USA* **89**: 5192-5196.
Schaller,M.D., Otey,C.A. Hildebrand,J.D. and Parsons,J.T. (1995) Focal adhesion kinase and Paxillin bind to peptides mimicking beta- integrin cytoplasmic domains. *J. Cell Biol.* **130**: 1181-1187.
Schlaepfer,D.D., Hanks,S.K., Hunter,T and van der Geer,P. (1994) Integrin-mediated signal transduction linked to Ras pathway by GRB2 binding to focal adhesion kinase. *Nature* **372**: 786-91.
Schlaepfer,D.D. and Hunter,T. (1996) Evidence for *in vivo* phosphorylation of the Grb2 SH2-domain binding site on focal adhesion kinase by Src-family protein-tyrosine kinases. *Mol. Cell Biol.* **16**: 5623-5633.
Schlaepfer,D.D. and Hunter,T. (1997) Focal adhesion kinase overexpression enhances ras-dependent integrin signaling to ERK2/mitogen-activated protein kinase through interactions with and activation of c-Src. *J. Biol. Chem.* **272**: 13189-13195.
Schneller,M., Vuori,K. and Ruoslahti,E. (1997) Alpha v beta 3 integrin associates with activated insulin and PDGF beta receptors and potentiates the biological activity of PDGF. *EMBO J.* **16**: 5600-5607.
Schwartz,M.A., Schaller,M.D., and Ginsberg,M.H. (1995) Integrins - emerging paradigms of signal-transduction. *Ann. Rev. Cell Dev. Biol.* **11**: 549-599.
Songyang,Z., Baltimore,D., Cantley,L.C. *et al.* (1997) Interleukin 3-dependent survival by the Akt protein kinase. *Proc. Natl. Acad. Sci. USA* **94**: 11345-11350.
Strange,R., Li,F., Saurer,S. *et al.* (1992) Apoptotic cell-death and tissue remodeling during mouse mammary-gland involution. *Development* **115**: 49-58.
Streuli,C.H., Bailey,N., and Bissell,M.J. (1991) Control of mammary epithelial differentiation - basement-membrane induces tissue-specific gene-expression in the absence of cell cell- interaction and morphological polarity. *J. Cell Biol.* **115**: 1383-1395.
Sympson,C.J., Talhouk,R.S., Alexander,C.M. *et al.* (1994) Targeted expression of stromelysin-1 in mammary-gland provides evidence for a role of proteinases in branching morphogenesis and the requirement for an intact basement-membrane for tissue-specific gene-expression. *J. Cell Biol.* **125**: 681-693.
Talhouk,R.S., Bissell,M.J. and Werb,Z. (1992) Coordinated expression of extracellular matrix-degrading proteinases and their inhibitors regulates mammary epithelial function during involution. *J. Cell Biol.* **118**: 1271-1282.
Toker,A. and Cantley,L.C. (1997) Signalling through the lipid products of phosphoinositide-3-OH kinase. *Nature.* **387**: 673-676.
Vachon,P.H., Loechel.,F., Xu.,H. *et al.*.(1996) Merosin and laminin in myogenesis; specific requirements for merosin in myotube stability and survival. *J. Cell Biol.* **134**: 1483-1497.
Vuori,K. and Ruoslahti,E. (1994) Association of insulin-receptor substrate-1 with integrins. *Science* **266**: 1576-1578.
Wang,H.G., Rapp,U.R. and Reed,J.C. (1996) Bcl-2 targets the protein-kinase Raf-1 to mitochondria. *Cell* **87**: 629-638.
Wary,K.K., F.Mainiero,F., Isakoff,S.J. *et al.* (1996) The adapter protein Shc couples a class of integrins to the control of cell-cycle progression. *Cell* **87**: 733-743.

Wen,L.-P., Fahrni,J.A., Troie,S. (1997) Cleavage of focal adhesion kinase by caspases during apoptosis. *J. Biol. Chem.* **272**: 26056-26061.

Xia,Z., Dickens,M., Raingeaud,J. *et al.* (1995) Opposing effects of ERK and JNK-p38 MAP kinases on apoptosis. *Science* **270**: 1326-1331.

Xing,Z., Chen,,H.C., Nowlen,J.K. *et al.* (1994) Direct interaction of v-Src with the focal adhesion kinase mediated by the Src SH2 domain. *Mol. Biol. Cell* **5**: 413-421.

Xu,L.H., Owens,L.V., Sturge,G.C. *et al.* (1996) Attenuation of the expression of the focal adhesion kinase induces apoptosis in tumor-cells. *Cell Growth Diff.* **7**: 413-418.

Zha,J.P., Harada,H., Yang,E. *et al.* (1996) Serine phosphorylation of death agonist Bad in response to survival factor results in binding to 14-3-3 not Bgl-X(L). *Cell* **87**: 619-628.

Zhang,Z.H., Morla,A.O., Vuori,K., *et al.*(1993) The alpha-V-beta-1-integrin functions as a fibronectin receptor but does not support fibronectin matrix assembly and cell-migration on fibronectin. *J. Cell Biol.* **122**: 235-242.

Zhang,Z.H., Vuori,K., Reed,J.C. and Ruoslahti,E. (1995) The alpha-5-beta-1 integrin supports survival of cells on fibronectin and up-regulates Bcl-2 expression. *Proc. Natl. Acad. Sci. USA* **92**: 6161-6165.

Zhu,X.Y. and Assoian,R.K. (1995) Integrin-dependent activation of Map Kinase - a link to shape-dependent cell-Pproliferation. *Mol. Biol. Cell* **6**: 273-282.

Zhu,X.Y., Ohtsubo,M., Bohmer,R.M. *et al.* (1996) Adhesion-dependent cell-cycle progression linked to the expression of cyclin D1, activation of cyclin E-Cdk2, and phosphorylation of the retinoblastoma protein. *J. Cell Biol.* **133**: 391-403.

Chapter 7

Death signalling in *C. elegans* and activation mechanisms of caspases

Asako Sugimoto and Masayuki Miura
Department of Biophysics and Biochemistry, Graduate School of Science, University of Tokyo and PRESTO, Japan Science and Technology Corporation, Japan [AS]. Department of Neuroanatomy, Osaka University Medical School and Core Research for Evolutional Science and Technology (CREST), Japan [MM].

1. 0 INTRODUCTION

Cell proliferation and cell death are two opposing sides of the same coin. The development and homeostasis of multicellular organisms requires the accurate control of both processes. The death that occurs normally as a process of development is called "programmed cell death" or "apoptosis". The phenomenon of developmentally programmed cell death has been known for more than 100 years, but it is only in the last decade that its molecular mechanisms have begun to be uncovered. The nematode *Caenorhabditis elegans* is one of the few experimental systems in which a genetic approach is available to dissect the processes of programmed cell death, and in fact, findings from studies using *C. elegans* have already played a very important role in elucidating these mechanisms. Here, we summarise the progress to date in understanding how cell death is controlled in *C. elegans*, and review the main machinery of programmed cell death/apoptosis with respect to its evolutionary conservation between *C. elegans* and other species.

2.0 PROGRAMMED CELL DEATH IN *C. ELEGANS*

2.1 Overview

2.1.1 C. elegans as a model organism

What makes the small worm an ideal model organism to study cell death? There are several advantages to using *C. elegans* to analyse the genetic, molecular, and cellular aspects of programmed cell death. First, since the egg shell and cells are transparent, the process of programmed cell death is readily observed in the live embryos and worms using differential interference contrast (DIC) microscopy. Second, the number of cells are small and each cell lineage in the whole organism is essentially invariant. As a fertilised egg becomes an adult, 1090 non-germ-cells are produced and 131 of these die by programmed cell death (Sulston and Horvitz, 1977; Sulston *et al.*, 1983) . Finally, the identification and characterisation of mutàtions that have defects in various processes of programmed cell death is a powerful tool to elucidate its genetic pathway. The existence of a nearly complete physical map and rapid progress in the genome sequencing project for *C. elegans* (expected to be completed in 1998) have accelerated the molecular cloning of the genes of interest (Sulston *et al.*, 1992) . The techniques for making transgenic worms are established and have been proved a useful method for testing the functions of cloned genes *in vivo* (Mello *et al.*, 1991) .

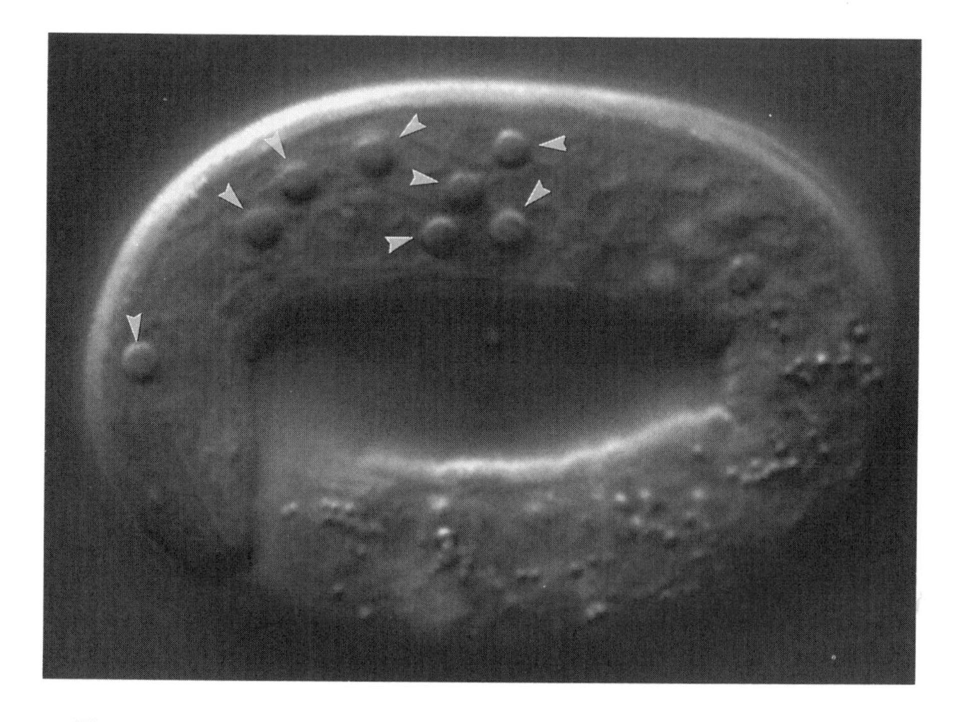

Figure 1: A DIC photograph of a ced-5 mutant embryo.
In ced-5 mutants, dying cells accumulate due to the defect in phagocytosis. The nuclei of dying cells are seen as 'flat buttons' (arrows).

2.1.2 Pattern of cell death in C. elegans

In *C. elegans*, a significant proportion of the total cells (12%) are removed by programmed cell death. Most of these deaths (113/131) occur during the 14 hours of embryogenesis, mostly in the short period between 250 and 450 minutes after fertilisation (Sulston *et al.*, 1983) . The 113 cells that undergo cell death correspond to one sixth of the total cells produced (671 cells) during embryogenesis. Post-embryonic development, which includes the four larval stages that precede the adult stage, produces 419 additional cells, but only 18 of these are removed by cell death (Sulston and Horvitz, 1977) . The majority of cells that die through apoptosis appear to be neuronal, although some hypodermal and muscle cells also die. There is no death of intestinal cells. Although not included in the numbers above, germ cell deaths occur in adult hermaphrodites, seemingly by the same molecular mechanism as somatic cells. In general, cell death occurs immediately after a cell divides and before it shows any sign of differentiation.

2.1.3 Morphological changes during cell death

The process of cell death in *C. elegans* takes approximately one hour (Sulston and Horvitz, 1977; Robertson and Thomson, 1982) , and appears to start even before the mother cell finishes dividing. The dying cells are smaller than their sisters and in some cases the engulfment of the dying cells begins even before cytokinesis finishes. At this step, however, no other obvious differences between the dying cells and their sisters can be detected. Next, the refractivity of the nuclei of the dying cells increases, and the dying cells are enclosed by the cytoplasm of the engulfing cells. By electron microscopy, swelling of the nuclear membrane and condensation of the cytoplasm are observed. The chromatin forms granular aggregates in the nucleus. These are electron-dense core particles, probably originating from the nucleolus. Parts of the cell also begin to be fragmented by the engulfing cells. The process takes approximately 30 minutes to get to this point. Next, the refractivity of the whole cell increases and the appearance of the dying cell becomes a "flat disc" by DIC optics. This step lasts ten to thirty minutes (Figure 1). Finally, the dying cells shrink and disappear. The nuclear membrane breaks down and chromatin-like structures scatter in the engulfing cells.

In early embryogenesis, some cell corpses are engulfed by their sister cells. In late embryogenesis, hypodermal cells are the major engulfing cells, but pharyngeal muscle cells and intestinal cells have also been observed to engulf the dying cells. In most cases, the cells destined to undergo cell death die within one to two hours after they are born. Only a fraction of the cells differentiate and function before they die. There are some common characteristics between the morphological changes of the cells undergoing programmed cell death in *C. elegans* and mammals (Wyllie *et al.*, 1980) . For example, condensation of cytoplasm, condensation of chromatin, fragmentation of dying cells, and engulfment by neighbouring cells are observed in both worms and mammals. However, the formation of a nucleosome-sized ladder by genomic DNA, which is often observed during apoptosis in higher animals, is not known in *C. elegans*, because it is difficult to collect dead cells and there are no cell lines from *C. elegans* available in which apoptosis can be induced.

2.1.4 Isolation of genes involved in cell death

The genetic analysis of programmed cell death in *C. elegans* was started in 1983 by the isolation of mutants defective in certain aspects of this process. Two of the cell death mutants isolated by Hedgecock *et al.* (1983), named *ced-1* and *ced-2* (ced=cell death abnormal), were isolated by their phenotype, in which cell corpses accumulate due to a defect in the engulfment process. In these mutants, the dying cells persist for many hours, in contrast to 10-30 minutes in wild type.

The *ced-3* gene was isolated by Ellis and Horvitz (1986) by screening for mutants in which the number of observable cell corpses in the *ced-1* mutant was reduced. The loss-of-function (lf) mutants of *ced-3* turned out to prevent all programmed cell death. The *ced-4(lf)* mutants and gain-of-function (gf) mutants of *ced-9* similarly showed no-cell death phenotypes (Ellis and Horvitz, 1986; Hengartner *et al.*, 1992) . This finding indicated that all the programmed cell death in *C. elegans* is controlled by a single genetic pathway that includes *ced-3*, *ced-4*, and *ced-9*. The functions of these genes will be discussed in more detail later in this review.

In *C. elegans*, programmed cell death is not essential for the viability of the animal, unlike in other higher organisms. The mutants that do not have any cell death have extra cells, but their movement, morphology, and brood size are unaffected (Ellis and Horvitz, 1986) . The question that therefore arises is why does *C. elegans* have cell death? Although the mutants can propagate as well as wildtype, their growth rates are slightly slower and their odorant response is worse than wildtype (Ellis *et al.*, 1991b) . Such slight disadvantages may not affect their viability in experimental (or, optimal) conditions, but are likely to cause significant selective disadvantages in the natural environment. The cells in the mutants that do not die as they should usually differentiate into the cell type of their sisters or near relatives (Ellis and Horvitz, 1986) , and in some cases have been shown to function as the differentiated cell type. For example, when a pharyngeal neuron is killed by laser ablation in a *ced-3* mutant, its surviving sister cell, which normally would undergo programmed cell death, can differentiate into a mature neuron and take over its sister's function (Avery and Horvitz, 1987) . The surviving cells do not proliferate.

2.1.5 Overview of the genetic pathway of cell death in C. elegans

Horvitz and his colleagues isolated more than 10 genes involved in cell death using mutants defective in various steps of the process. Genetic analyses revealed the cell death pathway shown in the Figure 2.

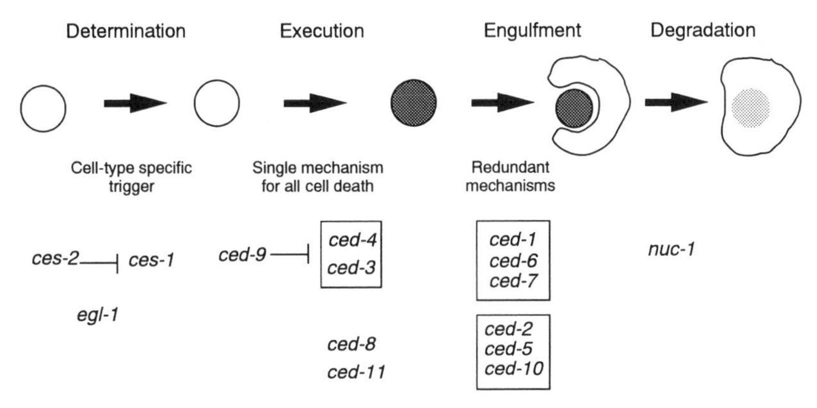

Figure 2. Genetic pathway of programmed cell death in C. elegans. (Adapted from Ellis et al. 1991b.)

Generally the process of programmed cell death is divided into four steps. The first step is the 'decision to live or die'. This step is likely to be regulated in a tissue/cell-type-specific manner. The *ces-1*, *ces-2,* and *egl-1* genes appear to be involved in this process. The second step is the 'execution of cell death' and, as mentioned above, *ced-3*, *ced-4,* and *ced-9* are classified into this category. In contrast to the cell-type-specific 'decision' step, the execution of all cell death is controlled by this single pathway. The third step is the 'engulfment of cell corpses'. At present two partially redundant pathways are known to regulate this process. One pathway includes *ced-1*, *ced-6*, and *ced-7*; the other includes *ced-2*, *ced-5,* and *ced-10*. When any of these genes are mutated, dead cells accumulate during embryogenesis, but eventually most corpses disappear, which suggests the presence of a second redundant pathway. The fourth and last step is the degradation of corpses. So far, only one mutant defective in degradation of DNA in the dead cells has been isolated. The genes that have been identified as having a role in cell death are described in more detail below (also see Table 1).

Table 1: Genes involved in programmed and pathological cell death in *C. elegans*.

Gene	Gene product (homologous gene in other organisms)	Phenotype of mutants (lf=loss-of-function, gf=gain-of-function)	Function
Execution of cell death			
ced-3	Caspase	lf: no cell death	Essential for all programmed cell death
ced-4	Apaf-1	lf: no cell death	Essential for all programmed cell death Activator of CED-3
Inhibition of cell death			
ced-9	bcl-2 family	lf: excess cell death (lethal) gf: no cell death	Inhibitor of CED-3
dad-1	DAD1 homologue a subunit of oligosaccharyltransferase?	overexpression: partial prevention of cell death	
Decision to die			
ces-1	Drosophila SNAIL-like Zn-finger protein	gf: NSM and I2 sisters survive lf: no phenotype	
ces-2	PAR family bZIP protein	lf: NSM sisters survive	
egl-1	not cloned	gf: HSNs undergo cell death lf: no cell death?	
Engulfment of corpses			
ced-1	not cloned	lf: persistent cell corpses	Phagocytosis of corpses
ced-2	not cloned	lf: persistent cell corpses	Phagocytosis of corpses
ced-5	not cloned	lf: persistent cell corpses	Phagocytosis of corpses
ced-6	not cloned	lf: persistent cell corpses	Phagocytosis of corpses
ced-7	not cloned	lf: persistent cell corpses	Phagocytosis of corpses
ced-10	not cloned	lf: persistent cell corpses	Phagocytosis of corpses
Degradation of cell corpses			
nuc-1	not cloned	lf: defect in degradation of DNA in cell corpses	Nuclease?
Death dependent on engulfment genes			
lin-24	novel	gf: Pn.P cells die or show abnormal lineages	
lin-33	not cloned	gf: Pn.P cells die or show abnormal lineages	
Necrosis			
deg-1	Degenerin	gf: necrotic death of a subset of neuronal cells	Na^+ channel
mec-4	Degenerin	gf: necrotic death of a subset of neuronal cells	Na^+ channel
mec-10	Degenerin	gf: necrotic death of a subset of neuronal cells	Na^+ channel

2.2 Main machinery of the execution of the cell death process

2.2.1 CED-3, a caspase prototype

In *C. elegans*, once a cell is fated to die, the genes in that cell that are responsible for the execution of cell death are activated. Two genes, *ced-3* and *ced-4*, are both required for all programmed cell deaths in *C. elegans* (Ellis and Horvitz, 1986) . If one of these genes is mutated, all programmed cell deaths are inhibited. Furthermore, these mutations are very cell-death-specific. The cell lineages of the *ced-3* or *ced-4* mutant animals are identical with those of wild-type animals except for the absence of all programmed cell deaths, suggesting that these two genes function only in cell death.

The *ced-3* gene encodes a 2.8 kb mRNA, which is most abundantly transcribed during embryogenesis when most programmed cell deaths occur. The level of *ced-3* mRNA is very high (comparable to that of actin). This abundance suggests that *ced-3* may not be transcribed only in dying cells, since there are usually no more than two or three cells dying at any given time during embryonic development. The CED-3 protein has significant homology with human and murine interleukin-1ß converting enzyme (ICE) (Cerretti *et al.*, 1992; Thornberry *et al.*, 1992) . ICE (caspase-1), a cysteine protease, cleaves the inactive 31kDa precursor of IL-1ß between Asp-116 and Ala-117 to release the carboxy-terminal 153 amino acid polypeptide from cells as the biologically active mature IL-1ß. Caspase (*c*ysteine *as*partate-specific protein*ase*), is the name for the family of proteases that includes caspase-1 and CED-3 (Alnemri *et al.*, 1996). All caspases are synthesised as inactive precursors (Figure 3), which contain prosequences that may be required for dimerisation of the precursor or interaction with other adaptor molecules. Active enzyme is generated by auto- and/or hetero-processing of the precursor and heterodimer formation of the p20 and p10 subunits.

The overall amino acid identity between the *C. elegans* CED-3 protein and human caspase-1 protein is 29%. In 5 out of 8 *ced-3* mutations identified, single amino acids that are conserved between caspase-1 and CED-3 were altered. Three other mutant alleles of *ced-3* have altered amino acids at the C-terminal end of CED-3 that are not conserved in caspase-1. Overexpression of the *caspase-1* gene or the *C. elegans ced-3* gene in Rat-1 fibroblast cells causes programmed cell death (Miura *et al.*, 1993) . Point mutations in a region conserved in the *caspase-1* and *ced-3* genes eliminated the cell death activities of these genes.

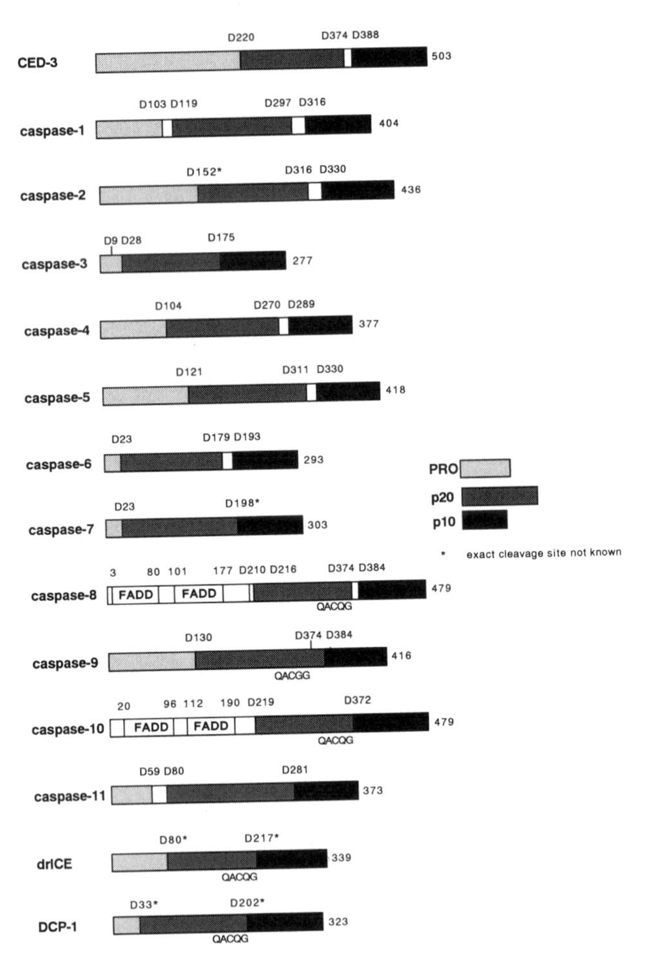

Figure 3. Structure of members of the caspase family.
All caspases are synthesised as long precursors and cleaved during activation into p20 and p10 subunits. The N-terminal prodomain is represented as PRO. The cleavage is dependent on the aspartic acid residue in the P1 position. Positions of the cleavage sites are indicated. Where the positions of cleavage sites have not been determined, it is marked by an asterisk. FADD represents the region that shares homology with the death effector domain of FADD/MORT1.

Full-length CED-3 expressed in *E. coli* undergoes proteolytic activation to generate an active cysteine protease. Active-site mutations of CED-3 abolish both *in vivo* cell-killing activities and *in vitro* protease activities (Xue and Horvitz, 1995). CED-3 can be categorised as a member of the caspase-3 subfamily based on its substrate specificities.

2.2.2 Anti-apoptotic gene, ced-9

The *ced-9* gene product inhibits the execution of the programmed cell death pathway. As described above, *ced-9* was initially identified as a gain-of-function mutant that, like *ced-3* and *ced-4* mutants, causes a complete loss of programmed cell death (Hengartner *et al.*, 1992) . In contrast, mutations that reduce or eliminate *ced-9* function cause cells that normally live to undergo programmed cell death, which eventually results in embryonic lethality (Hengartner *et al.*, 1992) . These observations indicate that *ced-9* acts to protect cells from programmed cell death. The lethality resulting from the reduction of *ced-9* function requires *ced-3* and *ced-4* gene activity, suggesting that *ced-9* acts by antagonising the actions of *ced-3* and *ced-4* (Hengartner *et al.*, 1992) . Cloning of the *ced-9* gene revealed that *ced-9* encodes a protein of 280 amino acids that is similar in sequence to the mammalian proto-oncogene *bcl-2* product (Hengartner and Horvitz, 1994b) . CED-9 and mouse Bcl-2 show 24% identity and 49% similarity, and contain several highly homologous regions that are conserved among the *bcl-2* family members (Figure 4). *bcl-2* had been shown to prevent apoptosis from being induced in a number of mammalian cells by a variety of stimuli.

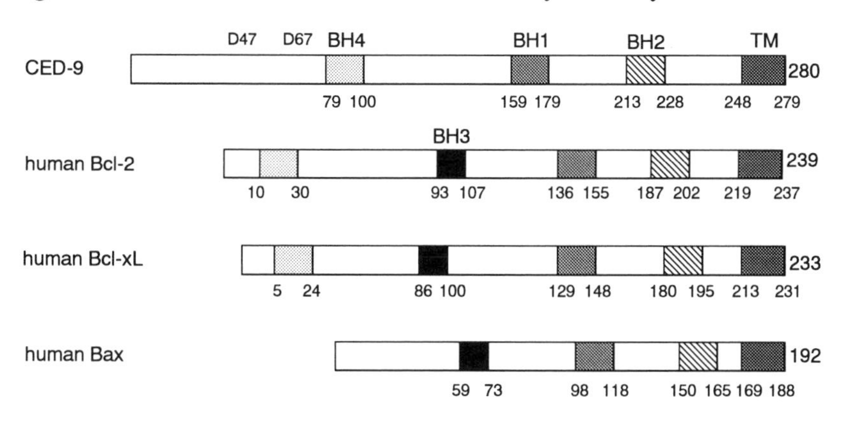

Figure 4. Structural comparison of CED-9 and other Bcl-2 families.
Conserved 'Bcl-2 homology (BH)' domains (BH1, BH2, BH3 and BH4) and N-terminal transmembrane domain (TM) are shown. CED-9 protein does not have apparent BH3 domain. D47 and D67 represent the cleavage sites of CED-9 by CED-3. (Adapted from Zha *et al.* 1996).

Strikingly, overexpression of *bcl-2* in *C. elegans* embryos using a heat-shock promoter prevented programmed cell death and could even complement the loss of *ced-9* function. These observations led to the conclusion that *ced-9* and *bcl-2* are evolutionarily conserved inhibitors of programmed cell death.In addition to *ced-9* and *bcl-2*, the '*bcl-2* family' includes many genes from vertebrates (e.g. *bax, bcl-x* , and *bad*) and viruses (e.g. Epstein-Barr virus BHRF1, and adenovirus E1B 19K). Vertebrate *bcl-2* family gene products have been shown to act as homo- and hetero-dimers and to possess either pro-apoptotic or anti-apoptotic activity. Interestingly, a gain-of-function allele of *ced-9* has a mutation in the region highly conserved among all *bcl-2* family members which is required for the heterodimerisation of Bcl-2 and Bax (Hengartner and Horvitz, 1994a) . In *C. elegans*, *ced-9* appears to be the only gene in the *bcl-2* family, and it is not known whether CED-9 functions as a dimer. Recently Xue and Horvitz showed that CED-9 can be cleaved by CED-3 at two sites near its amino terminus, and that at least one of these sites is important for CED-9 to have complete cell death inhibitor activity (Xue and Horvitz, 1997) . However, this site is not essential for CED-9 to have some inhibitor activity. The cleaved C-terminal product resembles *bcl-2*, and possesses cell death inhibitor activity. These findings led to the hypothesis that CED-9 protects against programmed cell death by two distinct mechanisms: First, by directly interacting with the CED-3 protease, possibly as a competitive inhibitor, and second, by acting through its C-terminal region using an unknown mechanism similar to that of mammalian *bcl-2*.

2.2.3 Activation of CED-3 by the cell death gene product CED-4

The *ced-4* gene has been cloned by transposon tagging using the Tc4 transposable element as a probe (Yuan and Horvitz, 1992) . Like *ced-3*, the *ced-4* gene is most abundantly expressed during embryogenesis. The *ced-4* gene is transcribed as a 2.2 kb mRNA that contains a 549 amino acid open reading frame. CED-4 amino acid residues 158-165 conform to a potential nucleotide-binding P-loop consensus sequence (Figure 5). CED-4 can bind to the ATP analogue 5'-fluorosulphonyl-benzoyladenosine (FSBA) whereas a deletion mutation of the CED-4 P-loop motif failed to bind to FSBA. Mutation of the CED-4 P-loop disrupts its ability to induce cell death in yeast, insect cells, and mammalian cells (Chinnaiyan *et al.*, 1997a; Seshagiri and Miller 1997; James *et al.*, 1997). The *C. briggsae* CED-4 sequence shows the highest similarity with the sequence of the flax rust-resistant protein L6, which also contains a P-loop motif (Lawrence *et al.*, 1995; Chinnaiyan *et al.*, 1997). The alignment of the *C. briggsase* CED-4 and L6 sequences shows 23% identity and 43% similarity over 322 amino acids. Part of the cell death-inducing function of both CED-4 and flax rust-resistant proteins may depend on their ATPase activity.

CED-4 interacts with CED-3, and several studies have elucidated the nature of this relationship. CED-3 is co-immunoprecipitated with CED-4 when 293T cells are transfected with CED-3 and CED-4 (Chinnaiyan *et al.*, 1997b; Irmler *et al.*, 1997a; Wu *et al.*, 1997a). A truncated CED-3 mutant (1-220) can be co-immunoprecipitated with CED-4, indicating that CED-4 associates with the prodomain of CED-3 (Wu *et al.*, 1997b). Expression of CED-4 does not induce cell death in 293T cells and transfection of CED-3 induces only a modest level of cell death in 293T cells. However, co-transfection of *ced-4* with *ced-3* greatly enhances the ability of *ced-3* to induce cell death (Wu *et al.*, 1997b; Seshagiri and Miller, 1997). Purified CED-4 extracted from 293T cells transfected with *ced-4* can stimulate CED-3 processing *in vitro* (Chinnaiyan *et al.*, 1997b). However, a P-loop mutant CED-4 fails to proteolytically activate CED-3 and FSBA prevents the CED-4-mediated activation of CED-3. These results support the idea that CED-4 induces CED-3 activation and cell death, and that this process depends on the ATP-ase activity of CED-4. It has been proposed that CED-4 might be a proCED-3-specific chaperone. ATP hydrolysis may cause conformational change and autocatalytic activation of CED-3 (Hengartner, 1997).

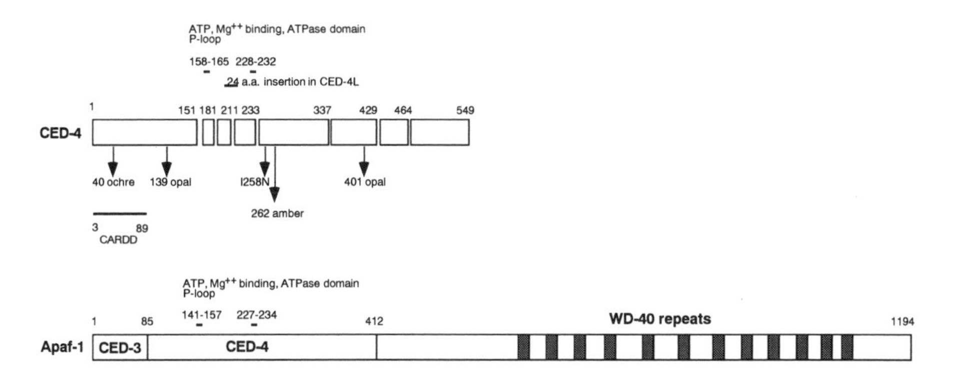

Figure 5. Structural comparison of CED-4 and Apaf-1.
In CED-4, all the exons are indicated by boxes and the positions of the exon boundaries are shown above the boxes. Both CED-4 and Apaf-1 contain P-loop nucleotide binding motifs (amino acid 158-165 and 228-232 in CED-4, and 141-157 and 227-234 in Apaf-1). The N-terminus of CED-4 contains a region called CARD (caspase-recruitment domain), which shares homology with sequences in caspase-2 and RAIDD. Apaf-1 contains a CED-3-homologous region in its N-terminus, a CED-4-homologous region in its middle, and 12 WD 40 repeats in its C-terminus.

Genetic studies show that *ced-9* prevents cell death by inhibiting the death- promoting activities of *ced-3* and *ced-4* (Shaham and Horvitz, 1996). Overexpression of *ced-4* in *C. elegans* ALM neurons causes cell death. *ced-4*-induced cell death requires *ced-3* activity for efficient killing, suggesting that *ced-4* acts upstream of *ced-3*. Protection against *ced-3*-induced cell death by *ced-9* requires *ced-4* activity, suggesting that *ced-9* inhibits *ced-3* by acting at least in part through *ced-4* (Shaham and Horvitz, 1996). Recently physical interaction of these cell-death gene products has been demonstrated. The use of yeast two-hybrid screening to search for proteins that bind CED-4 picked up proteins encoded by the entire CED-9 coding region. CED-4 is present in the cytosol when *ced-4* is overexpressed in 293T cells . CED-9 is localised in intracellular membranes and the perinuclear region in 293T cells. Expression of *ced-9* together with *ced-4* results in a relocation of *ced-4* from the cytosol to intracellular membranes in mammalian cells (Wu *et al.*, 1997a). Similar relocalisation of CED-4 resulting from co-expression with CED-9 has been observed in yeast *Schizosaccharomyces pombe* (James *et al.*, 1997).

A gene from mammalian cells that is homologous to CED-4 has been identified from biochemical studies of caspase activation (Zou *et al.*, 1997). A cell-free caspase-3 activation system was developed to identify factors that can stimulate caspases. Using this system, addition of dATP to extracts stimulates the proteolytic activation of caspase-3. Purification of the components in the extract required for this activity led to the identification of three apoptosis protease-activating factors (Apafs). Apaf-2 is cytochrome c. Upon the apoptotic stimulation of cells, cytochrome c is released from mitochondria into the cytosol. One of the important anti-apoptotic functions of *bcl-2* is to prevent the release of cytochrome c in response to apoptosis-inducing stimuli. Part of the sequence of Apaf-1 shows a striking similarity to that of CED-4, including the nucleotide-binding site (Figure 5). The amino terminal region of Apaf-1 shows some sequence similarities to the prodomain of CED-3. This region may bind directly to caspases and function as a caspase-recruitment domain (CARD). These structural features suggest Apaf-1 may be a mammalian homologue of CED-4. The carboxy terminus of Apaf-1 has a large domain containing 12 WD-40 repeats, which are also known to mediate protein-protein interactions. Apaf-1 can bind to cytochrome c, and WD-40 repeats may be involved in this interaction.

Like the physical interaction between CED-3 and CED-4 in worms, caspase-9 (Apaf-3) has been shown to interact with Apaf-1 through their N-terminal CARD domains in the presence of cytochrome c and dATP (Liu *et al.*, 1997b). *In vitro*, purified Apaf-1 in the presence of cytochrome c and dATP can induce the activation of caspase-9. Activated caspase-9 then cleaves caspase-3. DNA fragmentation factor (DFF) which induces DNA fragmentation *in vitro* can be activated by caspase-3 (Liu *et al.*, 1997a). There are more CARD domain-containing caspases in mammalian cells (caspase-1, -2, -8 and -10), so it might be possible that other caspases can be activated by Apaf-1 and cytochrome c.

2.3 Genes involved in engulfment and degradation of dead cells

2.3.1 Engulfment genes

Mutations in six genes (*ced-1*, *ced-2*, *ced-5*, *ced-6*, *ced-7*, and *ced-10*) prevent the efficient engulfment of dying cells (Ellis *et al.*, 1991a) . In the absence of functional engulfment genes, cells programmed to die can persist for hours in the stage where they are highly refractile by DIC optics, in contrast to 10-30 minutes in wild type cells. None of the mutations in the engulfment genes completely blocks phagocytosis of all dying cells. This is not because individual engulfment genes function only in a subset of cell corpses. Rather, analysis of double mutant combinations of the engulfment genes suggests that the engulfment of corpses is accomplished by more than one pathway. Thus, when one pathway is inactivated due to a loss of function of one of its components, the other pathway can still function (Ellis *et al.*, 1991a) .

Genetic analysis suggest that *ced-2*, *ced-5,* and *ced-10* comprise one pathway, and *ced-1*, *ced-6*, and *ced-7* the other (Ellis *et al.*, 1991a) . Double mutants within each group show the same degree of corpse persistence as single mutants in any one of the three genes. On the other hand, double mutant combinations of members of the two different sets exhibit significantly more persistence of dead cells, indicating that these two pathways act in parallel in the engulfment process. Interestingly, in addition to the engulfment defect, one group of mutants (*ced-2*, *ced-5*, and *ced-10*) have pleiotropic effects, including aberrant migration of the gonad arm (Nishiwaki; Hengartner; Horvitz personal communication). In some *ced-2* and *ced-10* mutants, sperm cannot stay in the spermatheca properly, suggesting decreased sperm motility or a reduced ability of sperm to interact with the matrix. These observations suggest that this group of genes may be involved not only in phagocytosis, but also in cell mobility or from within cell-cell interactions.

Some of the engulfment genes were recently cloned. *ced-5* encodes a protein similar to human DOCK180, which is implicated to function in the extension of cell surfaces (Wu and Horvitz, 1998a). CED-6 protein contains a phophotyrosine binding domain and a proline/serine-rich region, which suggest that it functions as an adaptor molecule (Liu and Hengartner, 1998). *ced-7* encodes a protein similar to ABC transporters (Wu and Horvitz,

1998b). CED-5 and CED-6 were shown to act within engulfing cells, and Ced-7 functions in both dying and engulfing cells.

2.3.2 Nuclease

The activity of *nuc-1* (abnormal nuclease) is required for the degradation of DNA from cells that have undergone programmed cell death. In the *nuc-1* mutant, cell death and engulfment proceed normally but the DNA of the engulfed corpses persists as a compacted mass of Feulgen-reactive material (Hedgecock *et al.*, 1983) . Thus, the activity of the *nuc-1* gene product is not required to initiate programmed cell death, but is needed to remove the debris of cell corpses. It is not known whether the nuclease is provided by the engulfing cells or within the dying cells. The *nuc-1* gene has not been cloned, thus it is not known whether it encodes an endonuclease or a positive regulator of the nuclease. Biochemical analysis has shown that the *nuc-1* gene encodes or controls the activity of a Ca^{2+}-, Mg^{2+}-independent deoxyribonuclease, which may be a lysosomal enzyme judging from its low pH optimum (pH 4.5) (Hevelone and Hartman, 1988) . The *nuc-1* nuclease does not function exclusively in programmed cell death, but is also required in the gut for the worm to digest the DNA of the bacteria that they have eaten.

2.4 Genes required for cell death in specific cell types

2.4.1 ces-1 and ces-2

Genetic evidence suggests that activation of the cell death program may be controlled in a cell-specific manner. Characterisation of mutations that specifically block cell death in the pharynx has resulted in the identification of two genes, *ces-1* and *ces-2* (cell death specification) that control the decision to live or die in only a few cells (Ellis and Horvitz, 1991) . Gain-of-function mutations in *ces-1* and loss-of-function mutations in *ces-2* lead to the survival of the NSM-neuron sister cells, which undergo programmed cell death during embryogenesis in the wildtype embryo. The *ces-1* gain-of-function mutations also lead to the survival of the sisters of the two pharyngeal I2 neurons. Loss of *ces-1* function causes no obvious abnormal phenotype. Genetic experiments suggest that *ces-2* acts as a negative regulator of *ces-1*. In double mutants of *ces-1(lf)* with *ced-3* or *ced-4*, the NSM sisters survive, indicating that *ces-1* acts upstream of the main cell death machinery, consistent with the proposed role of *ces-1* in controlling cell death specifically in the NSM and I2 sisters.

ces-2 encodes a bZIP (basic and zipper domain) family transcription factor (Metzstein *et al.*, 1996) . The bZIP domain of CES-2 is most similar to that of the PAR (proline- and acid-rich domain) family of transcriptional activators [hepatic leukemia factor (HLF), albumin D-element binding protein (DBP), and thyrotroph embryonic factor/vitellogenin gene binding protein (TEF/VBP)], as well as to that of transcriptional repressors encoded by the *Drosophila* gap gene *giant* and the mammalian gene E4BP4. Outside of the bZIP region, CES-2 does not have any sequence similarity to any other proteins in the database. Notably, although the other PAR family transcription factors possess a conserved PAR domain of unknown function, CES-2 does not have one. The CES-2 binding sequence was determined and found to be very similar to the PAR family target sequence.

A fusion protein containing the transactivation domain of the transcription factor E2A and the bZIP domain of the HLF, which is formed by a chromosomal translocation in acute lymphoblastic leukemia, has been shown to inhibit programmed cell death in mammalian cells (Inaba *et al.*, 1996) . These findings raise an attractive possibility that an evolutionarily conserved family of transcription factors plays an important role in determining whether a given cell lives or dies. Genetically, *ces-2* negatively regulates the activity of the *ces-1* gene (Ellis and Horvitz, 1991) , which in turn negatively regulates the activities of downstream pro-death genes. Such negative regulation of *ces-1* by *ces-2* might be mediated by the direct transcriptional repression of *ces-1*.

2.4.2 egl-1

In *egl-1* (egg-laying defective) gain-of-function mutants, a pair of neurons in the hermaphrodite inappropriately undergo programmed cell death (Desai and Horvitz, 1989) . These neurons called the hermaphrodite-specific neurons (HSN), normally innervate the vulval muscles and drive egg laying, thus the loss of HSNs results in the egg-laying defect (Egl phenotype). In males, the HSNs undergo programmed cell death. One hypothesis is that the death of HSNs is the result of a sexual transformation of cell fate. Recent results, however, suggest another, more intriguing possibility. Conradt and Horvitz isolated an *egl-1* loss-of-function mutation (Conradt and Horvitz 1998). Strikingly, *egl-1(lf)* not only suppresses the Egl phenotype, but also blocks most, if not all, programmed cell death during development. Genetic analysis suggests that *egl-1* acts upstream of *ced-9*, *ced-3*, and *ced-4*, and thus, is likely to be a new component of the general cell death machinery in *C. elegans*. The cloning of the *egl-1*gene revealed that it encodes a protein with a region similar to the BH3 domain of Bcl-2. It is proposed that EGL-1 activates cell death by binding to and directly inhibiting the activity of CED-9, perhaps by releasing CED-4 from a CED-9/CED-4 complex.

The genes described above were isolated from only a small fraction of all the cell types that undergo programmed cell death. It seems reasonable to assume that many more cell-type-specific cell death genes remain to be identified.

2.5 Pathological cell death in *C. elegans*

"Pathological death" occurs by different mechanism(s) from the programmed cell death described above. Pathological cell death in *C. elegans* can be categorised into three types: (1) murders, (2) death dependent on engulfment genes, and (3) necrotic death.

2.5.1 Murders

As described, most cells destined to die do not differentiate before they undergo cell death in *C. elegans*. Exceptions include a few cells in males that function to make the male reproductive system before they die. The death mechanism in these cells appears different from the normal programmed cell death - it is not 'suicide,' but 'murder' by neighbouring cells. The linker cell, one of the cells that is 'murdered,' is required to guide the developing male gonad to the tail, where the reproductive and digestive systems fuse (Kimble and Hirsh, 1979) . Once the destination is reached, the linker cell dies and is engulfed by either of the two cells (U.lp or U.rp) of the proctodeum, resulting in the opening of the channel between the vas deferens and the cloaca. The death of the linker cell is dependent on engulfment by the U.lp/U.rp cells. If the progenitor of the U.lp/U.rp cells is eliminated by laser ablation, or if the linker cell is prevented from reaching the proctodeum by mutation, the linker cell fails to die (Sulston and White, 1980) . Furthermore, the activities of the *ced-3* and *ced-4* genes are not necessary for linker cell death, supporting the hypothesis that this death occurs by a mechanism distinct from normal programmed cell death (Ellis and Horvitz, 1986) .

2.5.2 Death dependent on engulfment genes

The deaths induced by the gain-of-function mutations *lin-24* and *lin-33* are morphologically and genetically distinct from programmed cell deaths. These mutations can cause the death of Pn.p cells (a group of postembryonically derived precursor cells that generate hypodermal and vulval cells), resulting in a vulvaless phenotype (Ferguson and Horvitz, 1985; Ferguson *et al.*, 1987) . Under DIC optics, the nuclei of Pn.p cells in these mutants temporarily increase in refractivity for a few minutes to up to 3 hours (Kim, 1994) . Once the refractivity decreases, the nuclei become granular and some of the Pn.p cells die. The nuclei of the surviving Pn.p cells often remain abnormally small and fail to undergo normal division patterns. Electron microscopy has shown the necrotic nature of the refractile Pn.p cells, including swelling of nuclear and mitochondrial membranes and lysis of nuclei (Kim, 1994) .

The Pn.P cell deaths in *lin-24* and *lin-33* mutants do not depend on *ced-3* and *ced-4* activity, however, rather intriguingly, they require activity of *ced-2*, *ced-5*, and *ced-10*, a group of genes that participate in the engulfment of cell corpses as well as some cell migration (Kim, 1994) . One hypothesis is that the *lin-24(gf)* and *lin-33(gf)* mutations do not kill the Pn.p cells directly, but rather change the cells' conditions so that they are recognised as "dying cells" by neighbouring cells in a *ced-2/ced-5/ced-10*-dependent manner and are thus phagocytosed.

2.5.3 Necrosis

Degenerins are a family of Na^+ channels expressed in various neurons in *C. elegans*. Gain-of-function mutations in genes of the degenerin family, such as *mec-4, deg-1*, and *mec-10*, cause the degeneration of a subset of neuronal cells (Chalfie and Sulston, 1981; Chalfie and Au, 1989; Chalfie and Wolinsky, 1990) . Such deaths are distinct from programmed cell death in several ways. The degenerated cells are swollen and enlarged, in contrast to the compacted “button-like” corpses observed in normal programmed cell death. The process of programmed cell death takes less than an hour, whereas degenerative or necrotic deaths occur over several hours. Also, *ced-3* and *ced-4* are not required for degenerin-induced deaths. The identified mutation site in *mec-4(d)* mutations that cause necrosis is always an amino acid substitution at the alanine residue situated adjacent to the transmembrane domain (Driscoll and Chalfie, 1991) . One model for the initiation of necrosis by these alleles is that the degenerin channel cannot close effectively, thus resulting in inappropriate ion influx.

2.6 Approaches to studying cell death in *C. elegans*

The genetic screening for cell death mutants in *C. elegans* has already been very fruitful, with the identification of nearly 20 genes involved in various stages of cell death. Although the main cell death machinery consisting of CED-3, CED-4, and CED-9 has been identified, many questions have yet to be answered: How are the decisions to live or die made? What is the downstream event after the CED-3 protease is activated? What is the mechanism of engulfment? To answer these questions, the search for additional genes involved in this processes is necessary. Some of the approaches to identifying such genes are described below.

2.6.1 Screening of mutations with chromosomal deletions to identify genes involved in cell death

Since previous screens for cell death mutants have been biased against lethal mutations and the genes essential for both cell death and viability might have been overlooked. In addition, mutations that affect only a subset of cells are difficult to isolate because the assay method may not be sensitive enough to detect them. To circumvent these problems, Sugimoto and Rothman performed a screening using a collection of chromosomal deletions (deficiencies) in *C. elegans* (unpublished data). Animals homozygous for most deficiencies are arrested developmentally during embryogenesis, and these phenotypes usually reflect the null phenotype of one or many zygotic genes deleted by the deficiency. By observing the live embryos over time, and recording the number, morphology, and timing of cell deaths, abnormality in programmed cell death can be detected, which implicates the presence of a cell death gene in the deleted region of the genome. This assay would detect zygotic cell death genes whether or not they were essential for viability, as well as subtle or temporary defects that may be difficult to detect by the traditional screening process. More than 70% of the genome is represented by such deletion strains, thus it is possible to rapidly screen a substantial fraction of the genome for cell death genes. This type of deficiency screen for cell death genes in *Drosophila* was successful and resulted in the identification of *reaper, hid,* and *grim* (White *et al.*, 1994; Grether *et al.*, 1995; Chen *et al.*, 1996) .

By observing nearly 60 deficiencies within 70% of the genome, three phenotypic classes were observed: (1) no or significantly fewer cell corpses than wild type (8 loci), (2) excess cell corpses (16 loci), and (3) abnormally large corpses (two loci). The first class included the deficiency that deletes *ced-3*, supporting the validity of this screen. Most deficiencies in class 2 appeared to be engulfment defects, rather than a higher number of cell deaths. This screen indicated that there are genes that have an affect on the process of programmed cell death, in addition to those that have already been identified. Further characterisation of each locus is presently being carried out.

2.6.2 Transgenic approaches to studying cell death in C. elegans

Whereas mammals have multiple caspases and *bcl-2* family members that constitute a complex apoptotic pathway, *C. elegans* appears to have only one cell death pathway, consisting of *ced-3*, *ced-4*, and *ced-9,* which controls all the programmed cell death. The simpler pathway in *C. elegans* can be viewed as a prototype for the control mechanism of programmed cell death in higher organisms. Because of this simple and well characterised genetic pathway and because it readily lends itself to the construction of transgenic animals, *C. elegans* is a useful assay system for testing the activity of candidate cell death factors from various organisms. Baculovirus *p35* and the highly conserved *DAD1* are examples of factors whose cell death activities have been tested in *C. elegans*.

The infection of certain types of insect cells with mutant strains of *Autographa californica* multiply embedded nuclear polyhedrosis virus (AcMNPV), the prototype of the baculovirus family, results in apoptosis of the infected cells (Clem *et al.*, 1991) . These mutant viruses lack a functional *p35* gene. Sugimoto *et al.* constructed transgenic worms in which the *p35* gene from the baculovirus was put under the control of a *C. elegans* heat-shock promoter (Sugimoto *et al.*, 1994) . The expression of *p35* in *C. elegans* blocked programmed cell death very efficiently, thus clearly showing that *p35* itself, not the downstream gene in the virus or the infected cells, possesses the cell-death inhibitor activity. This was one of the first examples of heterologous gene activity testing in *C. elegans*. Later, *p35* was shown to be a very potent inhibitor of not only CED-3 but also of most known caspases (Bump *et al.*, 1995; Xue *et al.*, 1996) .

DAD1 (defender against apoptotic death) is a gene that was originally found as a temperature-sensitive mutation in a hamster cell line that undergoes apoptosis at the restrictive temperature, suggesting that DAD1 is required to protect these cells from apoptosis (Nakashima *et al.*, 1993) . To test whether DAD1 expression itself is sufficient to prevent cell death, human DAD1 as well as its *C. elegans* homologue (*dad-1*) was ectopically expressed in developing *C. elegans* embryos (Sugimoto *et al.*, 1995) . *DAD1/dad-1* expression partially prevented programmed cell death in *C. elegans*, showing that *DAD1/dad-1* has activity (though not as potent as *p35* or *ced-9*) as a cell death inhibitor. DAD1 homologues have also been identified in various organisms including plants and yeast. The yeast protein, Ost2p is a subunit of oligosaccharyltransferase (OST), an enzyme required for N-linked protein glycosylation in the endoplasmic reticulum (Silberstein *et al.*, 1995) . Similarly, human DAD1 was shown to be a subunit of the mammalian OST (Kelleher and Gilmore, 1997) . The inhibition of N-glycosylation by tunicamycin, however, does not necessarily induce apoptosis in mammalian cells (Makishima *et al.*, 1997) . In addition, unlike *C. elegans* or plant (rice and *Arabidopsis*) DAD1 homologues, the yeast Ost2 gene does not complement the mammalian DAD1^{ts} mutant cell line (Makishima *et al.*, 1997) , suggesting that DAD1 homologues with the exception of Ost2p might have functions in addition to their role as a subunit of OST. The function of *dad-1* in *C. elegans* is N-glycosylation, and its mode of action in interfering with the cell death machinery have yet to be characterised.

3.0 ACTIVATION MECHANISMS FOR CASPASES

3.1 Caspases

3.1.1 Caspases in Drosophila melanogaster

A genetic approach has been taken to screen a large fraction of the chromosomal regions that are involved in executing the cell death program in *Drosophila*. The chromosomal region 75C1,2 contains at least three pro-apoptotic genes, *rpr, hid,* and *grim* (White *et al* 1994; Grether *et al* 1995; Chen *et al.*, 1996). Overexpression of any one of these genes can induce cell death. All three may activate caspases that are inhibited by *p35*, suggesting that the regulation of caspase activity is crucial for cell death in *Drosophila* (Grether *et al* 1995; Chen *et al.*, 1996; White *et al.*, 1994; Pronk *et al.*, 1996; Hay *et al.*, 1995). *Drosophila* caspase-1 (DCP-1) was cloned and found to be structurally similar to caspase-3 and -6 (37% identity to both caspase-3 and -6) (Song *et al.*, 1997). Like caspase-3 and -6, DCP-1 has a short prodomain (33 amino acids). Most mammalian caspase family members have a QACRG pentapeptide in their active site, except for caspase-8 (QACQG), -9 (QACGG) and -10 (QACQG). DCP-1 contains a QACQG pentapeptide (Figure 4). Overexpression of an N-terminus truncated DCP-1 induces death in HeLa cells and the caspase inhibitor Ac-DEVD-CHO prevents PARP cleavage by DCP-1, whereas Ac-YVAD-CHO was ineffective. This result suggests that DCP-1 is a subfamily member of caspase-3 and CED-3. A loss-of- function mutation of DCP-1 was identified in P-element insertion mutants. The null mutation of DCP-1 is lethal during larval stages and melanotic tumours can be observed in the DCP-1 mutant, showing that *dcp-1* is essential for normal development (Song *et al.*, 1997).

Another caspase member, drICE was identified in *Drosophila* (Fraser and Evan, 1997). Like DCP-1, drICE contains the pentapeptide QACQG in its active site and shows highest identity (38.9%) with caspase-3 and -6. drICE contains a longer prodomain (80 amino acids) than DCP-1. The N-terminal region (S32-Y69) of drICE contains 30.8% Ser, a structural feature similar to that of CED-3 (36.5% Ser in S132-G206). Like DCP-1, the overexpression of N-terminal truncated drICE causes death in S2 cells. Cell-free lysates were prepared from *Drosophila melanogaster* S2 cells. Lysates prepared from cells undergoing cell death by *reaper* or cycloheximide contain a caspase activity with DEVD specificity and can trigger chromatin condensation in isolated nuclei. Immunodeprivation of drICE from this lysate is sufficient to eliminate most measurable apoptotic activity. These results suggest that drICE is an effector of apoptosis in S2 cells (Fraser *et al.*, 1997).

3.1.2 Function of worm genes in insect cells

To investigate whether CED-3 and caspase-1 elicit cell death in *Drosophila*, transgenic flies were created in which *ced-3* or *caspase-1* was ectopically expressed using the GAL4-UAS system. Ectopic expression of *ced-3* or *caspase-1* could induce death in *Drosophila*, which was rescued by coexpressing the viral anti-apoptotic gene *p35* (Shigenaga *et al.*, 1997). The ability of members of the *bcl-2* family to prevent the cell death induced by *ced-3* or *rpr* has also been tested in *Drosophila* cultured cells (Hisahara *et al.*, 1998). In a transient transfection assay in *Drosophlia* cells, *ced-9* prevented the *ced-3*-induced cell death more efficiently than *bcl-xL*. *ced-9*'s similarity in structure and subcellular localisation with *bcl-2* suggests that *ced-9* might function as an anti-apoptotic gene in mammals. However, the overexpression of *ced-9* could not prevent the cell death induced by *ced-3* in mammalian cells. This lack of anti-apoptotic function in mammalian cells suggests that *ced-9* requires other molecules to exert its anti-apoptotic functions (Hisahara *et al.*, 1998). Seshagiri *et al* (1997) reported that *ced-9* cannot prevent cell death in Sf-21 cells if *ced-3* is used alone to induce apoptosis but does work if *ced-3* and *ced-4* are used together. These results suggest that regulatory components are required for *ced-9* function, and that most likely a CED-4 homologue might be required for *ced-9*'s anti-apoptotic function. Such a factor may exist in SL2 cells but not in Sf-21 cells. In *C. elegans*, *ced-9*'s suppression of cell death due to *ced-3* overexpression is partly mediated via *ced-4* activity. CED-9 and CED-4 regulate the activation of CED-3 through physical interactions (Chinnaian *et al.*, 1997a; Wu *et al.*, 1997a, b, James *et al.*, 1997, Spector *et al.*, 1997). SL2 cells offer a unique and useful assay system for identifying additional genes, such as *ced-4* homologues, that are important in the *ced-9* regulation of *ced-3*-induced cell death.

SL2 cells are also useful for analysing to what extent cell death machinery has been conserved through evolution. Apaf-1, a human protein that participates in the cytochrome c-dependent activation of caspase-3 *in vitro,* has been cloned (Zou *et al.*, 1997). Part of its protein sequence shows significant similarities with the gene encoding CED-4 (Figure 5). Whether CED-3 and CED-9 physically bind to Apaf-1 has not yet been tested. It is possible that CED-9 cannot exert its function through this mammalian homologue of CED-4. In *Drosophila*, the overexpression of *ced-4* induces cell death and endogenous caspase activity, and *ced-4*-induced cell death can be prevented by the caspase inhibitory gene *p35* (Kanuka and Miura, unpublished observation). These observations strongly support the idea that the molecular mechanisms of caspase activation that lead to cell death have been conserved through evolution.

3.1.3 Interaction of caspases with death signalling receptors

The TNF receptor and Fas are receptors that are known to transduce death signals within the cytoplasm. Both receptors contain a death domain (DD) near their intracellular C-terminus. This DD (about 90 amino acid residues) is necessary to transduce death signals (Itoh and Nagata, 1993; Tartaglia *et al.*, 1993). Mutations in this region prevent cell death induction, and the expression of transfected cDNA encoding the DD is sufficient for inducing cell death (Boldin *et al.*, 1995). The DD exerts its function via interaction with other DD-containing proteins, which serve as adaptors in the signalling cascade. Fas recruits a DD-containing adaptor protein called MORT1/FADD (Chinnaiyan *et al.*, 1996; Hsu *et al.*, 1996). TNFR1 recruits a DD-containing adapter protein called TRADD (Hsu *et al.*, 1995). TRADD can bind to MORT1/FADD. The prodomains of caspase-8 and -10 can interact with MORT1/FADD through a motif called the death effector domain (DED), or the MORT motif (this motif is also found in the N-terminus of MORT1/FADD) (Muzio *et al.*, 1996; Boldin *et al.*, 1996). Caspase-8 was found to be one of the components of the death-inducing signalling complex (DISC), into which Fas/Apo-1 recruits a set of intracellular signalling proteins (Medema *et al.*, 1997). The binding of caspase-8 to MORT1/FADD is not sufficient to cleave and activate caspase-8. However, caspase-8 can be cleaved after binding to DISC. Caspase-8 activity can be inhibited by Ac-DEVD-CHO but not by Ac-YVAD-CHO *in vitro*, indicating that caspase-8 has a caspase-3-like substrate specificity (Medema *et al.*, 1997) (Figure 6). Overexpression of CrmA prevents cell death by Fas/Apo1, however it cannot inhibit the activation of caspase-8 (Medema *et al.*, 1997). This result suggests that CrmA-sensitive caspase-1-like protease must be located downstream of caspase-8 activation (Figure 3). An newly identified caspase, FLICE2, shares overall structural and functional similarity with caspase-8 (Vincenz and Dixit, 1997).

Caspase-2 has been found to interact with a molecule called RAIDD/CRADD, which has a homologous region with the prodomain of caspase-2 and CED-3 at its N-terminus (Duan and Dixit, 1997; Ahmad *et al.*, 1997) (Figure 6). RAIDD/CRADD has a dual-domain structure [N-terminus: CRAD (*c*aspase and *R*ip *ad*aptor) motif, C-terminus; death

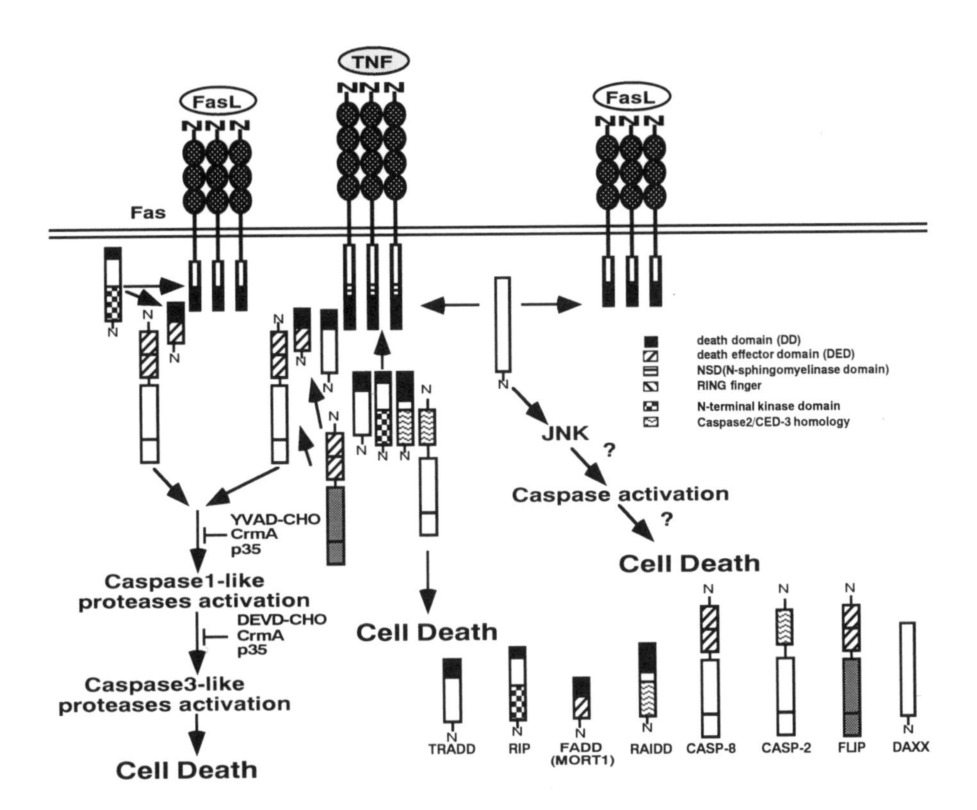

Figure 6. Signalling cascade of Fas and TNFR1.

Binding of TNF or FasL to their receptor induces receptor trimerisation. Adaptor molecules are recruited to the receptor death domain. At least three distinct death signals can be transduced into the cytoplasm: (1) activation of caspase-8 through FADD, (2) activation of caspase-2 through RAIDD, and (3) activation of JNK through DAXX.

domain]. The C-terminal death domain of RAIDD/CRADD binds to the homologous domain in RIP, a serine/threonine kinase component of the apoptosis pathway. Mutation of Leu 27 and Gly 65 in the prodomain of CED-3 has been found in the loss-of-function mutation allele of *ced-3* in *C. elegans* (*n1040* and *n718*, respectively). Mutations of residues corresponding to Leu 27 and Gly 65 in CED-3 were made in both RAIDD and caspase-2. These mutations eliminated the ability of RAIDD to bind CED-3 or caspase-2. These results suggest that these residues are important for the binding of these CRAD motif-containing proteins. RAIDD can function as an adaptor molecule in recruiting caspase-2 to the TNFR1 signalling complex. TNFR1 forms a complex with RAIDD and caspase-2 in the presence of TRADD and RIP. This complex does not transduce signals to activate NF-κβ (Duan and Dixit, 1997). The CARD motif (the same motif referred to as CRAD) is also found in CED-4, a death-regulatory protein in *C. elegans*. It is suggested that CED-4 can directly bind to CED-3 via its CARD motif. CED-4 also binds to caspase-8 and -1 (Chinnaiyan *et al*., 1997a). Deletion of the prodomain from CED-3, caspase-1, and caspase-8 does not prevent them from binding to CED-4 (Chinnaiyan *et al*., 1997a). These results suggest that CED-4 can bind to several caspases in a CARD motif- independent manner.

A novel protein, Daxx, can specifically bind to the Fas death domain. Overexpression of Daxx enhances Fas-induced cell death and activates the Jun N-terminal kinase pathway (Yang *et al*., 1997) (Figure 6). The Daxx apoptotic pathway is blocked by both *bcl-2* and a dominant-negative JNK pathway component and acts co-operatively with the FADD pathway. Fas-induced cell death may be mediated by at least two distinct apoptotic pathways such as Daxx and FADD. Daxx is sensitive to *bcl-2* but not to FADD. This different sensitivity may explain why the effects of *bcl-2* on Fas-mediated cell death varies depending on the cell lines used. Viruses have evolved anti-apoptotic genes to avoid causing the death of host cells. Both the cowpox virus anti-apoptotic gene *crmA* and baculovirus *p35* block the host's apoptotic response by inhibiting the caspase family (Xue and Horvitz, 1995) .

Virus FLICE inhibitory proteins (v-FLIPs) have been found in γ-herpesviruses and tumorigenic human molluscipoxviruses (Thome *et al*., 1997). v- FLIPs contain two death-effector domains that interact with the adaptor protein FADD. Expression of v-FLIPs inhibit the recruitment and activation of caspase-8, and prevent the cell death induced by Fas (Thome *et al*., 1997). It has been suggested that prevention of cell death in virus-infected cells by the expression of v-FLIPs during the lytic viral replication cycle may lead to higher virus production and contribute to the persistence and oncogenicity of several FLIP-encoding viruses. Cellular counterparts of FLIP (named FLAME-1, Casper, CASH, I-FLICE, CLARP and MRIT) have also been found (Irmler *et al.*, 1997b; Srinivasula *et al.*, 1997; Shu *et al.*, 1997; Hu *et al.*, 1997; Goltsev *et al.*, 1997, Inohara *et al.*, 1997; Han *et al.*, 1997). There are two splicing variants, designated $FLIP_S$ and $FLIP_L$ (Irmler *et al.*, 1997b). $FLIP_S$ contains two death-effector domains and is structurally similar to v-$FLIP_S$. $FLIP_L$ contains a caspase-like domain in which the active-site cysteine is replaced by a tyrosine residue. The overall structure is similar to that of caspase-8 except that it lacks several conserved amino acids found in all caspases (Figure 3). $FLIP_S$ and $FLIP_L$ can interact with FADD and caspase-8, and potently inhibit the apoptosis induced by Fas and TNF (Irmler *et al.*, 1997b; Srinivasula *et al.*, 1997; Hu *et al.*, 1997). In this case, $FLIP_L$ may act as a dominant-negative inhibitor of Fas-induced apoptosis. However, in one circumstance Casper can induce apoptosis. It interacts with FADD, caspase-8, caspase-3, TRAF-1, and TRAF-2 through distinct domains (Shu *et al.*, 1997). The overexpression of full-length Casper in HeLa cells induces apoptosis (Shu *et al.*, 1997; Goltev *et al.*, 1997), however, a C-terminus deletion mutant (Casper 1-435) prevents cell death induced by TNF (Shu *et al.*, 1997;). Casper interacts with caspase-8 and -3. Caspase-8 and -3 alone do not interact, however, caspase-8 can interact with caspase-3 in the presence of the protease-like domain of Casper. These results suggest that the protease-like domain of Casper can recruit caspase-3 to the protease domain of caspase-8 (Shu *et al.*, 1997).

3.1.4 Caspases and their substrates

Caspases have unique substrate specificities and show an absolute requirement for Asp in the P1. In case of caspase-1, the amino acids Cys-285 and His-237 involved in catalysis, and those involved in forming the P1 carboxylate binding pocket (Arg-179, Gln-283, Arg-341, and Ser-347), are conserved in all the other caspases (Wilson *et al.*, 1994). One exception is the conservative substitution of Thr for Ser-347 in caspase-10 (Fernades-Alnemri *et al.*, 1996). This structural conservation explains the absolute requirement for Asp in the P1 position. The residues that form the P2-P4 binding pocket are not well conserved, which suggests that these residues may determine the substrate specificities of different caspases. To define the individual substrate specificities of caspases, a new method has been developed that employs a positional scanning synthetic combinatorial library (PS-SCL). The results obtained divide the caspases into three major groups (Thornberry *et al.*, 1997) (Table 2).

Table 2. Subatrate specificities of caspases

Substrate specificities are determined by using PS-SCL (Thornberry et al., 1997).

Group I caspases are involved in inflammation and cell death. Group II caspases work at the effector phase of cell death. Group III caspases amplify the death signal.

	caspase	optimal sequence
Group I	caspase-1 (ICE)	WEHD
	caspase-4 (TX, ICH-2, ICErel-II)	(W/L)EHD
	caspase-5 (TY, ICErel-III)	(W/L)EHD
Group II	CED-3	DETD
	caspase-3 (CPP32, Yama, apopain)	DEVD
	caspase-7 (MCH-3, ICE-LAP-3, CMH-1)	DEVD
	caspase-2 (ICH-1, NEDD-2)	DEHD
Group III	caspase-6 (MCH-2)	VEHD
	caspase-8 (FLICE, MACH, MCH-5)	LETD
	caspase-9 (ICE-LAP6, MCH-6)	LEHD

The peptide substrate preferences within each group are remarkably similar. Members of group I (caspase-1, -4, and -5) prefer the tetrapeptide sequence WEHD. The optimal peptide sequence for group II (caspase-2, -3, -7, and CED-3) is DEXD. The caspases in group III (caspase-6, -8, and -9) prefer the sequence (L/V)EXD. Most of the endogenous substrates cleaved during apoptosis possess a sequence similar or identical to DEXD. These observations suggest that group II caspases work in the effector phase of apoptosis. For group III, the optimal sequence, (L/V)EXD, resembles the activation sites in effector caspase proenzymes. This suggests that group III enzymes have a role as upstream components in a protolytic cascade that amplifies the apoptotic signal. Group I enzymes may prefer substrates that are involved in inflammation, such as pro-interelukin-1β and pro-interferon-γ inducing factor. Multiple endogenous substrates of group II caspases have been reported (Table 3). Substrates can serve as competitive inhibitors for caspases. For example, baculovirus p35 gene product inhibits a broad range of caspases. The main target for pox virus protein CrmA, a viral serpin, is caspase-1.

Table 3. Substrates of caspases

Substrates
1. Cytokines pro-IL-1β, pro-IGIF
2. Kinases MEKK-1, PAK2, PKCδ, PITSLRE kinase, DNA-PKs
3. Structural proteins lamins, actin, fodrin
4. Oncogenes and tumor suppressor genes mdm2, APC, RB
5. Others PARP, Gas2, D4-GDI, U1-70kDa, cPLA2, DFF, SREBP, DSEB/RF-C140, presenilins, Huntingtin, gelsolin, FTase-α, CED-9, BCL-2, PP2A, β-catenin, FKBP45

It has been shown that CED-9 can be a substrate for CED-3 *in vitro* and mutations in the cleavage sites of CED-9 reduce its ability to protect against cell death. This suggests that the presence of CED-3 cleavage sites in CED-9 is important for CED-9 to function in death prevention in *C. elegans* (Xue and Horvitz, 1997). The structure of the cleaved product of CED-9 (68-280) is similar to mammalian Bcl-2. CED-9 (68-280) can still protect against cell death, and mutations in its BH1 and BH2 domains abolish this protective activity. These results suggest that CED-9 can prevent cell death by acting as a competitive inhibitor of CED-3, and also by a mechanism similar to that used by Bcl-2.

1. Proteins involved in nuclear and cell structure changes.

During the execution phase of apoptosis, structural changes are first observed in the nucleus. Lamins are the major structural proteins of the nuclear envelope, and are cleaved by caspases during apoptosis. This proteolysis of lamins may be responsible for some of the observed nuclear changes. This idea is supported by the observation that inhibition of lamin cleavage by a caspase inhibitor prevents some of these changes (Lazrbnik *et al.*, 1995; Neamati *et al.*, 1995). In addition to lamin, cleavage of other nuclear substrates may be involved in the morphological changes of nuclear apoptosis including chromatin condensation, fragmentation, and margination, and the internucleosomal cleavage of DNA (Kaufman, 1989).

The plasma membrane also changes dramatically during apoptosis. Important cytoskeletal proteins, including actin (Mashima *et al.*, 1995, 1997; Kayalar *et al.*, 1996) and fodrin, a membrane-associated cytoskeketal protein (Martin *et al.*, 1995; Cryns *et al.*, 1996; Vanags *et al.*, 1996), are cleaved. Regulatory proteins that are involved in cytoskeletal organisation are also cleaved by caspases. These include the proteins encoded by the growth arrest specific gene-2 (Brancolini *et al.*, 1995) and p21-activated kinases (PAKs), serine-threonine kinases whose activity is regulated by the small guanosine triphosphatases (GTPase) Rac and Cdc42. PAKs are thought to mediate actin cytoskeketon organisation acting as downstream effectors of the effects of Rac and Cdc42. Apoptosis induces the caspase-mediated proteolytic cleavage of PAK2. Cleavage occurs between the amino-terminal regulatory domain and the carboxyl-terminal catalytic domain. Cleaved PAK2 acts as a constitutively active kinase (Rundel and Bokoch, 1997). Another important actin-regulatory protein is gelsolin. Overexpression of gelsolin can prevent cell death induced by Fas, C2-ceramide, or dexamethasone, and it has been suggested that gelsolin acts upstream of the Caspase-3-like protease to inhibit apoptosis (Ohtsu *et al.*, 1997). Gelsolin was identified as a substrate for Caspase-3 by screening the translation products of small cDNA pools for sensitivity to cleavage by Caspase-3. Gelsolin is cleaved during the apoptosis induced by Fas and TNF. Caspase-3 cleaved-gelsolin severed actin filaments *in vitro* in a Ca^{2+}-dependent manner and expression of the gelsolin cleavage product caused apoptosis. Neutrophils isolated from gelsolin knockout mice exhibited a delayed onset of membrane blebbing and DNA fragmentation. These results suggest that cleaved gelsolin may be one of physiological effectors of the morphologic changes that take place during apoptosis (Kothakota *et al.*, 1997).

2. Proteins involved in cell proliferation.

The retinoblastoma (Rb) protein is phosphorylated by cyclin-dependent kinases. Phosphorylated Rb is inactive in its growth-suppressive function and drives cells through the cell cycle into mitosis. Hyperphosphorylated Rb is cleaved during apoptosis (Janicke *et al.*, 1996; An and Dou, 1996). Several cleavage products of Rb are generated during apoptosis, and one of its cleavage sites is located in the C-terminal DEADG sequence, suggesting that caspase-3 may be involved in this cleavage. Inactivation of Rb by caspases may lead to apoptosis by p53. The mdm2 oncogene product can bind to p53 and negatively regulate its functions in transcription, cell cycle arrest, and apoptosis. The mdm2 oncoprotein is cleaved by caspase-3-like protease during apoptosis. This proteolytic cleavage removes the C-terminal RING finger motif, resulting in the loss of RNA binding activity. However, the p53 binding and inhibition functions of mdm2 are not affected by this cleavage (Chen *et al.*, 1997). The physiological roles of mdm2 cleavage have yet to be elucidated.

3. Proteins involved in kinase cascades.

The Protein kinase C (PKC) family has been divided into classical (cPKC; α, β, γ), novel (nPKC; δ, ε, η, θ, μ), and atypical (aPKC; ξ, λ) groups. The $Ca2^+$-dependent cPKCs contain conserved regulatory regions and are activated by calpains. The $Ca2^+$-independent nPKC and aPKC isoforms lack the C2 regulatory domain and are activated by different means. Protein kinases δ and θ are specifically cleaved by caspase-3 during apoptosis to a catalytically active fragment (Emoto *et al.*, 1995; Ghayur *et al.*, 1996; Datta *et al.*, 1997). Overexpression of the cleaved kinase-active fragment, but not full-length kinase, results in the induction of chromatin condensation, nuclear fragmentation, and cell death.

Jun N-terminal kinases (JNKs) are activated in certain apoptotic situations. In some cases, such as loss of cell-matrix contact and Fas ligation, caspases must be active for the induction of JNK activity (Frisch *et al.*, 1996; Cahill *et al.*, 1996). The protein kinase MEKK initiates the JNK pathway, and MEKK is activated by the deletion of its N-terminal regulatory domain. A DEVD-sensitive caspase that cleaves MEKK-1 is specifically activated when cells undergo apoptosis by losing matrix contact (this phenomenon has been termed "anoikis"). MEKK-1 cleavage by caspase is required for the activation of the kinase activity. Overexpression of the MEKK-1 cleavage product stimulates cell death. A cleavage-resistant mutant of MEKK-1 partially protects cells from anoikis and also prevents the activation of caspase-7. These results suggest that caspases can induce apoptosis by activating MEKK-1, which in turn activates more caspase activity, comprising a positive feedback loop (Cardone *et al.*, 1997).

3.1.5 Involvement of caspases in neural cell death

In *C. elegans*, more than 80% of the total programmed cell death can be observed in the cells derived from neural lineages. In mammalian nervous systems, about 50% of the neurons and oligodendroglial cells that are generated during development die by apoptosis (Oppenheim 1991; Barres *et al.*, 1992). Adult neurons or oligodendroglial cells also undergo apoptosis in response to acute injury, such as ischemia and chronic degenerative disease. The neural tube develops from the neural plate, which invaginates along the length of the embryo. Dead cells are observed throughout the developing chick neural tube and are concentrated in the neural folds. When embryos at the 8-somite stage were cultured *in vitro* in the presence of the caspase inhibitor zVAD-fmk, cell death in the neural tube was drastically reduced, and closure of the neural tube was prevented (Weil *et al.*, 1997). This result suggests that programmed cell death is crucial during this early morphogenetic events.

A similar approach was taken to study the involvement of caspases in cell death in the developing motor neuron (Milligan *et al.*, 1995). When a caspase inhibitor was administered to chick embryos *in ovo*, the naturally occurring motoneuron cell death was inhibited. In addition, a caspase-3 knockout study provided more direct evidence for the importance of caspases in neural cell death. In this case, caspase-3 deficient mice were smaller than the other mice in their litters and died 1-3 weeks after birth. The brain development of caspase-3 deficient mice was profoundly affected, and discernible by embryonic day 12, resulting in a variety of hyperplasias and disorganised cell deployment (Kuida *et al.*, 1996). Pyknotic clusters at the site of major morphogenetic change during brain development were not observed in mutant embryos, suggesting decreased cell death in the absence of caspase-3.

Caspase inhibition has also been shown to protect against cell death from ischemic injury in mice (Friedlander *et al.*, 1997). Transgenic mice were generated that express dominant-negative caspase-1 (C285G) in the brain under the control of a neuron- specific enolase (NSE) promoter. Developmental cell death is not observed in these transgenic mice. After permanent focal ischemia by middle cerebral artery occlusion, the mutant caspase-1^{C285G}-expressing mice (M17Z mice) showed significantly reduced brain injury as well as behavioural deficits when compared to the wild-type controls (Friedlander *et al.*, 1997). To determine whether dominant negative caspase-1 might halt the progression of the amyotrophic lateral sclerosis (ALS)-like syndrome in mice expressing mutant superoxide dismutase [SOD (G93R)], M17Z mice were crossed with SOD (G93R) mice. The double-transgenic mice survived significantly longer following the onset of the disease than SOD (G93R) mice. These results suggest that the expression of dominant-negative caspase-1 in the neurons of mutant SOD mice can slow the symptomatic progression of this disease and delay mortality (Friedlander, 1997).

Oligodendrocytes are the myelin-forming cells in the mammalian central nervous system. About 50% of the oligodendrocytes (OLGs) undergo cell death during normal development. In addition, massive OLG cell deaths have been observed in multiple sclerosis (MS). Tumour necrosis factor (TNF) is thought to be one of the mediators responsible for the damage to oligodendrocytes (OLGs). The addition of TNF-α to primary cultures of OLGs significantly decreased the number of live OLGs. Caspase inhibitors Z-Asp-CH_2-DCB, Ac-YVAD-CHO (a specific inhibitor of caspase-1-like proteases), and Ac-DEVD-CHO (a specific inhibitor of caspase-3-like proteases) enhanced the survival of OLGs treated with TNF-α, indicating that the caspase-1- and the caspase-3-mediated cell-death pathways are activated in TNF-induced OLG cell death (Hisahara *et al.*, 1997). These results suggest that the inhibition of caspases may be a novel approach to treating neurodegenerative diseases such as MS.

3.2 Conservation of the cell death pathway from worm to mammals

The discovery of caspases as common mediators of cell death resulted from a fruitful marriage of the genetic studies of cell death in *C. elegans* and biochemical studies in mammals. Programmed cell death is an irreversible process of cell fate. To execute this process, multicellular organisms developed the proteolytic events that are required for all programmed cell death. The regulatory components of caspase activation in dying cells have yet to be elucidated. Furthermore, the steps in the cell death pathway that follow caspase activation, such as nuclear condensation, DNA fragmentation and engulfment are largely unknown. New genetic screens for identifying the genes involved in cell death in *C. elegans* will provide the components that regulate the execution of the programmed cell death and engulfment processes. The isolation and characterisation of mammalian homologues of such genes will provide a better understanding of the biochemical functions of cell death regulatory genes.

4.0 REFERENCES

Ahmad,M., Srinivasula,S.M., Wang,L. (1997) CRADD, a novel human apoptotic adaptor molecule for caspase-2, and FasL/tumor necrosis factor receptor-interacting protein RIP. *Cancer Res.* **57**: 616-619.

Alnemri,E.S., Livingston,D.J. Nicholson,D.W. *et al.* (1996) Human ICE/CED-3 protease nomenclature. *Cell* **87**: 171

An,B. and Dou,Q.P. (1996) Cleavage of retinoblastoma protein during apoptosis: An interleukin 1β-converting enzyme-like protease as candidate. *Cancer Res.* **56**: 438-442.

Avery,L. and Horvitz,H.R. (1987) A cell that dies during wild-type *C. elegans* development can function as a neuron in a *ced-3* mutant. *Cell* **51**: 1071-1078.

Barres,B.A., Hart,I.K. Coles,H.S.R. *et al.* (1992) Cell death and control of cell survival in oligodendrocytes lineage. *Cell* **70**: 31-46.

Brancolini,C., Benedetti,M. and Schneider,C. (1995) Microfilament reorganisation during apoptosis: the role of Gas2, a possible substrate for ICE-like proteases. *EMBO J.* **14**: 5179-5190.

Boldin,M.P., Mett,I.L., Varfolomeev,E.E. *et al.* (1995) Self-association of the "death domains" of the p55 tumor necrosis factor (TNF) receptor and Fas/APO1 prompts signalling for TNF and Fas/APO1 effects. *J. Biol. Chem.* **270**: 387-391.

Boldin,M.P., Goncharov,T.M., Goltsev,Y.V. *et al.* (1996) Involvement of MACH, a novel MORT1/FADD-interacting protease, in Fas/APO-1- and TNF receptor-induced cell death. *Cell* **85**: 803-815.

Bump,N.J., Hackett,M., Hugunin,M. *et al.* (1995) Inhibition of ICE family proteases by baculovirus antiapoptotic protein p35. *Science* **269**: 1885-1888.
Cardone,M.H., Salvesen,G.S., Widmann,C. *et al.* (1997) The regulation of anoikis: MEKK-1 activation requires cleavage by caspases. *Cell* **90**: 315-323.
Cerretti,D.P., Kozlosky,C.J., Mosley,B. *et al.* (1992) Molecular Cloning of the Interleukin-1β Converting Enzyme. *Science* **256**: 97-100.
Chahill,M., Peter,M. Kischkel,F. *et al.* (1996) CD95 (APO1/FAS) induced activation of SAP kinases downstream of ICE-like proteases. *Oncogene* **13**: 2087-2096.
Chalfie,M. and Au,M. (1989) Genetic control of differentiation of the *Caenorhabditis elegans* touch receptor neurons. *Science* **243**: 1027-1033.
Chalfie,M. and Sulston,J. (1981) Developmental genetics of the mechanosensory neurons of *Caenorhabditis elegans*. *Dev. Biol.* **82**: 358-370.
Chalfie,M. and Wolinsky,E. (1990) The identification and suppression of inherited neurodegeneration in *Caenorhabditis elegans*. *Nature* **345**: 410-416.
Chen,P., Nordstrom,W., Gish,B. *et al.* (1996) *grim*, a novel cell death gene in *Drosophila*. *Genes Dev.* **10**: 1773-1782.
Chen,L., Marechal,V., Moreau,J. *et al.* (1997) Proteolytic cleavage of the mdm2 oncoprotein during apoptosis. *J. Biol. Chem.* **272**: 22966-22973.
Chinnaiyan,A.M., Tepper,C.G., Seldin,M.F. *et al.* (1996) FADD/MORT1 is a common mediator of CD95 (Fas/APO-1) and tumor necrosis factor receptor-induced apoptosis. *J. Biol. Chem.* **271**: 4961-4965.
Chinnaiyan,A.M., O'Rourke, K., Lane,B.R. *et al.* (1997a) Interaction of CED-4 with CED-3 and CED-9: a molecular framework for cell death. *Science* **275**: 1122-1126.
Chinnaiyan,A.M., Chaudhary,D. O'Rourke,K. *et al.* (1997b) Role of CED-4 in the activation of CED-3. *Nature* **388**: 728-729.
Clem,R.J., Fechheimer,M. and Miller,L.K. (1991) Prevention of apoptosis by a baculovirus gene during infection of insect cells. *Science* **254**: 1388-90.
Conradt,B. and Horvitz,H.R. (1998). The *C. elegans* protein EGL-1 is required for programmed cell death and interacts with the Bcl-2-like protein CED-9. *Cell* **93**: 519-529.
Cryns,V.L., Bergeron,L. Zhu,H. *et al.* (1996) Specific cleavage of alpha-fodrin during Fas- and tumor necrosis factor-induced apoptosis is mediated by an interleukin-1β-converting enzyme/Ced-3 protease distinct from the poly(ADP-ribose) polymerase protease. *J. Biol.Chem.* **271**: 31277-31282.
Datta,R., Kojima,H. Yoshida,K. *et al.* (1997) Caspase-3-mediated cleavage of protein kinase Cθ in induction of apoptosis. *J. Biol. Chem.* **272**: 20317-20320.
Desai,C. and Horvitz,H.R. (1989) *Caenorhabditis elegans* mutants defective in the functioning of the motor neurons responsible for egg laying. *Genetics* **121**: 703-721.
Driscoll,M. and Chalfie,M. (1991) The *mec-4* gene is a member of a family of *Caenorhabditis elegans* genes that can mutate to induce neuronal degeneration. *Nature* **349**: 588-593.
Duan,H. and Dixit,V.M. (1997) RAIDD is a new 'death' adaptor molecule. *Nature* **385**: 86-89.
Ellis,H.M. and Horvitz,H.R. (1986) Genetic control of programmed cell death in the nematode *C. elegans*. *Cell* **44**: 817-829.
Ellis,R.E. and Horvitz,H.R. (1991) Two *C. elegans* genes control the programmed deaths of specific cells in the pharynx. *Development* **112**: 591-603.
Ellis,R.E., Jacobson,D.M. and Horvitz,H.R. (1991) Genes required for the engulfment of cell corpses during programmed cell death in *Caenorhabditis elegans*. *Genetics* **129**: 79-94.
Ellis,R.E., Yuan,J. and Horvitz,H.R. (1991b) Mechanisms and functions of cell death. *Ann. Rev. Cell. Biol.* **7**: 663-698.
Emoto,Y., Manome,Y. Meinhardt,G. *et al.* (1995) Proteolytic activation of protein kinase Cδ by an ICE-like protease in apoptotic cells. *EMBO J.* **14**: 6148-6156.
Ferguson, E.L. and Horvitz, H.R. (1985) Identification and characterisation of 22 genes that affect the vulval cell lineages of the nematode *Caenorhabditis elegans*. *Genetics* **110**: 17-72.
Ferguson,E.L., Sternberg,P.W. and Horvitz,H.R. (1987) A genetic pathway for the specification of the vulval cell lineages of *Caenorhabditis elegans* . *Nature* **326**: 259-267.
Fernandes-Alnemri,T., Armstrong,R.C. Krebs,J. *et al.* (1996) *In vitro* activation of CPP32 and Mch3 by Mch-4, a novel human apoptotic protein with homology to *Caenorhabditis elegans* cell death protein Ced-3 and mammalian interleukin-1β-converting enzyme. *J. Biol. Chem.* **269**: 30761-30764.
Fraser,A.G. and Evan,G.I. (1997) Identification of a *Drosophila melanogaster* ICE/CED-3-related protease, drICE. *EMBO J.* **16**: 2805-2813.
Fraser,A.G., McCarthy,N.J. and Evan,G.I. (1997) drICE is an essential caspase required for apoptotic activity in *Drosophila* cells. *EMBO J.* **16**: 6192-6199.
Friedlander,R.M., Gagliardini,V, Hara,H. *et al.* (1997) Expression of a dominant negative mutant of interleukin-1β-converting enzyme in transgenic mice prevents neuronal cell death induced by trophic factor withdrawal and ischemic brain injury. *J. Exp. Med.* **185**: 933-940.
Friedlander,R.M., Brown,R.H., Gagliardini,V. *et al.* (1997) Inhibition of ICE slows ALS in mice. *Nature* **388**: 31.
Frish,S., Vuori,K. Kelaita,D. *et al.* (1996) A role for Jun-N-terminal kinase in anoikis; suppression by *bcl-2* and *crmA*. *J. Cell Biol.* **135**: 1377-1382.
Goltsev,Y.V., Kovalenko,A.V. Arnold,E. *et al.* (1997) CASH, a novel caspase homologue with death effector domains. *J. Biol. Chem.* **272**: 19641-19644.

Ghayur,T., Hugunin,M. Talanian,R.V. *et al.* (1996) Proteolytic activation of protein kinase Cδ by an ICE/CED3-like protease induces characteristics of apoptosis. *J. Exp. Med.* **184**: 2399-2404.
Grether,M.E., Abrams,J.M., Agapite,J. *et al.* (1995) The *head involution defective* gene of *Drosophila* melanogaster functions in programmed cell death. *Gen. Dev.* **9**: 1694-1708.
Han,D.K.M., Chaudhary,P.M., Wright,M.E. *et al.* (1997) MRIT, a novel death-effector domain-containing protein, interacts with caspases and Bcl-XL and initiates cell death. *Proc. Natl. Acad. Sci. U.S.A.* **94**: 11333-11338.
Hay,B.A., Wasserman,D.A. and Rubin,G.M. (1995) *Drosophila* homolog of baculovirus inhibitor of apoptosis proteins function to block cell death. *Cell* **83**: 1253-1262.
Hedgecock,E.M., Sulston,J.E. and Thomson,J.N. (1983) Mutations affecting programmed cell deaths in the nematode *Caenorhabditis elegans*. *Science* **220**: 1277-1279.
Hengartner,M.O., Ellis,R.E. and Horvitz,H.R. (1992) *Caenorhabditis elegans* gene *ced-9* protects cells from programmed cell death. *Nature* **356**: 494-499.
Hengartner, M.O. and Horvitz, H.R. (1994a) Activation of *C. elegans* cell death protein CED-9 by an amino-acid substitution in a domain conserved in Bcl-2. *Nature* **369**: 318-320.
Hengartner,M.O. and Horvitz,H.R. (1994b) *C. elegans* cell survival gene *ced-9* encodes a functional homolog of the mammalian proto-oncogene *bcl-2*. *Cell* **76**: 665-676.
Hengartner,M.O. (1997) CED-4 is a stranger no more. *Nature* **388**: 714-715.
Hevelone, J. and Hartman, P.S. (1988) An endonuclease from *Caenorhabditis elegans*: partial purification and characterisation. *Biochem. Genet.* **26**: 447-461.
Hisahara,S., Shoji,S. Okano,H *et al.* (1997) ICE/CED-3 family executes oligodendrocyte apoptosis by tumor necrosis factor. *J. Neurochem.* **69**: 10-20.
Hisahara,S., Kanuka,H., Shoji,S. *et al.* (1998) *C. elegans* anti-apoptotic gene *ced-9* prevents *ced-3*-induced cell death in *Drosophila* cells. *J. Cell Sci.* **111**: 667-673.
Hsu,H., Xiong,J. and Goeddel,D.V. (1995) The TNF receptor-1 associated protein TRADD signals cell death and NF-κB activation. *Cell* **81**: 495-504.
Hsu,H., Shu,H.B. Pan,M.G. *et al.* (1996) TRADD-TRAF2 and TRADD-FADD interactions define two distinct TNF receptor-1 signal transduction pathways. *Cell* **84**: 299-308.
Hu,S., Vincenz,C. Ni,J. *et al.* (1997) I-FLICE, a novel inhibitor of tumor necrosis factor receptor-1- and CD-95-induced apoptosis. *J. Biol. Chem.* **272**: 17255-17257.
Inaba,T., Inukai,T., Yoshihara,T., *et al.* (1996) Reversal of apoptosis by the leukaemia-associated E2A-HLF chimeric transcription factor. *Nature* **382**: 541-544.
Inohara,N., Koseki,T., Hu,Y. *et al.* (1997) CLARP, a death effector domain-containing protein interacts with caspase-8 and regulates apoptosis. *Proc. Natl. Acad. Sci. U.S.A.* **94**: 10717-10722.
Irmler,M., Hofmann,K., Vaux,D. (1997a) Direct physical interaction between the *Caenorhabditis elegans* 'death proteins' CED-3 and CED-4. *FEBS Lett.* **406**: 189-190.
Irmler,M., Thome,M. Hahne,M. *et al.* (1997b) Inhibition of death receptor signals by cellular FLIP. *Nature* **388**: 190-195.
Itoh,N. and Nagata,S. (1993) A novel protein domain required for apoptosis. *J. Biol. Chem.* **268**: 10932-10937.
James,C., Gschmeissner,S. Fraser,A. *et al.* (1997) CED-4 induces chromatin condensation in *Schizosaccharomyces pombe* and is inhibited by direct physical association with CED-9. *Curr. Biol.* **7**: 246-252.
Janicke,R.U., Walker,P.A. Lin,X.Y. *et al.* (1996) Specific cleavage of the retinoblastoma protein by an ICE-like protease in apoptosis. *EMBO J.* **15**: 6969-6978.
Kaufmann,S.H. (1989) Induction of endonucleolytic DNA cleavage in human acute myelogenous leukemia cells by etoposide, camptothecin, and other cytotoxic anticancer drugs: a cautionary note. *Cancer Res.* **49**: 5870-5878.
Kayalar,C., Ord,T., Testa,M.P. *et al.* (1996) Cleavage of actin by interleukin 1β-converting enzyme to reverse DNase I inhibition. *Proc. Natl. Acad. Sci. U.S.A.* **93**: 2234-2238.
Kelleher,D.J. and Gilmore,R. (1997) DAD1, the defender against apoptotic cell death, is a subunit of the mammalian oligosaccharyltransferase. *Proc. Natl. Acad. Sci. U S A.* **94**: 4994-4999.
Kim,S. (1994) *Two C. elegans genes that can mutate to cause degenerative cell death"*, Massachusetts Institute of Technology, Cambridge.
Kimble,J. and Hirsh,D. (1979) The postembryonic cell lineages of the hermaphrodite and male gonads in *Caenorhabditis elegans*. *Dev. Biol.* **70**: 396-417.
Kothakota,S., Azuma,T., Reinhard,C. *et al.* (1997) Caspase-3-generated fragment of gelsolin: effector of morphological change in apoptosis. *Science* **278**: 294-298.
Kuida,K., Zheng,T.S., Na,S. *et al.* (1996) Decreased apoptosis in the brain and premature lethality in CPP32-deficient mice. *Nature* **384**: 368-372.
Lawrence,G.L., Finnegan,E.J., Ayliffe,M.A. (1995) The *L6* gene for flax rust resistance is related to the arabidopsis bacterial resistance gene *RPS2* and the tobacco viral resistance gene *N*. *The Plant Cell* **7**: 1195-1206.
Lazebnik,Y.A., Takahashi,A., Moir,R.D. *et al.* (1995) Studies of the lamin proteinase reveal multiple parallel biochemical pathways during apoptotic execution. *Proc. Natl. Acad. Sci. U.S.A.* **92**: 9042-9046.

Lui,Q.A. and Hengartner, M.O. (1998) Candidate adaptor protein CED-6 promotes the engulfment of apoptotic cells in *C. elegans*. *Cell* **93**: 961-972.
Liu,X., Kim,C.N. Yang,J. *et al.* (1996) Induction of apoptotic program in cell-free extracts: requirement for dATP and cytochrome c. *Cell* **86**: 147-157.
Liu,X., Zou,H. Slaughter,C. *et al.* (1997a) DFF, a heterodimeric protein that functions downstream of caspase 3 to trigger DNA fragmentation during apoptosis. *Cell* **89**: 175-184.
Liu,X., Nijhawan,D., Budihardjo,I. *et al.* (1997b) Cytochrome c and dATP-dependent formation of Apaf-1/caspase-9 complex initiates an apoptotic protease cascade. Cell **91**: 479-489.
Makishima,T., Nakashima,T., Nagata-Kuno,K. *et al.* (1997) The highly conserved DAD1 protein involved in apoptosis is required for N-linked glycosylation. *Genes Cells*, **2**: 129-141.
Martin,S.J., O'Brien,G.A. Nishioka,W.K. *et al.* (1995) Proteolysis of fodrin (non-erythroid spectrin) during apoptosis. *J.Biol. Chem.* **270**: 6425-6428.
Mashima,T., Naito,M., Fujita,N. *et al.* (1995) Identification of actin as a substrate of ICE and an ICE-like protease and involvement of an ICE-like protease but not ICE in VP-16-induced U937 apoptosis. *Biochem. Biophys. Res. Commun.* **217**: 1185-1192.
Mashima,T., Naito,M., Noguchi,K. *et al.* (1997) Actin cleavage by CPP-32/apopain during the development of apoptosis. *Oncogene* **14**: 1007-1012.
Medema,P.P., Scaffidi,C., Kischkel,F.C. (1997) FLICE is activated by association with the CD95 death-inducing signalling complex (DISC). *EMBO J.* **16**: 2794-2804.
Mello,C.C., Kramer,J.M., Stinchcomb,D. *et al.* (1991) Efficient gene transfer in *C.elegans*: extrachromosomal maintenance and integration of transforming sequences. *EMBO J.* **10**: 3959-3970.
Metzstein,M.M., Hengartner,M.O., Tsung,N. *et al.* (1996) Transcriptional regulator of programmed cell death encoded by *Caenorhabditis elegans* gene ces-2. *Nature* **382**: 545-547.
Milligan,C.E., Prevette,D. Yaginuma,H. *et al.* (1995) Peptide inhibitors of the ICE protease family arrest programmed cell death of motoneurons. *Neuron* **15**: 385-393.
Miura,M., Zhu,H., Rotello,R. *et al.* (1993) Induction of apoptosis in fibroblasts by IL-1ß-converting enzyme, a mammalian homolog of the *C. elegans* cell death gene *ced-3*. *Cell* **75**: 653-660.
Muzio,M., Chinnaiyan,A.M., Kischkel,F.C. *et al.* (1996) Flice, a novel FADD-homologous ICE/CED-3-like protease, is recruited to the CD95 (Fas/APO-1) death-inducing signalling complex. *Cell* **85**: 817-827.
Nakashima,T., Sekiguchi,T., Kuraoka,A. *et al.* (1993) Molecular cloning of a human cDNA encoding a novel protein, DAD1, whose defect causes apoptotic cell death in hamster BHK21 cells. *Mol Cell Biol*, **13**: 6367-6374.
Neamati,N. Fernadez,A., Wright,S. *et al.* (1995) Degradation of lamin B1 precedes oligonucleosomal DNA fragmentation in apoptotic thymocytes and isolated thymocyte nuclei. *J. Immunol.* **154**: 3788-3795.
Ohtsu,M., Sakai,N. Fujita,H. *et al.* (1997) Inhibition of apoptosis by the actin-regulatory protein gelsolin. *EMBO J.* **16**: 4650-4656.
Oppenheim,R.W. (1991) Cell death during development of the nervous system. *Annu. Rev. Neurosci.* **14**: 453-501.
Pronk,G.J., Ramer,K. Amiri,P. *et al.* (1996) Requirement of an ICE-like protease for induction of apoptosis and ceramide generation by REAPER. *Science* **271**: 303-310.
Robertson,A.M.G. and Thomson,J.N. (1982) Morphology of programmed cell death in the ventral cord of *Caenorhabditis elegans*. *J. Embryol. Exp. Morphol.* **67**: 89-200.
Rundel,T. and Bokoch,G.M. (1997) Membrane and morphological changes in apoptotic cells regulated by caspase-mediated activation of PAK2. *Science* **276**: 1571-1574.
Seshagiri,S. and Miller,L.K. (1997) *Caenorhabditis elegans* CED-4 stimulates CED-3 processing and CED-3-induced apoptosis. *Curr. Biol.* **7**: 455-460.
Shaham,S. and Horvitz,H.R. (1996) Developing *Caenorhabditis elegans* neurons may contain both cell-death protective and killer activities. *Gen. Dev.* **10**: 578-591.
Shigenega,A., Funahashi,Y. Kimura,K.-I. *et al.* (1997) Targeted expression of *ced-3* and *Ice* induces programmed cell death in *Drosophila*. *Cell Death Diff.* **4**: 371-377.
Silberstein,S., Collins,P.G., Kelleher,D.J. *et al.* (1995) The essential *OST2* gene encodes the 16-kD subunit of the yeast oligosaccharyltransferase, a highly conserved protein expressed in diverse eukaryotic organisms. *J. Cell Biol.* **131**: 371-383.
Shu,H.-B., Halpin,D.R. and Goeddel,D.V. (1997) Casper is a FADD- and caspase-related inducer of apoptosis. *Immunity* **6**: 751-763.
Song,Z., McCall,K. and Steller,H. (1997) DCP-1, a *Drosophila* cell death protease essential for development. *Science* **275**: 536-540.
Spector,M.S., Desnoyers,S., Hoeppner,D.J. and Hengartner,M.O. (1997) Interaction between the *C.elegans* cell-death regulators CED-9 and CED-4. *Nature* **385**: 653-656.
Srinivasula, S.M., Ahmad, M. Ottilie, S. *et al.* (1997) FLAME-1, a novel FADD-like anti-apoptotic molecule that regulates Fas/TNFR1-induced apoptosis. *J. Biol. Chem.* **272**: 18542-18545.
Sugimoto,A., Friesen,P.D. and Rothman,J.H. (1994) Baculovirus p35 prevents developmentally programmed cell death and rescues a *ced-9* mutant in the nematode *Caenorhabditis elegans*. *EMBO J.* **13**: 2023-2028.
Sugimoto,A., Hozak,R.R., Nakashima,T. *et al.* (1995) *dad-1*, an endogenous programmed cell death suppressor in *Caenorhabditis elegans* and vertebrates. *EMBO J.* **14**: 4434-4441.

Sulston,J., Du,Z., Thomas,K. *et al.* (1992) The *C. elegans* genome sequencing project: a beginning. *Nature* **356**: 37-41.
Sulston,J. and Horvitz,H. (1977) Post-embryonic cell lineages of the nematode, *C. elegans*. *Dev. Biol.* **56**: 110-156.
Sulston,J.E., Schierenberg,E., White,J.G. *et al.* (1983) The embryonic cell lineage of the nematode *Caenorhabditis elegans*. *Dev. Biol.* **100**: 64-119.
Sulston,J.E. and White,J.G. (1980) Regulation and cell autonomy during postembryonic development of *Caenorhabditis elegans*. *Dev. Biol.* **78**: 577-597.
Tartaglia,L.A., Ayres,T.M. Wong,G.H.W. *et al.* (1993) A novel domain with the 55 kd TNF receptor signals cell death. *Cell* **74**: 845-853.
Thome,M., Schneider,P. Hofmann,K. *et al.* (1997) Viral FLICE-inhibitory proteins (FLIPs) prevent apoptosis induced by death receptors. *Nature* **386**: 517-521.
Thornberry,N.A., Bull,H.G., Calaycay,J.R. *et al.* (1992) A novel heterodimeric cysteine protease is required for interleukin -1β processing in monocytes *Nature* **356**: 768-774.
Thornberry,N.A., Rano,T.A. Peterson,E.P. *et al.* (1997) A combinational approach defines specificities of members of the caspase family and granzyme B. *J. Biol. Chem.* **272**: 17907-17911.
Vanags,D.M., Porn-Ares,M.I. Coppola,S. *et al.* (1996) Protease involvement in fodrin cleavage and phosphatidylserine exposure in apoptosis. *J. Biol. Chem.* **271**: 31075-31085.
Vincenz,C. and Dixit,V.M. (1997) Fas-associated death domain protein interleukin-1β-converting enzyme 2 (FLICE2), an ICE/Ced-3 homologue, is proximally involved in CD95- and p55-mediated death signalling. *J. Biol. Chem.* **272**: 6578-6583.
Weil,M., Jacobson,M. and Raff.M. (1997) Is programmed cell death required for neural tube closure? *Curr. Biol.* **7**: 281-284.
White,K., Grether,M.E., Abrams,J.M. *et al.* (1994) Genetic control of programmed cell death in *Drosophila*. *Science* **264**: 677-683.
Wilson,K.P., Black,J.-A.F., Thomson,J.A. (1994) Structure and mechanism of interleukin-1β-converting enzyme. *Nature* **370**: 270-275.
Wu,D., Wallen,H.D. and Nunez,G. (1997a) Interaction and regulation of subcellular localisation of CED-4 by CED-9. *Science* **275**: 1126-1129.
Wu,D., Wallen,H.D., Inohara,N. *et al.* (1997b) Interaction and regulation of the *Caenorhabditis elegans* death protease CED-3 by CED-4 and CED-9. *J. Biol. Chem.* **272**: 21449-21454.
Wu,Y.C. and Horvitz,H.R. (1998a). *C. elegans* phagocytosis and cell migration protein CED-5 is similar to human DOCK180. *Nature* **392**: 501-504.
Wu,Y.C. and Horvitz,H.R. (1998b). The *C. elegans* cell corpse engulfment gene *ced-7* encodes a protein similar to ABC transporters. *Cell* **93**: 951-960.
Wyllie,A.H., Kerr,J.F. and Currie,A.R. (1980) Cell death: the significance of apoptosis. *Int. Rev. Cytol.* **68**: 251-306.
Xue,D. and Horvitz,H.R. (1995) Inhibition of the *Caenorhabditis elegans* cell-death protein CED-3 by a CED-3 cleavage site in baculovirus p35 protein. *Nature* **377**: 248-251.
Xue,D., Shaham,S. and Horvitz,H.R. (1996) The *Caenorhabditis elegans* cell-death protein CED-3 is a cysteine protease with substrate specificities similar to those of human CPP32 protease. *Gen. Dev.* **10**: 1073-1083.
Xue,D. and Horvitz,H.R. (1997) *Caenorhabditis elegans* CED-9 protein is a bifunctional cell-death inhibitor. *Nature* **390**: 305-308.
Yang,X., Khosravi-Far,R. Chang,H.Y. *et al.* (1997) Daxx, a novel Fas-Binding protein that activates JNK and apoptosis. *Cell* **89**: 1067-1076.
Yuan,J. and Horvitz,H.R. (1992) The *Caenorhabditis elegans* cell death gene *ced-4* encodes a novel protein and is expressed during the period of extensive programmed cell death. *Development* **116**: 309-320.
Zha,H., Aime-Sempe,C., Sato,T. *et al.* (1996) Proapoptotic protein Bax heterdimerises with Bcl-2 and homdimerises with Bax via a novel domain (BH3) distinct from BH1 and BH2. *J. Biol. Chem.* **271**: 7440-7444.
Zou,H., Henzel,W.J., Liu,X. *et al.* (1997) Apaf-1. a human protein homologous to *C. elegans* CED-4, participates in cytochrome c-dependent activation of caspase-3. *Cell* **90**: 405-413.

Chapter 8

Apoptosis in *Drosophila*

John P. Wing and John R. Nambu
Biology Department and Molecular and Cell Biology Program, University of Massachusetts at Amherst, Amherst MA 01003.

1.0 THE *DROSOPHILA* MODEL SYSTEM

Holometabolous insects, such as the dipteran *Drosophila*, exhibit two distinct life forms and undergo a complete metamorphosis. The *Drosophila* embryo develops over the course of one day and hatches into a motile and feeding first instar larva, which grows and subsequently moults twice (Figure 1). The first and second larval instars each last about a day, while the third instar lasts approximately three days. At the end of the third larval instar, the animal ceases to feed, climbs up a suitable substrate, and initiates pupariation. The pupal stage lasts approximately four days, during which time the larva metamorphoses into the adult fly. The ease of rearing *Drosophila*, their fecundity, and short life cycle have all contributed to their use as a genetic model system (see Ashburner, 1989). In addition, the *Drosophila* genome is relatively small ($\sim 1.6 \times 10^8$ base pairs per haploid genome), and is organised into only four chromosomes which are all linked by a common chromocentre. The large polytene chromosomes of the larval salivary gland greatly facilitate the cytological mapping of chromosomal rearrangements as well as the localisation of cloned genes via *in situ* hybridisation. In addition, the use of balancer chromosomes, which repress recombination and provide dominant phenotypic markers, greatly facilitates genetic screens and the stable maintenance of homozygous lethal chromosomes.

While the fruit fly *Drosophila melanogaster* has long been a classic genetic system for studying the mechanisms of inheritance (see Kohler, 1994), it is really over the last twenty years that *Drosophila* has emerged as a pre-eminent model system in developmental biology. The *Drosophila* system has two crucial qualities: the availability of powerful molecular genetic tools to identify and analyse genes that are required for specific developmental processes, and a balance between complexity and accessibility. The ascendance of *Drosophila* developmental biology is due largely to the pioneering work of Christianne Nüsslein-Volhard and Eric Wieschaus, who performed large scale saturation screens that identified dozens of mutations which disrupt the normal body plan of the *Drosophila* larva (Nüsslein-Volhard and Wieschaus, 1980; Jürgens *et al.*, 1984; Nüsslein-Volhard *et al.*, 1984; Wieschaus *et al.*, 1984). These studies were strongly complemented by the prescient work of Ed Lewis to analyse homeotic mutations that alter normal segmental identities (Lewis, 1978). Lewis, Nüsslein-Volhard, and Wieschaus were all honoured for their work with the Nobel Prize for Physiology or Medicine in 1995. The subsequent isolation and analysis of the genes corresponding to these mutations, undertakings facilitated by the chromosome walking techniques developed

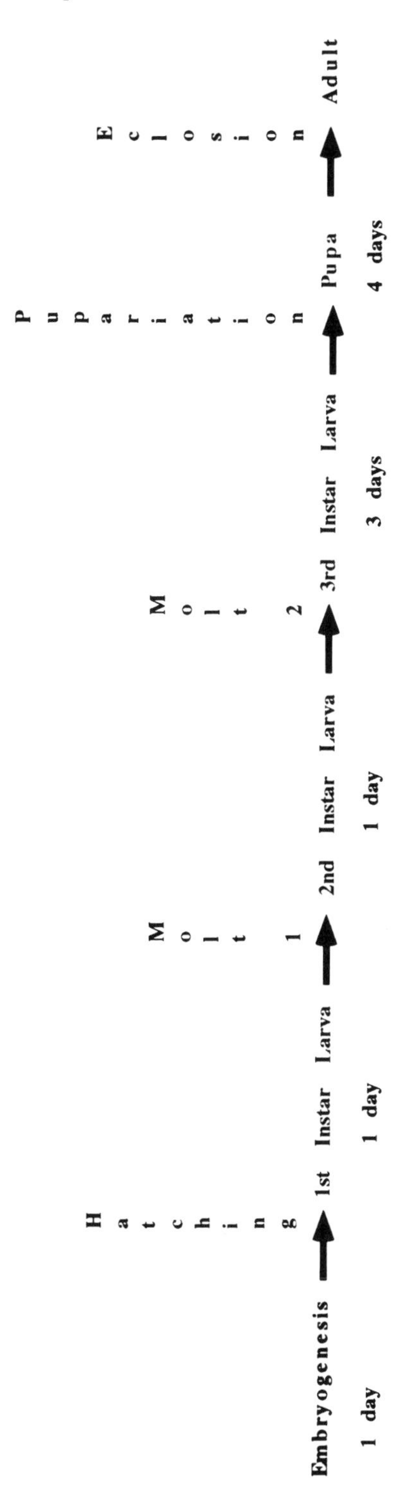

Figure 1. The life cycle of *Drosophila melanogaster*.

by Welcome Bender and David Hogness (Bender *et al.*, 1983a,b), has provided tremendous insight into the mechanisms underlying pattern formation and organogenesis.

Another milestone in *Drosophila* molecular genetics was the development of the P element mediated germ line transformation system by Alan Spradling and Gerald Rubin (Spradling and Rubin, 1982). This technique, which utilises a *D. melanogaster* transpososable DNA element, has served as a cornerstone for facilitating detailed analyses of gene function and regulation. The facility of inducing the transposition or "hopping" of different types of P element constructs has permitted screens for insertions that yield mutant phenotypes and/or specific expression patterns (Bellen *et al.*, 1989; Bier *et al.*, 1989; Wilson *et al.*, 1989), and these insertions readily permit cloning of the corresponding gene. Use of P elements has also permitted development of several other important methodologies, such as targeted gene expression (Brand and Perrimon, 1993) and directed mitotic recombination (Golic and Lindquist, 1989; Xu and Rubin, 1993).

Taken together, these advances have resulted in a system that permits high resolution molecular genetic studies of a wide array of biological processes. The ensuing concerted assault on the biological problems of pattern formation, cell determination, differentiation, and organogenesis, have led to great progress in our understanding of the development of not only this small insect, but many other organisms as well. Several of the most significant findings that have emerged include: molecular determination of a morphogen; elucidation of anterior/posterior and dorsal/ventral patterning mechanisms; the identification of the homeobox and the HOX gene clusters; and the elucidation of several fundamental cell signalling pathways (see Lawrence, 1992; Gerhart and Kirschner, 1997). Significantly, it has become apparent that many of the molecules and pathways used to mediate specific developmental processes in *Drosophila* are conserved across a wide phylogenetic spectrum. Thus, many of the pathways used to regulate cell differentiation and organisation appear to have arisen very early in evolution, and have been largely preserved during the radiation of disparate animal phyla. These findings have had strong implications for understanding normal human development, as well as cancer and disease (Lee *et al.*, 1995; van Leeuwen and Nusse, 1995; Johnson *et al*, 1996; Roessler *et al.*, 1996; Alman *et al.*, 1997; Capobianco *et al.*, 1997; Fan *et al.*, 1997; Oro *et al.*, 1997; Xie *et al.*, 1997), and have caused a re-examination of issues that include the evolution of eyes (reviewed in Halder *et al.*, 1995), and the establishment of dorsal/ventral axes in chordates and arthropods (e.g. DeRobertis and Sasai, 1996). The current availability of conserved gene probes is greatly facilitating examinations of the evolutionary origins and divergence of specific developmental pathways (see Gerhart and Kirschner, 1997).

As we will describe in this review, *Drosophila* is also making significant contributions to our understanding of another fundamental biological process, programmed cell death (PCD). Programmed cell death was a term originally coined to describe the death of insect intersegmental muscles after eclosion (Lockshin and Williams, 1965). These cells exhibit an endocrine controlled pattern of death that is dependent upon new protein synthesis (Lockshin, 1969; Schwartz *et al.*, 1990). The death of these cells appears to occur via pathways that may be distinct from those of apoptosis (Schwartz *et al.*, 1993), a common form of PCD that occurs in many cell types, and is characterised by stereotypic changes in cellular morphology and chromatin degradation (Kerr *et al.*, 1972; Wyllie *et al.*, 1980). In this review, unless otherwise referred to, we will concentrate on a discussion of apoptotic cell deaths in *Drosophila*.

Cell death permits the controlled elimination of unwanted or deleterious cells in a number of developmental, homeostatic, and pathological contexts. While cell death has been recognised as a developmental process for over a hundred years, only recently has it received widespread experimental attention. Great progress has been made over the last decade in identifying and analysing the components of a conserved cellular death machinery (reviewed in Steller, 1995; Chinnaiyan and Dixit, 1996; Vaux and Strasser, 1996; White E., 1996; Yuan, 1997). In addition, it has become more evident that a variety of diseases, such as cancer, AIDS, neurodegenerative and autoimmune disorders, are associated with aberrant patterns of cell death (reviewed in Bellamy *et al.*, 1995; Fisher *et al.*, 1995; Thompson, 1995; Rudin and Thompson, 1997). Thus it is of great clinical significance to understand the mechanisms of cell death and devise approaches to effectively manipulate them.

Despite the powerful genetic tools available, *Drosophila* has not traditionally been a key model system for studying cell death. Instead, due to the pioneering work of Robert Horvitz, the nematode *C. elegans* emerged as the most significant genetic system for the identification of key cell death regulators (reviewed in Ellis *et al.*, 1991; Driscoll, 1992; Hengartner and Horvitz, 1994; Horvitz *et al.*, 1994; Osborne, 1996). As discussed by Miura and Sugimoto in another chapter in this book, genetic screens for cell death mutants in *C. elegans* led to the identification of several *cell death abnormal* (*ced*) genes, including *ced-3*, *ced*-4, and *ced*-9, that play essential and evolutionarily conserved roles in mediating cell death (Yuan and Horvitz

1990; Hengartner *et al.*, 1992; Vaux *et al.*, 1992; Yuan *et al.*, 1993; Hengartner and Horvitz, 1994; Zou *et al.*, 1997). However, as the overall importance of cell death has been more widely recognised, *Drosophila* has also begun to demonstrate its strong experimental utility in exploring this area. Several important and novel cell death regulators have been identified in flies, and this system promises to provide great insight into the function of cell death during normal developmental processes. In addition, as discussed later, *Drosophila* may be a useful system for bridging the gap between the relatively simple *C. elegans* and more complex animals, such as mammals.

1.1 Identification of dying cells in *Drosophila*

An important aspect of studying cell death is the ability to identify dead and dying cells. There are a number of different methods to identify dying cells that rely on corresponding changes in morphological properties, membrane permeability, chromatin degradation, and gene expression. While most of these techniques are generally applicable to a diverse set of experimental systems, we will focus here on several that have been commonly used to analyse dying cells in *Drosophila* (reviewed in Abrams *et al.*, 1993). Traditionally, examination at the light and electron microscope level were used to identify dying cells in both embryonic and postembryonic tissues (e.g. Demerec 1950; Campos-Ortega and Hartenstein, 1997). The dying cells are generally intermingled with living cells, but can be distinguished due to their overall shrinkage, separation from neighbouring cells, fragmentation into membrane bound bodies, and electron dense nuclei. In addition, dying cells often appear pyknotic, and are flattened and rounded with a cytoplasmic clear area around the periphery. The mitochondria of these cells are generally intact. The cellular fragments or apoptotic bodies generated via apoptosis are quickly phagocytosed by neighbouring cells or circulating macrophages (Abrams *et al.*, 1993; Tepass *et al.*, 1994; Campos-Ortega and Hartenstein, 1997).

There are a number of histological reagents that facilitate the identification of dying cells in a heterogeneous cell population. Toluidine blue can be used in fixed and sectioned tissue; it labels the cytoplasm of both isolated and engulfed dying cells. The dying cells are intensely stained compared to their surrounding surviving neighbours. It is also possible to label dying cells in live tissues using several vital dyes that are generally excluded from living cells. These dyes include the DNA intercalating agent acridine orange (AO), as well as nile blue. AO has been widely used to assess the patterns of cell death in many *Drosophila* tissues. This agent yields a high signal-to-noise level, although there is significant autofluorescence of the yolk in embryos. Significantly, AO staining was found to correspond well with the labelling pattern obtained via other histological markers for PCD and does not appear to label cells undergoing necrotic cell death, such as that induced by hypoxia (Abrams *et al.*, 1993). Both individual dying cells as well as engulfed apoptotic bodies present in macrophages are labelled with AO, and time lapse studies suggest that the label is present in unengulfed cells for 30-45 minutes while labelling of engulfed corpses can persist up to 2 hours (Abrams *et al.*, 1993). Due to the strong labelling of engulfed corpses, some care must be taken to distinguish between labelled macrophages and isolated dying cells. This is particularly true since it is not clear what the stoichiometry is between the number of apoptotic bodies generated from a dying cell, and the number of macrophages that engulf them.

Other techniques that have been used for labelling dying cells in *Drosophila* are those that rely on the generation of fragmented chromosomal DNA during apoptosis (Wyllie *et al.*, 1984). These include both TUNEL and *in situ* nick translation. TUNEL relies on the enzyme terminal deoxynucleotidyl transferase to incorporate labelled nucleotides in a non-template dependent fashion from a double stranded DNA break (Gavrieli *et al.*, 1992). The *in situ* nick translation technique utilises the enzyme DNA polymerase I to incorporate labelled nucleotides from single-stranded DNA breaks into an extended DNA chain. Various histochemical techniques can then be used to detect the presence of labelled nucleotides in the nuclei of dying cells (Figure 2). Similar to AO labelling, these techniques detect both isolated and engulfed dying cells. These techniques can be used in sectioned or whole mount material, and due to the fixation of tissues in these preparations, they can provide very high levels of cellular resolution.

1.2 Patterns of cell death in *Drosophila*

1.2.1 Postembryonic cell death is regulated by hormonal cues

Much of the study of cell death in insects has traditionally focused on descriptions of the massive histolysis of larval tissues that occur during pupariation, where there is a dramatic reorganisation of the larval body plan to give rise to the adult animal. Postembryonic cell deaths in insects are largely governed by hormonal cues that also influence the timing and types of moults. In *Drosophila*, changes in the circulating levels of ecdysone (20-hydroxyecdysteroid) play a crucial role in regulating cell death. An increase in ecdysone titres at the end of the third instar larval stage triggers cell death in many tissues during pupariation (e.g. Truman *et al.*, 1992; Truman *et al.*,

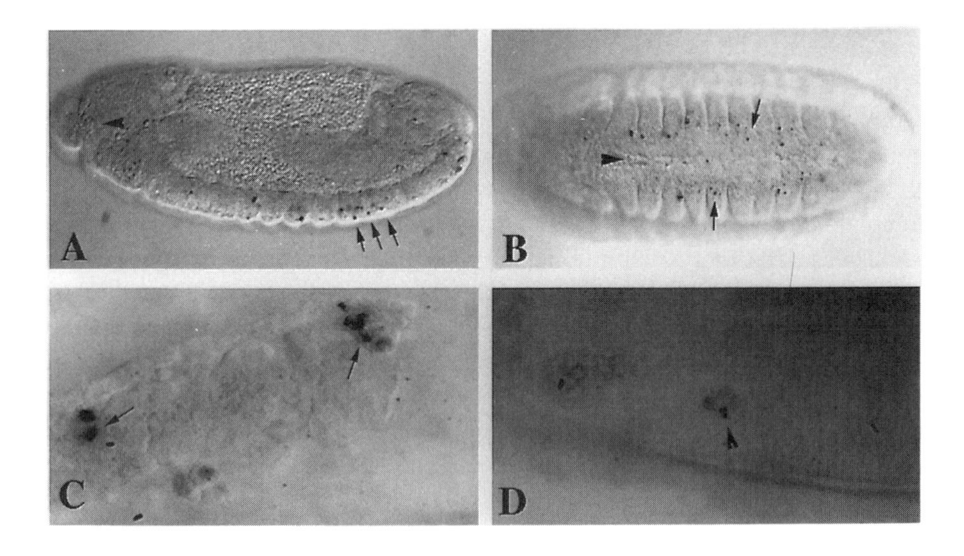

Figure 2: Detection of dead and dying cells in Drosophila embryos via in situ nick translation.

(A) Sagittal view of a late stage 12 wild type embryo. Note prominent cell death in the retracting germ band (arrows). Significant cell death is also detected in the cephalic region of the embryo (arrowhead). Anterior is to left and dorsal is up. (B) Ventral view of a late stage 12 embryo. Note asymmetric pattern of corpses (arrows) in the neuroectoderm on both sides of the CNS midline (arrowhead). Anterior is to left. (C) Section of tissue showing multiple labelled apoptotic bodies within phagocytic macrophages (arrows). (D) Ventral region of stage 16 embryo indicating an apoptotic body (arrowhead) engulfed by a circulating macrophage.

1993), while a drop in ecdysone levels precipitates a second set of cell deaths following adult eclosion (e.g. Kimura and Truman, 1990; Robinow *et al.* 1993).

Ecdysone acts by entering cells and complexing with DNA binding receptor proteins that directly regulate gene transcription (Koelle *et al.*, 1991). The ecdysone receptor (EcR) is a member of the steroid receptor superfamily of zinc finger proteins, and appears to function as a heterodimeric complex with the ULTRASPIRACLES protein, which encodes a homologue of the vertebrate transcription factor, Retinoid X Receptor (Yao *et al.*, 1992). In *Drosophila*, there are three isoforms of the ecdysone receptor, EcR-A, EcR-B1, and EcR-B2, which derive from a common gene and share identical DNA and ligand binding domains (Talbot *et al.*, 1993). They differ, however in their NH2-terminal regions, which has permitted the generation of isoform specific antisera. Work by Jim Truman, David Hogness and colleagues have shown that each isoform displays a distinctive tissue and temporal specific pattern of expression at the onset of metamorphosis (Robinow *et al.*, 1993; Talbot *et al*, 1993).

Among the tissues that undergo cell death during metamorphosis are the larval salivary gland and midgut. Recent work by Carl Thummel and colleagues has investigated the synchronous death of the larval salivary gland cells (Jiang *et al.*, 1997). These cells die within 15 hours after puparium formation. The salivary gland imaginal ring cells then proliferate to give rise to adult salivary gland. The larval midgut begins to die between 2-4 hours after puparium formation; it significantly shortens as the gastric caeca disappears and the proventriculus is reduced in size. A layer of adult midgut cells then surround the condensing larval midgut and elongate. Cell death in both these tissues can be detected via AO staining and TUNEL labelling and was shown to be ecdysone-dependent. The use of cultured mid-prepupae salivary glands indicated that that in the absence of ecdysone, no cell death was detected within 7 hours of culture (Jiang *et al.*, 1997). In contrast, cultured cells treated with ecdysone began to die within 5 hours of treatment. Premature midgut cell death was induced upon injection of ecdysone into intact third instar larvae.

Ecdysone regulates a wave of cell death that occurs after emergence of the pharate adult from the pupal case, as muscle and neuronal tissues which were required specifically for adult emergence or wing inflation rapidly die (Kimura and Truman 1990; Truman *et al.*, 1992; 1993). Within the nervous system, large numbers of dying neurons can be detected during the first 6 hours after eclosion. For example, within the CNS there exist a distinctive set of Type II neurons that exhibit high expression levels of the EcR-A isoform (Robinow *et al.*, 1993). Of the roughly 300 type II neurons, about 90% of them are located in the third thoracic and abdominal ganglia, regions of prominent cell death. Indeed, within 6 hours after eclosion there is a dramatic reduction in the number of type II neurons, although the doomed neurons continued to express high levels of EcR-A. The dependence of type II neuron cell death on the decreased levels of ecdysone after eclosion was demonstrated via injection of ecdysone into newly eclosed adults. These injections were able to delay the death of type II neurons for up to 12 hours (Robinow *et al.*, 1993). These injections were only effective during a short period after eclosion, as within two hours the doomed neurons became refractile to rescue by ecdysone, indicating that they had become committed to die.

1.2.2 Stage specific embryonic cell death

Analyses of the pattern of programmed cell death during embryogenesis have also revealed a dynamic and highly regulated process. *Drosophila* undergoes a long germ band pattern of embryonic development where the entire germ band is specified concurrently. Cell death is not observed during early embryonic stages in the blastoderm or gastrulating embryos. It is not until stage 11 (stages defined by Campos-Ortega and Hartenstein, 1997) when the embryo is fully germ band extended, that dying cells begin to be detected (Abrams *et al.*, 1993; Campos-Ortega and Hartenstein, 1997). This is approximately 7 hours after fertilisation, about a third of the way through embryogenesis. The dying cells are first detected in head segments located in the ventral anterior region of the embryo. In addition, dying cells are also detected in the posterior tip of the extended germ band.

During germ band retraction the levels of cell death increase and dying cells are detected in the lateral epidermis, the central nervous system, and dorsal cephalic regions. Cell death in the head region may facilitate the dramatic morphogenic movements that occur during head invagination. Dying cells also come to be detected in the dorsal epidermis which extends over the surface of the embryo during dorsal closure. Within the CNS, the pattern of deaths include cells located at symmetric and asymmetric positions; the precise positions of dying cells appear largely stochastic, with significant variability between individual embryos. This suggests that the processes through which individual cells in the *Drosophila* embryo are chosen to die are less stereotypic and invariant than those in *C. elegans*. In both *Drosophila* and *C. elegans* the highest levels of cell death occur in the nervous system. In *C. elegans*, approximately 80% of all deaths occur in neural lineages (Sulston and Horvitz, 1977; Horvitz *et al.*, 1982; Sulston *et al.*, 1983) while in *Drosophila* 50-70% of all the cells formed in the CNS ultimately die (White *et al.*, 1994; Grether *et al.*, 1995).

The dying embryonic cells do not persist, but are generally phagocytosed by circulating macrophages. These specialised cells differentiate from haemocytes that derive from anterior mesodermal cells (Tepass *et al.*, 1994). At the onset of germ band retraction these cells begin to migrate posteriorally along both dorsal and ventral pathways. They secrete basement membrane components that insulate all the internal organs (see Fessler and Fessler 1989), and are all capable of engulfing and removing dead and dying cells (Tepass *et al.*, 1994). Macrophage migration can be directed by signals from dying cells (Tepass *et al.*, 1994), however, there exist cell death-independent migration signals as well (Zhou *et al.*, 1995). The macrophages appear to utilise scavenger receptors similar to those found in vertebrate macrophages to detect dying cells (Abrams *et al.*, 1992; Krieger *et al.*, 1992; Pearson *et al.*, 1995). While macrophages are found throughout the body cavity, they are not present within the mature CNS, where dying cells are engulfed by neighbouring glia (Sonnenfeld and Jacobs, 1995a). Interestingly, in *Drosophila*, the *glial cells missing* gene appears to influence the development of embryonic glial and haemocyte lineages (Bernadoni *et al.*, 1997), and in the vertebrate CNS, microglia also phagocytose debris from dying cells (reviewed in Davis *et al.*, 1994).

2.0 GENES THAT REGULATE APOPTOSIS IN *DROSOPHILA*

The identification of cellular death genes has been accomplished using genetic, immunological, biochemical, and molecular approaches (reviewed in Ellis *et al.*, 1991; Driscoll, 1992; Nagata and Golstein, 1995; Osborne *et al.*, 1996; McCall and Steller, 1997). Such studies have identified both tissue-specific cell death regulators, such as the vertebrate *fas* and *nur77* genes (Itoh *et al.*, 1991; Oehm *et al.*, 1992; Liu *et al.*, 1994; Woronicz *et al.*, 1994), as well as evolutionarily conserved families of genes, such as caspases and *ced-9/bcl-2* homologues, that appear to play general roles in the death of many different tissues (reviewed in Steller, 1995; Chinnaiyan and Dixit, 1996; Schwartz and Milligan, 1996; Vaux and Strasser, 1996; White, 1996; Reed, 1997; Yuan, 1997). In *Drosophila*, it is clear that developmental mutants often exhibit altered patterns of cell death, typically with increased numbers of dying cells in specific tissues (e.g. Magrassi and Lawrence, 1988; Klingensmith *et al.*, 1989; Namba *et al.*, 1997). Indeed, the ultimate phenotypes of many mutants are likely to be as much the result of ectopic cell deaths as defects in cell fate determination and differentiation. The relationship between cell death and differentiation is intricate and intimate, and it appears that cells which fail to differentiate properly often commit suicide. Much like the functions of trophic factor support, this may provide a selective mechanism to help ensure proper organogenesis. In the next sections we will focus our discussion on what is known about several genes which appear to play direct and general roles in regulating global patterns of apoptosis in *Drosophila*. We will place particular emphasis on mechanisms where *Drosophila* is providing novel contributions to our understanding of cell death.

2.1 The *grim-reaper* locus

A seminal study in the regulation of PCD in *Drosophila* was carried out by Hermann Steller's laboratory. To identify genes essential for cell death, Steller's group analysed the patterns of AO staining in embryos homozygous for a series of chromosomal deletions. They screened 129 deletion strains which together covered approximately 50% of the genome (White *et al.* 1994). Strikingly, 3 strains, Df(3L)WR4, Df(3L)WR10, and Df(3L)Cat DH104, which each eliminated the 75C1,2 region of chromosome 3, exhibited an absence of essentially all cell death. Subsequently, they identified another 75C1,2-deficiency mutant, Df(3L)H99, which also exhibited a blockade of embryonic cell death, as assayed via both AO staining and transmission electron microscopy. In addition, no apoptotic corpses were detected in circulating macrophages and there was a 2-3 fold increase in the normal number of cells present in the ventral nerve cord (White *et al.*, 1994). Thus, one or more genes in the 75C1,2 interval are essential for the normally occurring embryonic cell deaths.

75C1,2 genes are also required for ectopic cell death induced by environmental insults. Thus, exposure of wild type embryos to 500 rads of X-irradiation induced high levels of cell death, whereas this dosage failed to induce deaths in Df(3L)H99 mutants (White *et al.*, 1994). Significantly, very high levels of X-irradiation did induce some cell death in the mutants and the dying cells exhibited normal apoptotic morphologies. This finding suggested that the cell death defect in Df(3L)H99 mutant embryos occurs upstream of the actual cell death machinery itself, and that it is the activation of cell death pathways that is disrupted. This hypothesis was substantiated by subsequent molecular and genetic studies.

After identification of the 75C1,2 region as an essential locus for cell death, a series of studies were performed to identify the corresponding genes. To date three 75C1,2 cell death genes (Figure 3), *reaper, head involution defective* (*hid*) and *grim*, have been identified (White *et al.*, 1994; Grether *et al.*, 1995; Chen *et al.*, 1996a). Each of these genes is expressed in dying cells and, upon over-expression, can induce some cells to die. In the following sections we will discuss the methods by which these genes were identified and their roles in regulating cell death.

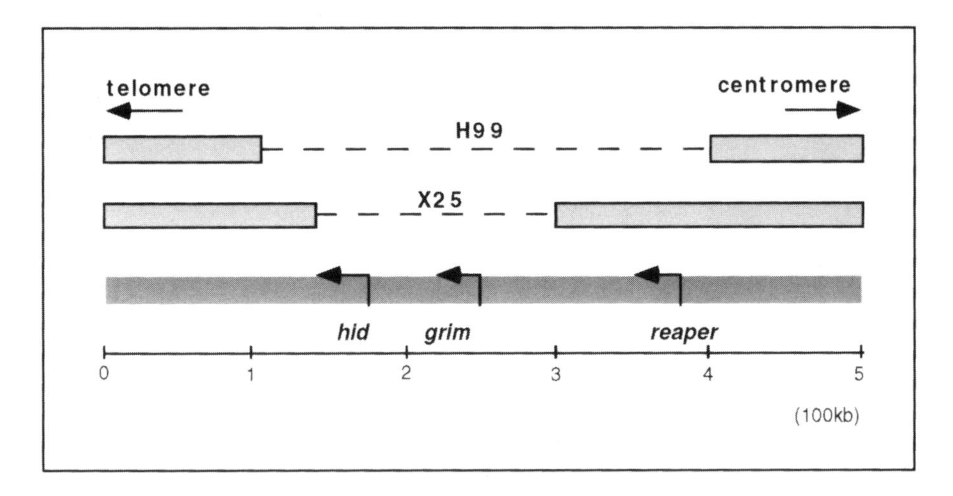

Figure 3: Organisation of the grim-reaper locus.
hid, grim, and *reaper* are all located at the 75C 1,2 genomic interval. The Df(3L)H99 deficiency strain deletes all three of these genes, whereas, the Df(3L)X25 strain lacks only *hid* and *grim.* Note that all 3 genes are transcribed in the same direction, towards the telomere. (adapted from White *et al.*, 1994).

2.2 *reaper*

To initially identify a cell death gene in the 75C1,2 interval, a chromosomal walk was performed to isolate 85 kb of overlapping genomic DNA (White *et al.*, 1994). Several cosmids containing DNA from this region were then tested for their ability to restore cell death in the Df(3L)H99 mutants. One cosmid, NT1B1, restored cell death to the mutant embryos, though not to wild type levels (White *et al.*, 1994). Only a single transcription unit was contained in NT1B1, and this gene, named *reaper*, gives rise to a 1300 bp transcript that encodes a small and novel 65 amino acid protein (White *et al.*, 1994). The predicted protein did not exhibit any strong homologies, although it was later noted to have weak similarity to the death domains present in several vertebrate cell death signalling proteins (Golstein *et al.*, 1995). Embryonic expression of *reaper* mRNA was detected in a pattern which generally resembled that of embryonic cell death. Over-expression of a *reaper* cDNA under the control of the Hsp70 promoter induced cell death in wild type and Df(3L)H99 mutant embryos, confirming that *reaper* is sufficient to activate cell death pathways. The P[Hsp70-*reaper*] strain was subsequently used to examine the developmental specificity of *reaper* action. The induction of ectopic *reaper* expression resulted in lethality during all stages of development, with the exception of late embryos and late pupae (White *et al.*, 1996). While the basis for this temporal specificity is uncertain, it probably reflects the modulation of *reaper* activity by other positive or negative apoptosis regulators.

A P[GMR-*reaper*] strain, where *reaper* expression is driven in the developing eye imaginal disc via multimerised DNA binding sites of the GLASS transcription factor, exhibited a dosage-sensitive ablation of eye tissue (Hay *et al.* 1995; White *et al.*, 1996). Two copies of the GMR-*reaper* transgene resulted in an incomplete loss of ommatidia and a "rough" eye phenotype, whereas, three to four copies resulted in an eyeless fly. The eye ablation induced by ectopic *reaper* expression was blocked by co-expression of the baculovirus p35 protein (Hay *et al.*, 1995; White *et al.*, 1996), a caspase inhibitor (Bump *et al.*, 1995; Xue and Horvitz, 1995). These results indicated that: 1) the loss of eye tissue was the direct result of ectopic cell death and not secondary to differentiation defects, and 2) *reaper* acts through one or more members of the conserved family of pro-apoptotic caspases.

reaper transcription is strongly influenced by cell death signals. The levels of *reaper* mRNA rapidly increased when embryos were exposed to high levels of ionising radiation, and *reaper* transcription was also up-regulated in *crumbs* mutant embryos, which exhibit extensive epidermal cell death (White *et al.*, 1994). In addition, *reaper* expression is strongly modulated by the levels of ecdysone in several dying postembryonic tissues (Jiang *et al.*, 1997; Robinow *et al.*, 1997). These findings suggest that *reaper* transcription is influenced by both extrinsic and intrinsic cell death-inducing stimuli, and that regulation of *reaper* gene expression is a key aspect of controlling cell death activation. Interestingly, *reaper* transcripts are first detected in blastoderm stage embryos, several hours before any cell death is detectable (Nordstrom *et al.*, 1996). This suggests that *reaper* may also function in signalling pathways independent of its functions in cell death.

The mechanisms regulating *reaper* gene transcription have been analysed through the generation of P element constructs that contain the bacterial lacZ gene fused to *reaper* genomic DNA fragments (Nordstrom *et al.*, 1996). An 11 kb fragment of *reaper* upstream DNA conferred lacZ expression which resembled that of the native *reaper* gene, indicating that *reaper* regulatory DNA elements had been isolated. This 11 kb fragment is likely to still lack some *reaper* regulatory elements, as the transgene exhibited reduced levels of expression in the CNS when compared to endogenous *reaper* expression (Nordstrom *et al.*, 1996). Nonetheless, this approach should provide significant insight into the mechanisms through which cell death transcriptional pathways are regulated, as the transgene did exhibit strong enhanced expression upon exposure to ionising radiation, as well as in *crumbs* mutant backgrounds. It will now be of interest to define the *cis* DNA sequences required to confer these transcriptional properties.

2.3 *head involution defective*

The inability of the *reaper* gene to fully rescue the cell death blockade in Df(3L)H99 mutants suggested there may be other genes in the 75C1,2 interval involved in regulating cell death activation. To test this hypothesis, Steller and colleagues carried out a screen for mutations which produced lethal or visible phenotypes over the Df(3L)H99 deficiency. Five lethal mutations were identified, all of which belonged to a single complementation group (Grether *et al.*, 1995). These alleles all failed to complement *head involution defective* (*hid*) mutants, which are characterised by a pronounced defect in the morphogenic movements of embryonic head involution (Abbot and Lengyel, 1991).

Analysis of *hid* mutant embryos indicated reduced levels of cell death, particularly in the head region prior to involution (Grether *et al.*, 1995). For example, in *hid* mutants there was an approximately 50% increase in the number of larval photoreceptor cells. Since head involution is associated with high levels of cell death (Abrams *et al.*, 1993; Campos-Ortega and Hartenstein, 1997), at least part of the *hid* morphogenic defect may result from a reduction in cell death. Overall, the extent of cell death blockade in *hid* mutants was lower than that observed for the Df(3L)H99 mutants, suggesting that the loss of *reaper* also contributes to the Df(3L)H99 cell death phenotype. The *hid* gene was isolated via a 75C1,2 P element insertion into the *hid* locus; it gives rise to a single 4.2 kb transcript that encodes a novel 410 amino acid protein (Grether *et al.*, 1995). The predicted HID protein does not possess any recognisable functional domains, however, the NH2-terminal 14 amino acids of HID exhibits 55% identity to the corresponding region of REAPER.

The functions of the *hid* gene in activating cell death pathways were also analysed through use of a P[Hsp70-*hid*] transformant fly strain (Grether *et al.*, 1995). Over-expression of *hid* resulted in ectopic cell death in both wild type and Df(3L)H99 mutant embryos, indicating that *hid* can induce cell death in the absence of *reaper* function. In addition, it was determined that *hid* does not activate *reaper* expression, indicating that *reaper* is not directly downstream of *hid*. Expression of *hid* in the eye imaginal disc via a P[GMR-*hid*] strain resulted in severe loss of ommatidia. Interestingly, the eye mechanosensory bristles were left intact, suggesting that these cells may be less susceptible to *hid*-induced killing. Like that of *reaper*, the *hid*-induced eye cell death was blocked by co-expression of p35.

The *hid* gene is expressed throughout embryogenesis in regions where cell death occurs (Grether *et al.*, 1995). The overall expression of *hid* is less restricted than that of *reaper*, as *hid* mRNA is also present in cells which normally survive. For example, *hid* mRNA is detected throughout the entire optic lobe primordium, where only some cells undergo cell death. Like that of *reaper*, this expression pattern suggests that *hid* may act in pathways distinct from those that mediate cell death. Interestingly, both *hid* and *reaper* transcripts are detected in apoptotic bodies engulfed by phagocytic macrophages (White *et al*, 1994; Grether *et al.*, 1995; Zhou *et al.*, 1995). This is not generally observed for other mRNAs and suggests that *reaper* and *hid* mRNAs may be specifically stabilised in dying cells.

2.4 *grim*

Additional genetic analyses of the 75C1,2 region suggested that a third cell death gene may reside within the 75C1,2 interval. Thus, two different deficiency mutants, Df(3L)X14 and Df(3L)X25, which delete *hid* yet leave *reaper* intact, yield distinct cell death phenotypes when placed in *trans* with Df(3L)H99. Df(3L)X25/Df(3L)H99 trans-heterozygotes exhibit a severe cell death phenotype whereas Df(3L)X14/Df(3L)H99 trans-heterozygotes yield a subtle cell death defect (Chen *et al.*, 1996a). To test whether a third cell death gene may exist between the Df(3L)X14 and Df(3L)X25 breakpoints, John Abrams and colleagues generated transformant fly strains which contained a cosmid that spans this breakpoint interval and does not contain *reaper* or *hid*. This cosmid was able to restore some cell death in Df(3L)H99 mutant embryos.

The cosmid was shown to contain a single gene, named *grim*, which gives rise to a 1.6 kb mRNA that encodes a novel 138 amino acid protein (Chen *et al.*, 1996a). The predicted GRIM protein shares the conserved NH2-terminal region with REAPER and HID. This 14 amino acid region is most similar between REAPER and GRIM, as the two exhibit 71% identity. GRIM and HID share the least similarity in this region, with 21% identity. This similarity suggests conserved functional roles for this region (discussed in a section 3.2). Due to this sequence similarity, we will refer to this region as the RHG domain, for REAPER, HID, and GRIM (Wing *et al.*, 1998).

The expression and actions of *grim* share many similarities with that of *reaper* and *hid*. *grim* mRNA is detected in dying cells and within apoptotic bodies engulfed by phagocytic macrophages (Chen *et al.*, 1996a). Over-expression of *grim* is capable of inducing ectopic, caspase-dependent cell death in both the developing embryo and adult eye. *grim* can induce cell death independently of *reaper* or *hid*, and there do not appear to be any cross-regulatory interactions between these genes; ectopic expression of one gene does not activate expression of any other. *grim* was able to induce cell death in the embryo at an earlier stage than either *reaper* or *hid*, suggesting that *grim* may access the cell death machinery more efficiently, and/or be regulated in a distinct manner from either *reaper* or *hid*.

The closely linked *reaper*, *hid*, and *grim* genes all function to activate apoptosis and they all share conserved RHG domains, exhibit transcription in the same direction, and are expressed in overlapping patterns (White *et al.*, 1994; Grether *et al.*, 1995; Chen *et al.*, 1996a). It will be of interest to determine whether the genes share any common gene regulatory elements and whether the conserved RHG domain has a conserved exonic structure, suggesting that these genes may share sequences from a common ancestral progenitor. Because of these similarities, we have proposed referring to this genomic interval as the *grim-reaper* locus (Wing *et al.*, 1998). One important issue is whether these genes may have overlapping functions to regulate the normal patterns of cell death. While single gene mutations in *reaper* or *grim* have not been yet been published, it is likely that similar to *hid* mutants, they will exhibit only a partial blockade of cell death. The initial characterisations of *reaper*, *hid*, and *grim* indicated that these genes do not function in a strict hierarchy, and may act either in parallel or in converging, overlapping pathways. Recent experiments have also suggested that, at least within some cell lineages, these genes can act synergistically (Zhou *et al.*, 1997; Wing *et al.*, 1998). Resolution of the mechanisms through which these genes act is a crucial issue, and one that we will revisit in section 3.

2.5 *Drosophila caspases*

A series of elegant genetic screens carried out by Robert Horvitz and colleagues in *C. elegans* resulted in the identification of several central players in the conserved apoptotic machinery (reviewed in Ellis *et al.*, 1991; Hengartner and Horvitz, 1994; Horvitz *et al.*, 1994; see also Miura and Sugimoto, this book). This included the pro-apoptotic genes *ced-3* and *ced-4*, as well as the negative regulator, *ced*-9. These three archetypal genes have structural and functional homologues in vertebrate species and appear to act together via conserved mechanisms (Vaux *et al.* 1992; Yuan *et al.*, 1993; Zou *et al.*, 1997). In particular, recent studies indicate that CED-4-related proteins promote cell death by directly binding to and activating CED-3-related proteins (Chinnaiyan *et al.*, 1997; Seshagiri and Miller 1997; Wu *et al.*, 1997; Xue and Horvitz, 1997; Zou *et al.*, 1997). This interaction is repressed by several CED-9-related proteins, which in turn act to inhibit apoptosis.

The *ced-3* gene encodes a member of a family of at least 10 related cysteine proteases, termed caspases (Alnemri *et al.*, 1996), that play crucial roles in the degradation of specific proteins during the cell death process (reviewed in Schwartz and Milligan, 1996; Salvesen and Dixit, 1997; Thornberry, 1997; Villa *et al.*, 1997). These endopeptidases are generated as zymogens which are proteolytically processed to generate a pro-domain, a linker region, and a large and a small subunit that assemble into an active tetramer (Cerretti *et al.*, 1992; Thornberry *et al.*, 1992). This processing can be triggered by other caspases (e.g. Muzio *et al.*, 1997), or through auto-processing (Yang *et al.*, 1998). Caspases all cleave after aspartate residues found within related tetrapeptide sequences, and exhibit distinct yet overlapping substrate specificities (see Thornberry *et al.*, 1997). These cleavages can either activate [e.g. pro-interleukin-1ß, DNA fragmentation factor (DFF), and pro-caspases] or disrupt [e.g. Poly(ADP-ribose) polymerase (PARP), nuclear lamins, Gas2, and gelsolin] substrate functions (Thornberry *et al.*, 1992; Brancolini *et al.*, 1995; Lazebnik *et al.*, 1995; Kothakota *et al.*, 1997; Liu *et al.*, 1997). Interestingly, the apoptosis inhibitor BCL-2 appears to be cleaved by CASPASE-3, resulting in a product which acts instead to promote cell death (Cheng *et al.* 1997). Ectopic expression of caspases can induce cells to die (e.g. Miura *et al.*, 1993; Wang *et al.*, 1994; Lazebnik *et al.*, 1995), and mouse knockout mutants of the *caspase-1* and *caspase-3* genes exhibit distinct tissue specific cell death defects (Kuida *et al.*, 1995; Kuida *et al.*, 1996; Bergeron *et al.*, 1998).

In *Drosophila*, ectopic expression of both the *C. elegans ced-3* gene as well as the vertebrate *interleukin-1ß converting enzyme* (*Ice*) gene can induce cell death in embryonic and imaginal tissues (Shigenaga *et al.*, 1997). These results suggest that *Drosophila* possess an endogenous and conserved caspase-activated cell death machinery. In this regard, two *Drosophila* caspase genes, *dcp-1* (Song *et al.*, 1997) and *drIce* (Fraser and Evan, 1997) were recently identified via their sequence homology to conserved regions of other caspases. Each predicted fly protein contains a conserved pentapeptide sequence essential for caspase function. DCP-1 possesses a QACRG sequence also found in CASPASE-3 and CED-3, while DRICE contains a QACQG sequence present in CASPASE-8. DRICE also contains a NH2-terminal Ser-Gly rich region similar to that of CED-3.

In vitro biochemical assays indicated that while full length DCP-1 exhibits little proteolytic activity, a truncated version lacking the prodomain was able to cleave the caspase substrates PARP and p35 into stereotypic proteolytic fragments (Song *et al.*, 1997). In addition, the CASPASE-3 inhibitor, Ac-DEVD-CHO, completely inhibited DCP-1 activity, while an ICE inhibitor, Ac-YVAD-CHO was ineffective. *dcp-1* mRNA is expressed in a uniform pattern in the blastoderm embryo and appears to be maternally deposited into the oocyte (Song *et al.*, 1997; McCall and Steller, 1998). This ubiquitous expression pattern is maintained during embryogenesis, although the expression becomes enhanced in the developing CNS. The *dcp-1* gene was mapped cytologically to the 59F region of the second chromosome, and P element insertions were identified that disrupt the first exon of the *dcp-1* gene. Analysis of these *dcp-1* mutants indicated that the pattern of embryonic cell death was essentially normal (Song *et al.*, 1997). This finding suggests either that the maternal component of *dcp-1* expression is sufficient to rescue embryonic cell deaths, or that other (embryonically-expressed) caspases have redundant functions. *dcp-1* mutants die during larval stages and the *dcp-1* mutant larvae lack imaginal discs and contain internal melanotic tumours (Song *et al.*, 1997). While it is not clear how these defects may correspond to disruptions in cell death, the melanotic tumours might result from an overproliferation of blood cells, or the survival of abnormal cells which elicit an aberrant immune response.

The importance of the maternal component of *dcp-1* expression was investigated through additional genetic studies to generate germ line clones of *dcp-1* mutant cells (McCall and Steller, 1998). During normal oogenesis, a set of 15 polyploid nurse cells extrude or "dump" their cytoplasm through a series of ring canals into the growing oocyte, delivering large amounts of RNAs and proteins that will be used during embryogenesis (see Spradling, 1993). After cytoplasmic dumping has occurred, the nurse cells undergo cell death which can be monitored via DNA fragmentation and AO staining (McCall and Steller, 1998; Foley and Cooley, 1998). *dcp-1* mRNA is expressed in the nurse cells and deposited into the developing oocyte. Germ line clones of *dcp-1* mutant egg chambers exhibited a "dumpless" phenotype where the nurse cells failed to extrude their contents into the oocyte, and nurse cell death was delayed (McCall and Steller, 1998). There was also an abnormal breakdown of a DCP-1 cleavage substrate, nuclear lamin B, as well as defects in cytoskeletal reorganisation that occurs during nurse cell degeneration. These defects may contribute to the inability of the nurse cells to properly expel their contents and may also slow the morphological changes associated with cell death.

Interestingly, recent work from Lynn Cooley's laboratory has shown that other maternal effect mutants which yield a "dumpless" phenotype, including *chickadee* and *kelch*, also exhibit delayed nurse cell death (Foley and Cooley, 1998). These investigators further demonstrated that the *reaper*, *hid*, and *grim* genes are all transcribed in the developing nurse cells, suggesting they might regulate nurse cell death. However, analysis of germ line clones of Df(3L)H99 mutant egg chambers revealed normal nurse cell cytoplasmic extrusion and DNA fragmentation, suggesting that the *grim-reaper* genes are not essential for nurse cell death (Foley and Cooley 1998). In addition, *chickadee* and *kelch* mutations exhibit normal expression of *grim-reaper* mRNAs, suggesting that the presence of these gene products is not sufficient for nurse cell death.

The cell killing properties of DRICE have been assayed in *Drosophila* S2 tissue culture cells. Expression of full length DRICE did not induce significant cell death, although, it did sensitise these cells to other apoptotic stimuli (Fraser and Evan 1997). In contrast, removal of the pro-domain of DRICE resulted in efficient induction of cell death and cleavage of both p35 and nuclear lamin, effects which were blocked by caspase inhibitors. As examined via developmental Northern blots, the *drIce* gene is expressed during all developmental stages, with highest levels found during early to mid-embryogenesis. Thus, expression of the *drIce* and *dcp-1* genes appear to overlap extensively. Interestingly, like *dcp-1*, *drIce* is expressed embryonically several hours before the onset of cell death (Fraser and Evan, 1997; Song *et al.*, 1997), raising the possibility that similar to some vertebrate caspases, they may also act in non-cell death associated processes. To date, two distinct *Drosophila* caspases have been identified, and biochemical evidence suggests that others exist (Kondo *et al.*, 1997). Thus, unlike *C. elegans*, where there is a single caspase, CED-3, that is required for all cell deaths, *Drosophila* appears more akin to mammals, where the existence of multiple caspases provide genetic redundancy, and elimination of a single gene only affects a subset of normally occurring cell deaths (Kuida *et al.*, 1995; Kuida *et al.*, 1996; Bergeron *et al.*, 1998).

2.6 *Drosophila Iaps*

Many viruses express anti-apoptotic proteins which serve to repress the ability of the host cell to die and block viral reproduction. Lois Miller and colleagues have identified several such anti-apoptotic proteins expressed by insect baculoviruses (Clem *et al.*, 1991; Crook *et al.*, 1993; Birnbaum, *et al.*, 1994; Clem and Miller 1994). This includes p35, which blocks cell death in a wide range of organisms, including mammals, *C. elegans*, and *Drosophila* (Clem and Miller, 1994; Rabizadeh *et al.*, 1993; Hay *et al.*, 1994; Sugimoto *et al.*, 1994). p35 serves as a substrate for the pro-apoptotic caspase enzymes and blocks their ability to cleave normal cellular substrates during cell death (Bump *et al.*, 1995; Xue and Horvitz, 1995). As yet no cellular homologues of p35 have been identified. Another class of baculovirus cell death inhibitors are the Inhibitor-of-Apoptosis proteins (IAPs) (Crook *et al.*, 1993; Birnbaum *et al.*, 1994; Clem and Miller, 1994). These viral proteins define an extended family that also include cellular homologues from both vertebrates and invertebrates (Hay *et al.*, 1995; Rothe *et al.*, 1995; Liston *et al.*, 1996). IAPs contain one to three BIR (Baculovirus IAP Repeat) motifs located at the NH2-terminal and centre portions of the protein, and most also contain a COOH-terminal RING finger. The RING finger is thought to constitute a zinc binding domain which may participate in protein/protein interaction. IAPs have also been found to block both naturally occurring and induced cell deaths in several different species (Crook *et al.*, 1993; Birnbaum *et al.*, 1994; Hay *et al.*, 1995; Rothe *et al.*, 1995; Liston *et al.*, 1996). In humans, the neuronal IAP (NAIP) has been implicated as a candidate gene underlying some forms of the neuorodegenerative disorder, spinal muscular atrophy (Roy *et al.*, 1995).

In *Drosophila*, a cellular IAP gene, *Diap1* was identified by Bruce Hay and Gerald Rubin (Hay *et al.*, 1995) in a genetic enhancer/supressor screen for modifiers of *reaper* activity. They screened a series of deficiencies in *trans* to P[GMR-*reaper*] chromosomes and found that several deletions in the 72D1,2 region significantly enhanced the P[GMR-*reaper*] eye ablation phenotype. This suggested that one or more genes in this interval may repress *reaper* function. Several lethal P element insertions at 72D1,2 also enhanced the P[GMR-*reaper*] phenotype, and these insertions all corresponded to alleles of the *thread* gene. *thread* hypomorphic mutants exhibit adult appendage defects which include thin antennal arristae that lack side branches as well as defects in leg tarsal claws (Lindsley and Zimm, 1992). In addition, some *thread* alleles also exhibit behavioural deficits in mating and geotaxis. While it is not clear how these mutant phenotypes may be the result of defects in cell death, molecular analyses of the *thread* gene revealed that it encodes a protein which shares strong homology to baculovirus IAPs (Hay *et al.*, 1995). This protein, DIAP1, contains two BIR domains in the NH2-terminal two-thirds of the protein and a single RING finger domain in the COOH-terminal region. A second *Drosophila* IAP gene, *Diap2* was identified via homology searches of expressed sequence tags (ESTs) in the Berkeley *Drosophila* Genome Project. The *Diap2* gene resides at position 52D on the second chromosome and encodes a protein containing 3 BIR domains and a single RING finger. Sequence comparisons indicated that DIAP1 is most closely related to the baculovirus OpIAP and CpIAP proteins (46% amino acid identity) which all contain two BIR domains and one RING finger, and is less related (36% identity) to DIAP2. DIAP2 resembles mammalian IAPs that contain three BIR domains and one RING finger.

Diap1 or *Diap2* also moderately suppressed the ectopic eye cell death in P[GMR-*hid*] strains. It was determined that the NH2-terminal portion of DIAP1, including the two BIR domains but not the RING finger, was sufficient to block both naturally occurring eye cell deaths, as well as deaths induced by X-ray irradiation and P[GMR-*reaper*] and P[GMR-*hid*]. The level of activity displayed by this truncated protein was greater than that of full length DIAP1, suggesting a potential modulatory function of the RING finger. Consistent with this, ectopic expression of the DIAP1 RING finger domain resulted in ectopic eye cell death, possibly by acting as a dominant-negative to block endogenous DIAP functions. As described in section 3.4, recent studies have extended these observations to reveal that: 1) DIAP1 can directly associate with REAPER, HID, and GRIM proteins (Vucic *et al.*, 1997a; Vucic *et al.*, 1998), and 2) *Diap2* blocks both *reaper*- and *hid*-induced cell death *in vivo*, but fails to block *grim*-induced cell death (Wing *et al.*, 1998).

The functional roles of the *Diaps* in regulating normal developmental cell deaths are not entirely clear, and it is not yet known whether they have distinct functions in repressing apoptosis. The *Diap1* gene is widely expressed during embryogenesis and eye imaginal disc development (Hay *et al.*, 1995). Lethal *diap-1* mutations do not exhibit excessive embryonic cell death, suggesting that the gene is not absolutely essential for inhibiting cell death. However, in somatic recombination experiments in the adult eye, homozygous mosaic clones of *diap-1* mutant cells failed to be detected, suggesting potential ectopic cell death. There are no point mutations for the *diap2* gene, and deletions that remove *diap2* failed to enhance the P[GMR-*reaper*] phenotype. However, P[GMR-*diap1*] and P[GMR-*diap2*] strains both exhibited a partial block of normally occurring eye cell death, although with less efficiency than a P[GMR-p35] strain (Hay *et al.*, 1995).

IAPs can block the functions of a wide variety of pro-apoptotic proteins, and have been shown to associate with the vertebrate FAS Associated Death Domain (FADD) protein (Rothe *et al.*, 1995), CASPASE-3 and CASPASE-7 (Deveraux *et al.*, 1997; Roy *et al.*, 1997) and REAPER, HID, and GRIM (Vucic *et al.*, 1997a; Vucic *et al.*, 1998). Another *Drosophila* protein, DOOM, was found to interact with IAPs in a yeast 2-hybrid screen for OpIAP-interacting clones (Harvey *et al.*, 1997). DOOM is one of 4 distinct proteins encoded by the *modifier of mdg-4* (*mod*(*mdg4*)) gene. It contains both a leucine-zipper containing motif, the BTB domain, shared with 2 other products of the *mod*(*mg4*) gene, as well as a novel DOOM-Specific-Domain (DSD) not present in any of the other *mod*(*mg4*) gene products. Expression of full length DOOM or the DSD domain can induce some cell death in lepitodpteran Sf-21 cells, suggesting DOOM may normally act as a cell death regulator (Harvey *et al.*, 1997). Other proteins from the *mod*(*mg4*) locus have been shown to bind to many distinct chromosomal sites, and appear to modulate changes in chromatin structure associated with position effect variegation, transvection, and genetic imprinting (Dorn *et al.*, 1993; Gerasimova *et al.*, 1995). It will be of interest to determine whether DOOM influences chromatin structure to regulate gene expression of cell death regulators.

3.0 MECHANISMS OF *GRIM-REAPER* FUNCTIONS

The use of *Drosophila* to study programmed cell death has led to the identification of the *grim-reaper* genes, which encode a set of novel and essential activators of cell death. Defining the mechanisms through which these genes act is of keen interest, especially given that the sequences of the corresponding proteins do not exhibit strong similarity to any well defined functional domains, and as yet no homologues have been identified from any organism outside the *Drosophila* genus (White *et al.*, 1994). Thus, although these genes all act upstream of one or more caspases, it remains unclear whether they are components of a general and conserved cell death pathway. Several studies using different experimental approaches have begun to shed light on the actions of these proteins, and are discussed below.

3.1 Does REAPER act as a death domain?

One intriguing observation regarding the potential function of REAPER was made by Pierre Golstein (Golstein *et al.*, 1995), who noted that REAPER exhibits limited homology to the death domain, a protein/protein interaction motif present in the cytoplasmic regions of several vertebrate cell death signalling proteins, including TNFR1 and FAS (Cleveland and Ihle 1995; Nagata and Golstein, 1995). These transmembrane proteins act via multimerisation and subsequent formation of an intracellular complex that includes other death domain containing adapter proteins, and ultimately results in the activation of downstream caspases (reviewed in White E., 1996; Nagata, 1997; Villa *et al.*, 1997; Yuan, 1997). As REAPER appears to form multimers *in vitro* (Cerinda Carboy-Newcomb, Chia-Lin Wei, and Hermann Steller, unpublished), it may function analogously to the death domains of FAS or TNFR1, acting as a cytoplasmic adapter between externally derived cell death signals and downstream components of the cell death machinery.

A number of recent studies have begun to test this possibility. Shigekazu Nagata and colleagues directly compared the cell death inducing functions of REAPER and the cytoplasmic region of the vertebrate FAS (FAS-C) protein (Kondo *et al.*, 1997). They demonstrated that while FAS-C and

REAPER were both able to induce apoptosis of *Drosophila* S2 cells, REAPER-induced death was more rapid than FAS-C-induced death. In addition, experiments using specific caspase substrates and inhibitors suggested that FAS-C-induced S2 cell death is mediated through CASPASE-1 and CASPASE 3-like proteases, while REAPER-induced death involved only a CASPASE-3-like protease. A different study by Lois Miller's lab compared the actions of REAPER and the vertebrate FADD protein, which contains a death domain and mediates signalling through TNFR1 (Vucic *et al.*, 1997b). FADD recruits the FLICE/MACH protein to the cytoplasmic region of TNFR1 (see White E., 1996; Villa *et al.*, 1997; Yuan, 1997). While both REAPER- and FADD-induced apoptosis in transfected insect cells, the REAPER-induced killing occurred with a much greater rapidity, and exhibited differential sensitivity to inhibition by BCL-2 family members (Vucic *et al.*, 1997b). Taken together, these data suggest that REAPER may not function analogously to vertebrate death domains. However, it should be noted that it was not determined in these assays whether distinct vertebrate death domain proteins would induce apoptosis with identical or similar time courses.

Finally, the number of identified death domain containing proteins has increased, re-inspection of the REAPER sequence does not appear to support the original proposed death domain homology (Vucic *et al.*, 1997b). In addition, mutations of specific residues conserved in the original sequence alignments, including those essential for activity of FAS or TNFR1, did not significantly reduce the cell killing ability of REAPER (Chen *et al.*, 1996b; Vucic *et al.*, 1997b). In contrast, mutations of two residues, Phenylalanine-5 or Tyrosine-6, conserved between the RHG domains of REAPER and HID, resulted in a two-fold reduction in REAPER cell killing activity, suggesting these residues are important, although not essential, for cell death activity (Vucic *et al.*, 1997b). Mutation of Proline-8, a residue conserved between the RHG domains of REAPER, HID, and GRIM, did not alter REAPER activity. It thus remains to be determined which residues in REAPER might be absolutely essential for its normal activities.

3.2 Functions of the RHG domain

An important step in deciphering the functions of REAPER, HID, and GRIM lies in functional analyses of the conserved RHG domain. In tissue culture assays, a truncated REAPER protein specifically lacking the RHG domain exhibited reduced levels of cell killing compared to wild type REAPER; however, this may have been due to its lower levels of accumulation (Chen *et al.*, 1996b; Vucic *et al.*, 1997b). This result indicated that the RHG domain is not absolutely required for REAPER cell killing activity. This conclusion was supported by recent *in vivo* studies (Wing *et al.*, 1998) where expression of a similarly truncated REAPER protein was driven using the P[Gal4]/P[UAS] system of Brand and Perrimon (1993). In these studies the truncated REAPER protein was sufficient to induce cell death in the developing adult eye, but its activity was reduced compared to full length REAPER. Interestingly, when expression of the truncated REAPER protein was driven in the embryonic CNS midline, it was able to function co-operatively with both HID and GRIM (see following section) to induce high levels of cell death. Indeed, the levels of midline cell killing were greater than those observed for similar experiments using full length REAPER. This suggests that the RHG domain may act in part to modulate REAPER activity. Other experiments have indicated that the RHG domains of GRIM can induce cell death when it is stabilised by fusion to a heterologous polypeptide (Vucic *et al.*, 1998). Thus, the RHG domain may mediate multiple aspects of GRIM-REAPER protein functions (see also below).

3.3 Synergistic functions of *grim-reaper* genes

Genetic studies have indicated that the *grim-reaper* genes do not act in a strict hierarchy, and may function in parallel or overlapping pathways (Grether *et al.*, 1995; Chen *et al.* 1996a; Zhou *et al.*, 1997). Analyses of the cell killing abilities of these genes in the adult eye further indicated that REAPER, HID, and GRIM proteins are all sufficient to independently induce cell death, and no synergistic effects were detected (White *et al.*, 1994; 1996; Grether *et al.*, 1995; Hay *et al.*, 1995; Chen *et al.* 1996a). Similar effects were seen with ectopic *reaper* and *hid* expression in specific pupal neurosecretory cells (McNabb *et al.*, 1997; Robinow *et al.* 1997). However, recent studies on the embryonic CNS midline have revealed that these genes can act synergistically in some tissues (Zhou *et al.*, 1997). During normal development, about two-thirds of the midline glial cells which are born, ultimately die, while few if any of the Ventral Unpaired Median (VUM) neurons undergo cell death (Sonnenfeld and Jacobs, 1995b; Zhou *et al.*, 1995). It was found that: 1) *hid* null mutants exhibit a 1.5-fold increase in the number of surviving midline glia, 2) Df(3L)X25 mutants, which lack *hid* and *grim* but leaves *reaper* intact, exhibit a 2-fold increase in midline glia, and 3) Df(3L)H99 mutants which eliminates all three genes, exhibit a 3-fold increase in midline glia (Zhou *et al.*, 1997). Thus, the normal pattern of cell death in the developing embryonic CNS midline was shown to require the functions of multiple *grim-reaper* genes. In addition, while targeted CNS midline expression of *reaper* or *hid* was not sufficient to induce ectopic cell death, co-expression of both genes together resulted in significant cell killing (Figure 4; Zhou *et al.*, 1997). These synergistic effects were dosage-sensitive, as lower levels of *reaper* and *hid* co-expression resulted in the excess death of midline glia only, whilst higher levels also resulted in death of the VUM neurons. Thus, different cell types exhibit distinct sensitivities to the effects of ectopic *reaper* and *hid* expression, suggesting that in earlier experiments where ubiquitous embryonic expression of these genes was induced (White *et al.*, 1994; Grether *et al.*, 1995), potentially different subsets of cells were induced to die. These CNS midline experiments revealed synergistic effects of the ectopic expression of *reaper* and *hid* and suggested that *reaper* and *hid* may function co-operatively.

Similar experiments have also been carried out to test the actions of *grim*. Unlike *reaper* or *hid*, *grim* can act alone to efficiently induce death of the midline glia that normally survive (Figure 4; Wing *et al.* 1998). Along with previous results indicating that *grim* can induce embryonic cell death at an earlier stage than either *reaper* or *hid* (Chen *et al.*, 1996a), these data suggest that either *grim* is more efficient at accessing the cell death machinery, or that its activity is regulated in a distinct manner (see following section). Co-expression of *grim* with *reaper* or *hid* also resulted in synergistic cell killing effects of CNS midline cells. Taken together, these data suggest that the *grim-reaper* genes could act in a combinatorial fashion to regulate the patterns of cell death in distinct tissues or cell types.

3.4 Interactions between *grim-reaper* and *Diap* genes

Genetic studies indicated that *reaper* activity can be negatively regulated by the *Diaps* (Hay *et al.*, 1995). Lois Miller and colleagues analysed the mechanism of this repression and showed that REAPER protein can directly associate with OpIAP, DIAP1, and DIAP2 (Vucic *et al.*, 1997a). Interestingly, this interaction was mediated through the BIR domain of the IAPs, the same region that also is able to interact with DOOM, as well as TRAF-1 and TRAF-2, two mammalian proteins important for mediating signal transduction from TNFR (Uren *et al.*, 1996), and CASPASE-3 and CASPASE-7 (Devereaux *et al.*, 1997). This may suggest a common mechanism through which IAPs bind to and repress cell death activators, although interestingly, none of these pro-apoptotic proteins share significant sequence similarity. IAPs may also repress REAPER activity by altering its normal subcellular localisation or degradation, as co-expression of OpIAP with REAPER in Sf21 cells blocked REAPER-induced cell death and induced a cytoplasmic to perinuclear re-localisation and enhanced persistence of REAPER protein (Vucic *et al.*, 1997a).

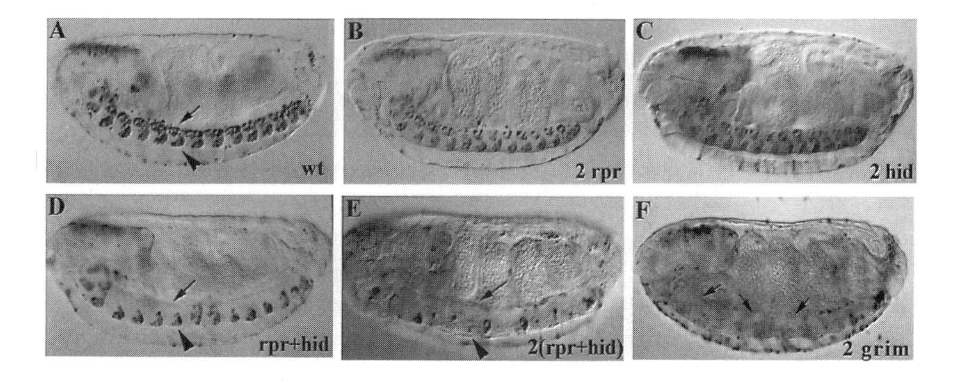

Figure 4. grim-reaper genes induce cell death of the embryonic CNS midline cells.

Each panel presents a stage 16 embryo where one or more grim-reaper genes was specifically expressed in the CNS midline using the Gal4/UAS system of Brand and Perrimon (1993). The midline glia (arrows) and VUM neurons (arrowheads) were detected via anti-ß-gal staining (A). Expression of reaper (B) or hid (C) alone was not sufficient to induce ectopic midline cell death, however, reaper and hid co-expression results in a synergistic killing of midline glia and/or VUM neurons (D, E). Ectopic expression of grim alone is capable of inducing midline cell death (F), suggesting it may function or be regulated in a distinct manner. All views sagittal with anterior to left. (adapted from Zhou *et al.*, 1997; Wing *et al.*, 1998*)*

A set of recent data suggest that the inhibitory effects of the DIAPs may be mediated through the RHG domains of GRIM-REAPER proteins. Thus, it was found using the P[Gal4]/P[UAS] system of Brand and Perrimon (1993) that DIAP2 does not block cell killing induced by a truncated REAPER protein lacking the RHG domain as effectively as it inhibits the activity of full length REAPER (Wing *et al.*, 1998). *Diap2* also failed to block *grim*-induced killing in both the adult eye and CNS midline, but was able to block *reaper*- or *hid*-induced deaths (Wing *et al.*, 1998). Because recent *in vitro* studies have indicated that DIAP1 can directly associate with the RHG domain of GRIM, and an extended NH2-region of HID (Vucic *et al.*, 1998), it will be of much interest to decipher the potential functional distinctions between the RHG domains of REAPER, HID, and GRIM.

3.5 Conserved pathways of *reaper* signalling?

An important insight into *reaper* action was provided by a study from Gijsbertus Pronk and colleagues which indicated that *reaper*-induced death of *Drosophila* S2 cells resulted in an increase in ceramide production (Pronk *et al.*, 1996). Ceramide, a hydrolysis product of the membrane phospholipid sphingomyelin, acts as a second messenger in the cell death signalling pathways initiated by the vertebrate FAS and TNFR proteins, as well as ionising radiation (reviewed in Hannun and Obeid, 1995; Haimovitz-Friedman *et al.*, 1997). Thus, *reaper* may function through a conserved signal transduction pathway.

Potentially conserved functions of REAPER were also provided by studies of Sally Kornbluth and colleagues to test the ability of REAPER to induce apoptosis in vertebrate cells (Evans *et al.*, 1997). They used a cell-free extract of *Xenopus* eggs that can support apoptotic events including nuclear fragmentation and nucleosomal laddering of genomic DNA (Newmeyer *et al.*, 1994). The ability of the extracts to undergo apoptosis requires the presence of a fraction enriched in mitochondria, which play a key role in apoptosis through the release of the caspase activator cytochrome C (Liu *et al.*, 1996). The cytosolic release of cytochrome C is regulated at least in part by members of the BCL-2 family of proteins (reviewed in Reed, 1997; Mignotte and Vayssiere, 1998; see Chittenden, this book). REAPER accelerated the release of cytochrome C and activation of caspases in the *Xenopus* extracts (Evans *et al.* 1997). This effect required the presence of other cytosolic factors, indicating that REAPER was capable of functioning with endogenous vertebrate proteins. These results suggest that REAPER can access the vertebrate cell death machinery, and may act through stimulation of mitochondrial cytochrome C release. In this regard, addition of BCL-2 exhibited dose-sensitive repression of REAPER-induced apoptosis. Whether this effect can be generalised to other cell types is unclear, as a similar study using insect cells indicated that several BCL-2 family members, including CED-9, BCL-2, BCL-XL, and adenovirus E1B19K did not effectively block REAPER-induced cell death (Vucic *et al.*, 1997b).

4.0 INSIGHTS INTO CELL DEATH FROM *DROSOPHILA*

Studies on programmed cell death in *Drosophila* have contributed to a rapidly growing body of knowledge on the regulation and mediation of apoptosis. In particular, the identification of the *grim-reaper* genes have revealed a novel set of key cell death regulators. Given the central role these genes play in *Drosophila* apoptosis, and the widely conserved mechanisms of cell death, it seems likely that homologues of these genes will be found in other species. The recent identification of caspases in flies strongly suggests that the mechanisms of apoptosis in *Drosophila* will share central features with those in other organisms, and it should only be a matter of time before *Drosophila ced-9/bcl-2* and *ced-4/apaf-1* homologues are identified.

A great advantage of the *Drosophila* system is the availability of powerful genetics and molecular biology that permit identification of key regulatory genes and subsequent *in vivo* functional analyses. Yet, similar techniques are available in the nematode *C. elegans*, the system which pioneered the genetic analysis of programmed cell death (reviewed in Ellis *et al.*, 1991; Hengartner and Horvitz, 1994; Horvitz *et al.*, 1994; Osborne, 1996), and continues to be a pre-eminent system for elucidating cell death mechanisms (e.g. Chinnaiyan *et al.*, 1997; Seshagiri and Miller, 1997; Spector *et al.*, 1997; Wu *et al.*, 1997; Xue and Horvitz, 1997). In addition, the potent transgenic capabilities now available in mice also provide powerful methods for analysing the *in vivo* functions of cell death genes (e.g. Veis *et al.*, 1993; Knudson *et al.*, 1995; Kuida *et al.*, 1995; 1996; Knudson and Korsmeyer, 1997; Bergeron *et al.*, 1998). So, the question may be asked as to what specific contributions to our understanding of cell death are likely to be gleaned from the use of *Drosophila* as a model system. One answer is that *Drosophila* may provide a very useful intermediary system that offers a balance between cellular and molecular complexity, and experimental accessibility. For example, compared to vertebrates, *Drosophila* and *Caenorhabditis* share stronger similarities in respective physical sizes, cellular complexity, developmental rates and early embryonic cleavage patterns, as well as genome sizes and estimated numbers of genes (see Gerhart and Kirschner, 1997). However, *Drosophila* also shares several features with vertebrates that are not found in *C. elegans*, including segmented tissues and limbs, complex eyes, paired brain lobes and bilaterally symmetric nervous systems, and obligate heterosexualism. Germane to this discussion is the observation that cell death in *Drosophila* appears to be mediated by more complex sets of effectors than present in *C. elegans*. Thus, *C. elegans* possesses a single caspase (*ced-3*), and a single *ced-4/apaf-1* and a single *ced-9/bcl-2* family member, each of which plays an essential role in all apoptotic deaths. On the other hand, vertebrates possess several related genes in each of these families and the individual family members exhibit overlapping though distinct functions and patterns of expression. In *Drosophila*, the situation may be more similar to vertebrates than to *C. elegans*, as there are multiple caspase genes that likey have overlapping functions. It will be of keen interest to determine whether this will also be true for the *ced-4/apaf-1* and *ced-9/bcl-2* families. Thus, although *Drosophila* is phylogenetically more closely related to *Caenorhabditis*, its level of cellular complexity may require a cell death machinery more similar to that in vertebrates. In *C. elegans*, only approximately 10% of all cells born ultimately undergo programmed cell death and these deaths are not essential for the organismal viability or fertility (Ellis and Horvitz, 1986). This is unlikely to be the case in vertebrates, where massive amounts of cell death occur in organs such as the thymus and brain, and de-regulation of cell death patterns may ultimately lead to autoimmune disorders or tumourigenesis (see Thompson, 1995; Rudin and Thompson, 1997). While the question of whether cell death is essential for organismal viability in *Drosophila* is not yet resolved, *hid* null mutants as well as several relatively small 75C1,2 deficiency mutants are all lethal (Abbott and Lengyel, 1991; White *et al.*, 1994; Grether *et al.*, 1995; Chen *et al.*, 1996a).

Drosophila may also prove to be an important and unique system for studying the transcriptional regulation of cell death gene expression. Thus, in many species, tissues require *de novo* gene expression to synthesise proteins essential for cell death (e.g. Lockshin, 1969; Martin *et al.*, 1988; Oppenheim *et al.*, 1990; Schwartz *et al.*, 1990). This property has been exploited in differential and subtractive cloning strategies to identify genes that are activated in dying cells (Schwartz *et al.*, 1990; Liu *et al.*, 1994; Schwartz *et al.*, 1994; Osborne *et al.*, 1996). In *Drosophila*, transcription of the *grim-reaper* genes appears to be tightly regulated and generally restricted to dying cells. In addition, inappropriate expression of these genes can result in aberrant cell death. The availability of P element transformation systems provide a powerful means for analysing important *cis*-regulatory elements as well as facilitating identification of *trans*-acting regulatory proteins that may play crucial roles in the gene expression mechanisms underlying cell death activation. This approach has already begun to be productively used, as key regulatory regions of the *reaper* gene have been identified (Nordstrom *et al.*, 1996). The identification of sequences that control *reaper*, *hid*, and *grim* expression will permit detailed analyses of the regulation via various developmental events, such as changes in ecdysone levels (Jiang *et al.*, 1997; Robinow *et al.*, 1997) or MAP kinase signalling (Stemerdink and Jacobs, 1997; Sawamoto *et al.*, 1998; K. White, personal communication), as well as by environmental cell death signals such as irradiation (Nordstrom *et al.*, 1996). It will be of interest to determine whether the linked *grim-reaper* genes share any common regulatory elements. Finally, the *Diap* genes also appear to be strongly transcriptionally regulated (Jiang *et al.*, 1997; Robinow *et al.*, 1997) and one interesting question is whether there are mutually antagonistic mechanisms controlling transcription of the *Diaps* and *grim-reaper* genes.

As discussed in the outset of this review, there are many compelling examples of evolutionarily conserved developmental and physiological pathways between flies and vertebrates, and many instances where identification of key genes using fly genetics have ultimately provided insight into vertebrate developmental disorders and cancer. There are also potential examples of de-regulation of cell death pathways that may have similar effects in flies and humans. For example, in both flies and humans there are light-induced retinal degeneration syndromes associated with rhodopsin mutations (e.g. Dryja *et al.*, 1990; Leonard *et al.*, 1992). The defects in Rhodopsin protein structure ultimately lead to aberrant photoreceptor cell physiology and a gradual death of these cells. These defects, which include the dominant *ninaE* mutations in *Drosophila* and retinitis pigmentosa in humans, appear to result from ectopic activation of cell death pathways. In *Drosophila* the photoreceptor cell loss in *ninaE* mutants was recently shown to be blocked by p35 (Davidson and Steller, 1998), and significantly, functional assays indicated that this blockade resulted in a rescue of the mutant electrophysiological and behavioural defects in visual function (Davidson and Steller, 1998). These results suggest that use of cell death inhibitors may ultimately be useful to treat human photoreceptor degeneration disease. Two other recent studies provide further evidence for the potential utility of *Drosophila* in studying neurodegeneration. First, Nancy Bonini and colleagues found that specific neurodegeneration was induced in *Drosophila* by targeted expression of a human, expanded glutamine-repeat protein associated with Machado Joseph disease (Warrick *et al.*, 1997). Second, Seymour Benzer and Colleagues found that *spongecake* and *eggroll* mutant flies, which undergo premature and sudden death, also exhibit defects in brain morphology resembling those observed in human Tay Sach's and Creutzfeld-Jacob diseases (Min and Benzer, 1997). Thus, *Drosophila* may prove to be a useful system for modelling neuronal death during disease or trauma. While these areas are as yet relatively unexplored, it is likely that research on cell death in *Drosophila* will continue to provide insights into our growing understanding of the conserved mechanisms of cell death, and the roles this process plays in development, physiology, and disease.

Acknowledgements

The authors would like to thank Lois Miller for sharing unpublished data, and Larry Schwartz, Emily Niemitz, and Mousumi Mutsuddi for comments on the manuscript.

5.0 REFERENCES

Abbott,M.K. and Lengyel,J.A. (1991) Embryonic head involution and rotation of male terminalia require the *Drosophila* locus *head involution defective*. *Genetics* **129**: 783-789.

Abrams,J.M., Lux,A., Steller,H. and Krieger,M. (1992) Macrophages in *Drosophila* embryos and L2 cells exhibit scavenger receptor-mediated endocytosis. *Proc. Natl. Acad. Sci. USA* **89**: 10375-10379.

Abrams,J.M., White,K., Fessler,L. I. and Steller,H. (1993) Programmed cell death during *Drosophila* embryogenesis. *Development* **117**: 29-43.

Alman,B.A., Li,C., Pajerski,M.E., *et al.* (1997) Increased beta-catenin protein and somatic APC mutations in sporadic fibromatoses. *Am. J. Pathol.* **151**: 329-334.

Alnemri,E.S., Livingston,D.J., Nicholson,D.W., *et al.* (1996) Human ICE/CED-3 protease nomenclature. *Cell* **87**: 171.

Ashburner,M. (1989) Drosophila a laboratory handbook. Cold Spring Harbor Laboratory Press, Cold Spring Harbor.

Bellamy,C.O., Malcolmson,R.D., Harrison,D.J. and Wyllie,A.H. (1995) Cell death in health and disease: the biology and regulation of apoptosis. *Semin. Cancer Biol.* **6**: 3-16.

Bellen,H.J., O'Kane,C.J., Wilson,C., *et al.* (1989) P-element-mediated enhancer detection: a versatile method to study development in *Drosophila. Genes Dev.* **3**: 1288-1300.

Bender,W., Spierer,P. and Hogness,D.S. (1983a) Chromosomal walking and jumping to isolate DNA from the ACE and rosy loci and the Bithorax complex in *Drosophila melanogaster. J. Mol. Biol.* **168**: 17-33.

Bender,W., Akam,M.E., Karch,F., *et al.* (1983b) Molecular genetics of the Bithorax complex in *Drosophila melanogaster. Science* **221**: 23-29.

Bergeron,L., Perez,G.I., MacDonald G., *et al.* (1998). Defects in regulation of apoptosis in caspase-2-deficient mice. *Genes Dev.* **12**, 1304-1314.

Bernardoni,R., Vivancos,B. and Giangrande,A. (1997) *glide/gcm* is expressed and required in the scavenger cell lineage. *Dev. Biol.* **191**: 118-130.

Bier,E., Vassin,H., Shepherd,S., *et al.* (1989) Searching for pattern and mutation in the *Drosophila* genome with a P-lacZ vector. *Genes Dev.* **3**: 1273-1287.

Birnbaum,M.J., Clem,R.J. and Miller,L.K. (1994) An apoptosis-inhibiting gene from a nuclear polyhedrosis virus encoding a peptide with Cys/His sequence motifs. *J. Virol.* **68**: 2521-2328.

Brancolini,C., Benedetti,M. and Schneider,C. (1995) Microfilament reorganization during apoptosis: the role of Gas2, a possible substrate for ICE-like proteases. *EMBO J.* **14**: 5179-5190.

Brand,A.H. and Perrimon,N. (1993) Targeted gene expression as a means of altering cell fates and generating dominant phenotypes. *Development* **118**: 401-415.

Bump,N.J., Hackett,M., Hugunin,M., *et al.* (1995) Inhibition of ICE family proteases by baculovirus antiapoptotic protein p35. *Science* **269**: 1885-1888.

Campos-Ortega,J.A. and Hartenstein,V. (1997) in The Embryonic Development of *Drosophila melanogaster*, Springer-Verlag.

Capobianco,A.J., Zagouras,P., Blaumueller,C.M., *et al.* (1997) Neoplastic transformation by truncated alleles of human NOTCH/TAN1 and NOTCH2. *Mol. Cell Biol.* **17**: 6265-6273.

Cerretti,D.P., Kozlosky,C.J., Mosley,B., *et al.* (1992) Molecular cloning of the interleukin-1 beta converting enzyme. *Science* **256**: 97-100.

Chen,P., Nordstrom,W., Gish,B. and Abrams,J.M. (1996a) *grim*, a novel cell death gene in *Drosophila. Genes Dev.* **10**: 1773-1782.

Chen,P., Lee,P., Otto,L. and Abrams,J. (1996b) Apoptotic activity of REAPER is distinct from signaling by the tumor necrosis factor receptor 1 death domain. *J. Biol. Chem.* **271**: 25735-25737.

Cheng,E.H.-Y., Kirsch,D.G., Clem,R.J., *et al.* (1997) Conversion of Bcl-2 to a Bax-like death effector by caspases. *Science* **278**: 1966-1968.

Chinnaiyan,A.M. and Dixit,V.M. (1996). The cell-death machine. *Curr. Biol.* **6**: 555-62.

Chinnaiyan,A.M., O'Rourke,K., Lane,B.R. and Dixit, V.M. (1997) Interaction of CED-4 with CED-3 and CED-9: a molecular framework for cell death. *Science* **275**: 1122-1126.

Clem,R.J., Fechheimer,M. and Miller,L.K. (1991) Prevention of apoptosis by a baculovirus gene during infection of insect cells. *Science* **254**: 1388-1390.

Clem,R.J. and Miller,L.K. (1994) Control of programmed cell death by the baculovirus genes p35 and iap. *Mol. Cell Biol.* **14**: 5212-5222.

Cleveland,J.L. and Ihle,J.N. (1995) Contenders in Fas/TNF death signalling. *Cell* **81**: 479-482.

Crook,N.E., Clem,R.J. and Miller,L.K. (1993) An apoptosis-inhibiting baculovirus gene with a zinc finger-like motif. *J. Virol.* **67**: 2168-2174.

Davidson,F.F. and Steller,H. (1998) Blocking apoptosis prevents blindness in *Drosophila* retinal degeneration mutants. *Nature* **391**: 587-591.

Davis,E.J., Foster,T.D. and Thomas,W.E. (1994) Cellular forms and functions of brain microglia. *Brain Res. Bull.* **34**: 73-78.

Demerec,M. (1950) in Biology of *Drosophila*. John Wiley & Sons, Inc. New York; Chapman & Hall, Limited, London; (1994) re-issued by Cold Spring Harbor Laboratory Press, Cold Spring Harbor.

DeRobertis,E.M. and Sasai,Y. (1996) A common plan for dorsoventral patterning in Bilateria. *Nature* **380**: 37-40.

Devereaux,Q.L., Takahashi,R., Salvesen,G.S. and Reed,J.C. (1997) X-linked Iap is a direct inhibitor of cell-death proteases. *Nature* **388**: 300-304.

Dorn,R., Krauss,V., Reuter,G. and Saweber,H. (1993) The enhancer of position-effect variegation of *Drosophila E(var)*3-93D codes for a chromatin protein containing a conserved domain common to several transcription factors. *Proc. Natl. Acad. Sci. USA* **90**: 11376-11380.

Driscoll,M. (1992) Molecular genetics of cell death in the nematode *Caenorhabditis elegans*. *J. Neurobiol.* **23**: 1327-1351.

Dryja,T.P., McGee,T.L., Reichal.,E., *et al.* (1990) A point mutation of the rhodopsin gene in one form of retinitis pigmentosa. *Nature* **343**: 364-366.

Ellis,R., Yuan,J. and Horvitz,H.R. (1991) Mechanisms and functions of cell death. *Ann. Rev. Cell Biol.* **7**: 663–698.

Ellis,H.M. and Horvitz,H.R. (1986). Genetic control of programmed cell death in the nematode *C. elegans*. *Cell* **44**: 817-29

Evans,E.K., Kuwana,T., Strum,S.L., *et al.* (1997) Reaper-induced apoptosis in a vertebrate system. *EMBO J.* **16**: 7372-7381.

Fan,H., Oro,A.E. Scott,M.P. and Khavari,P.A. (1997) Induction of basal cell carcinoma features in transgenic human skin expressing Sonic Hedgehog. *Nature Med.* **3**: 788-792.

Fessler,J.H. and Fessler,L.I. (1989) *Drosophila* extracellular matrix. *Ann. Rev. Cell Biol.* **5**: 309-339.

Fisher,G.H., Rosenberg,F.J., Straus,S.E., *et al.* (1995). Dominant interfering Fas gene mutations impair apoptosis in a human autoimmune lymphoproliferative syndrome. *Cell* **16**: 935-46.

Foley,K. and Cooley,L. (1998) Apoptosis in late stage *Drosophila* nurse cells does not require genes within the H99 deficiency. *Development* **125**: 1075-1082.

Fraser,A.G. and Evan,G.I. (1997) Identification of a *Drosophila melanogaster* ICE/CED-3-related protease, drICE. *EMBO J.* **16**: 2805-2813.

Gavrieli,Y., Sherman,Y. and Ben Sassoon,S. (1992) Identification of programmed cell death *in situ* via specific labeling of nuclear DNA fragmentation. *J. Cell Biol.* **119**: 493-501.

Gerasimova,T.I., Gdula,D.A., Gerasimov,D.V., *et al.* (1995) A *Drosophila* protein that imparts directionality on a chromatin insulator is an enhancer of position effect variegation. *Cell* **82**: 587-597.

Gerhart,J. and Kirschner,M. (1997) in Cells, Embryos, and Evolution. Blackwell Science Inc., Malden.

Golstein,P., Marguet,D. and Depraetere,V. (1995) Homology between Reaper and the cell death domains of Fas and TNFR1. *Cell* **81**: 185.

Golic,K.G. and Lindquist,S. (1989) The FLP recombinase of yeast catalyzes site-specific recombination in the *Drosophila* genome. *Cell* **59**: 499-509.

Grether,M.E., Abrams,J.M., Agapite,J., *et al.* (1995) The *head involution defective* gene of *Drosophila melanogaster* functions in programmed cell death. *Genes Dev.* **9**: 1694-1708.

Haimovitz-Friedman,A., Kolesnick,R.N. and Fulks,Z. (1997) Ceramide signalling in apoptosis. *Br. Med. Bull.* **53**: 539-553.

Halder,G., Callaerts,P. and Gehring,W. J. (1995) New perspectives on eye evolution. *Curr. Opin. Gene. Dev.*.**5**: 602-609.

Hannun,Y.A. and Obeid,L.M. (1995) Ceramide: an intracellular signal for apoptosis. *Trends Biochem. Sci.* **20**: 73-77.

Harvey,A.J., Bidwai,A.P. and Miller,L.K. (1997) Doom, a product of the *Drosophila* mod(mgd4) gene, induces apoptosis and binds to baculovirus inhibitor-of-apoptosis. *Mol. Cell Biol.* **17**: 2835-2843.

Hay,B.A., Wolff,T. and Rubin,G.M. (1994) Expression of baculovirus P35 prevents cell death in *Drosophila. Development* **120**: 2121-2129.

Hay,B.A., Wassarman,D.A. and Rubin,G.M. (1995) *Drosophila* homologs of baculovirus inhibitor of apoptosis proteins function to block cell death. *Cell* **83**: 1253-1262.

Hengartner,M.O., Ellis,R. and Horvitz,H.R. (1992) *Caenorhabditis elegans* gene *ced–9* protects cells from programmed cell death. *Nature* **356**: 494–501.

Hengartner,M.O. and Horvitz,R.H. (1994) Programmed cell death in *Caenorhabditis elegans*. *Curr. Opin. Gen. Dev.* **4**: 581-586.

Horvitz,H.R., Ellis,H.M. and Sternberg,P.W. (1982) Programmed cell death in nematode development. *Neurosci. Comm.* **1**: 56-65.

Horvitz,H.R., Shaham,S. and Hengartner,M.O. (1994) The genetics of programmed cell death in the nematode *Caenorhabditis elegans*. *Cold Spring Harbor Symp. Quant. Biol.* **59**: 377-385.
Itoh,N., Yonehara,S., Ishii,A., *et al.* (1991) The polypeptide encoded by the cDNA for human cell surface antigen Fas can mediate apoptosis. *Cell* **66**: 233-243.
Jiang,C., Baehrecke,E.H. and Thummel,C. (1997) Steroid regulated programmed cell death during *Drosophila* metamorphosis. *Development* **124**: 4673-4683.
Johnson,R.L., Rothman,A.L., Xie,J., *et al.* (1996) Human homolog of *patched*, a candidate gene for basal cell nevus syndrome. *Science* **272**: 1668-1671.
Jürgens,G., Wieschaus,E, Nüsslein-Volhard,C. and Kluding,H. (1984) Mutations affecting the pattern of the larval cuticle in *Drosophila melanogaster*. II. Zygotic loci on the third chromosome. *Roux Arch. Dev. Biol.* **193**: 283-295.
Kerr,J.F.R., Wyllie,A. and Currie,A.R. (1972) Apoptosis: a basic biological phenomenon with wide–ranging implications in tissue kinetics. *Br. J. Cancer* **26**: 239–257.
Kimura,K.I. and Truman,J.W. (1990) Postmetamorphic cell death in the nervous and muscular systems of *Drosophila melanogaster*. *J. Neurosci.* **10**: 403–411.
Klingensmith,J., Noll,N. and Perrimon,N. (1989) The segment polarity phenotype of *Drosophila* involves differential tendencies toward transformation and cell death. *Dev. Biol.* **134**: 130-145.
Kluck,R.M., Bossy-Wetzel,E., Green,D.R. and Newmeyer,D.D. (1997) The release of cytochrome c from mitochondria: a primary site of Bcl-2 regulation of apoptosis. *Science* **275**: 1132-1136.
Knudson,C.M. and Korsmeyer,S.J. (1997) Bcl-2 and Bax function independently to regulate cell death. *Nature Genet.* **16**: 358-363.
Knudson,C.M., Tung,K.S.; Tourtellotte,W.G., *et al.* (1995) Bax-deficient mice with lymphoid hyperplasia and male germ cell death. *Science* **270**: 96-99.
Koelle,M.R., Talbot,W.S., Segraves,W.A., *et al.* (1991) The *Drosophila* EcR gene encodes an ecdysone receptor, a new member of the steroid receptor superfamily. *Cell* **67**: 59-77.
Kohler,R.E. (1994) In: Lords of the Fly. The University of Chicago Press, Chicago.
Kondo,T., Yokokura,T. and Nagata,S. (1997) Activation of distinct caspase-like protease by fas and reaper in *Drosophila* cells. *Proc. Natl. Acad.Sci. USA* **94**: 11951-11956.
Kothakota,S., Azuma,T., Reinhard,C., *et al.* (1997) Caspase-3-Generated Fragment of Gelsolin: Effector of Morphological Change in Apoptosis. *Science* **278**: 294-298.
Krieger,M., Abrams,J.M., Lux,A. and Steller,H. (1992) Molecular flypaper, athlerosclerosis, and host defense: structure and function of the macrophage scavenger receptor. *Cold Spring Harbor Symp. Quant. Biol.* **57**: 605-609.
Kuida,K., Lippke,J.A., Ku,G., *et al.* (1995) Altered cytokine export and apoptosis in mice deficient in interleukin-1ß converting enzyme. *Science* **267**: 2000-2003.
Kuida,K., Zheng,T.S., Na,S., *et al.* (1996) Decreased apoptosis in the brain and premature lethality in CPP32-deficient mice. *Nature* **28**: 368-372.
Lawrence P.A. (1992) The making of a fly. Blackwell Scientific Publications, Oxford.
Lazebnik,Y.A., Takahashi,A., Moir,R.D., *et al.* (1995) Studies of the lamin proteinase reveal multiple parallel biochemical pathways during apoptotic execution. *Proc. Natl. Acad. Sci. USA* **92**: 9042-9046.
Lee,F.S., Lane,T.F., Kuo,A., *et al.* (1995) Insertional mutagenesis identifies a member of the Wnt gene family as a candidate oncogene in the mammary epithelium of int-2/Fgf-3 transgenic mice. *Proc. Natl. Acad. Sci. USA* **92**: 2268-2272.
Leonard,D.S., Bowman,V.D., Ready,D.F. and Pak,W.L. (1992) Degeneration of photoreceptors in rhodopsin mutants of *Drosophila*. *J. Neurobiol.* **23**: 605-626.
Lewis,E.B. (1978) A gene complex controlling segmentation in *Drosophila*. *Nature* **276**: 565-570.
Lindsley,D.L. and Zimm,G.G. (1992) In: The genome of *Drosophila Melanogaster*. Academic Press, San Diego.
Liston,P., Roy,N., Tamai,K., *et al.* (1996) Suppression of apoptosis in mammalian cells by NAIP and a related family of IAP genes. *Nature* **379**: 349-353.
Liu,X., Zou,H., Slaughter,C. and Wang,X. (1997) DFF, a heterodimeric protein that functions Downstream of Caspase-3 to trigger DNA Fragmentation during Apoptosis. *Cell* **89**: 175-184.
Liu,X., Kim,C.N., Yang,J., *et al.* (1996). Induction of apoptotic program in cell-free extracts: requirement for dATP and cytochrome c. *Cell* **86**: 147-57.
Liu,Z.G., Smith,S., McLaughlin,K.A., *et al.* (1994) Apoptotic signals delivered through the T cell receptor require the immediate early gene *nur77*. *Nature* **367**: 281–284.
Lockshin,R.A. and Williams,C.M. (1965) Programmed cell death: Cytology of degeneration in the intersegmental muscles of the silkmoth. *J. Insect Physiol.* **11**: 123–33.
Lockshin,R.A. (1969) Activation of lysis by a mechanism involving the synthesis of protein. *J. Insect Physiol.* **15**: 1505–16.
Magrassi,L. and Lawrence,P.A. (1988) The pattern of cell death in *fushi tarazu*, a segmentation gene of *Drosophila*. *Development* **104**: 447-451.
Martin,D.P., Schmidt,R.E., DiStefano,P.S., *et al.* (1988) Inhibitors of protein synthesis and RNA synthesis prevent neuronal death caused by nerve growth factor deprivation. *J. Cell Biol.* **106**: 829-844.
McCall,K. and Steller,H. (1997) Facing death in the fly: Genetic analysis of apoptosis in *Drosophila*. *Trends Genet.* **13**: 222-226.

McCall,K. and Steller,H. (1998) Requirements for DCP-1 caspase during *Drosophila* oogenesis. *Science* **279**: 230-234.
McNabb,S., Baker,J.D., Agapite,J., *et al.* (1997) Disruption of a behavioral sequence by targeted death of peptidergic neurons in *Drosophila. Neuron* **19**: 813-823.
Mignotte,B. and Vayssiere,J.-L. (1998) Mitochondria and apoptosis. *Eur. J. Biochem.* **252**: 1-15.
Min,K.T. and Benzer,S. (1997) Spongecake and Eggroll: Two Hereditary diseases in *Drosophila* resemble patterns of human brain degeneration. *Curr. Biol.* **7**: 885-888.
Miura,M., Zhu,H., Rotello,R., *et al.* (1993) Induction of apoptosis in fibroblasts by IL-1 beta-converting enzyme, a mammalian homolog of the *C. elegans* cell death gene *ced-3*. *Cell* **75**: 653-60.
Muzio,M., Salvesen,G.S. and Dixit,V.M. (1997) FLICE-induced apoptosis in a cell-free system. Cleavage of caspase zymogens. *J. Biol. Chem.* **272**: 2952-2956.
Nagata,S. (1997) Apoptosis by death factor. *Cell* **88**: 355-365.
Nagata,S. and Golstein,P. (1995) The Fas death factor. *Science* **267**: 1449-1456.
Namba,R., Pazdera,T.M., Cerrone,R.L. and Minden,J.S. (1997) *Drosophila* embryonic pattern repair: how embryos respond to *bicoid* dosage alteration. *Development* **124**: 1393-1403.
Newmeyer,D.D., Farschon,D.M. and Reed,J.C. (1994) Cell-free apoptosis in Xenopus egg extracts: inhibition by Bcl-2 and requirement for an organelle fraction enriched in mitochondria. *Cell* **79**: 353-364.
Nordstrom,W., Chen,P., Steller,H. and Abrams,J.M. (1996) Activation of the *reaper* gene during ectopic cell killing in *Drosophila. Dev. Biol.* **180**: 213-226.
Nüsslein-Volhard,C. and Wieschaus,E. (1980) Mutations affecting segment number and polarity in *Drosophila. Nature* **287**: 795-801.
Nüsslein-Volhard,C., Wieschaus,E. and Kluding,H. (1984) Mutations affecting the pattern of the larval cuticle in *Drosophila Melanogaster.* I. Zygotic loci on the second chromosome. *Wilhelm Roux's Arch. Dev. Biol.* **193**: 267-282.
Oehm,A., Behrmann,I., Falk, W., *et al.* (1992) Purification and molecular cloning of the APO-1 cell surface antigen, a member of the tumor necrosis factor/ nerve growth factor superfamily. Sequence identity with the Fas antigen. *J. Biol. Chem.* **267**: 10709-10715.
Oppenheim,R.W., Prevette,D., Tytell,M. and Homma,S. (1990) Naturally occurring and induced neuronal death in the chick embryo *in vivo* requires protein and RNA synthesis: evidence for the role of cell death genes. *Dev. Biol.* **138**: 104-113.
Oro,A.E., Higgins,K.M., Hu,Z., *et al.* (1997) Basal carcinomas in mice overexpressing soinc the hedgehog. *Science* **276**: 817-821.
Osborne,B.A. (1996) Cell death in vertebrates: lessons from the worm. *Trends Genet.* **12**: 489-491.
Osborne,B.A., Smith,S.W., McLaughlin,K.A., *et al.* (1996) genes that regulate apoptosis in the mouse thymus. *J. Cell Biochem.* **60**: 18-22.
Pearson,A., Lux,A. and Krieger,M. (1995) Expression cloning of dSR-CI, a class C macrophage-specific scavenger receptor from *Drosophila melanogaster. Proc. Natl. Acad. Sci. USA* **92**: 4056-4060.
Pronk,G.J., Ramer,K., Amiri,P. and Williams,L.T .(1996) Requirement of an ICE-like protease for induction of apoptosis and ceramide generation by REAPER. *Science* **271**: 808-810.
Rabizadeh,S., LaCount,D.J., Friesen,P.D. and Bredesen,D.E. (1993) Expression of the baculovirus p35 gene inhibits mammalian neural cell death. *J. Nuerochem.* **61**: 2318-2321.
Reed,J.C. (1997) Double identity for proteins of the BCL-2 family. *Nature* **387**: 773-776.
Robinow,S., Talbot,W.S., Hogness,D.S. and Truman,J.W. (1993) Programmed cell death in the *Drosophila* CNS is ecdysone-regulated and coupled with a specific ecdysone receptor isoform. *Development* **119**: 1251-1259.
Robinow,S., Draizen,T.A. and Truman,J.W. (1997) Genes that induce apoptosis: Transcriptional regulation in identified, doomed neurons of the *Drosophila* CNS. *Dev. Biol.* **190**: 206-213.
Roessler,E., Belloni,E., Gaudenz,K., *et al.* (1996) Mutations in the human Sonic Hedgehog gene cause holoprosencephaly. *Nature Genet.* **14**: 357-360.
Rothe,M. Pan,M-G., Henzel,W.J., *et al.* (1995) The TNFR2-TRAF signaling complex contains two novel proteins related to baculoviral inhibitor of apoptosis proteins. *Cell* **83**: 1243-1252.
Roy,N., Mahadevan,M.S., McLean,M., *et al.* (1995) The gene for neuronal apoptosis inhibitory protein is partially deleted in individuals with spinal muscular atrophy. *Cell* **80**: 167-78.
Roy,N, Devereaux,Q.L., Takahashi,R., *et al.* (1997) The c-IAP-1 and c-IAP-2 proteins are direct inhibitors of specific caspases. *EMBO J.* **16**: 6914-6925.
Rudin,C.M. and Thompson,C.B. (1997) Apoptosis and disease: regulation and clinical relevence of programmed cell death. *Ann. Rev. Med.* **48**: 267-281.
Salvesen,G.S. and Dixit,V.M. (1997) Caspases: intracellular signaling by proteolysis. *Cell* **14**: 443-446.
Sawamoto,K., Taguchi,A., Hirota,Y., *et al.* (1998) Argos induces programmed cell death in the developing *Drosophila* eye by inhibition of the Ras pathway. *Cell Death Diff.* **5**: 262-270.
Schwartz,L.M., Kosz,L. and Kay,B.K. (1990) Gene activation is required for developmentally programmed cell death. *Proc. Natl. Acad. Sci. USA* **87**: 6594–6598.
Schwartz,L.M., Smith,S.W., Jones,M.E. and Osborne,B.A. (1993) Do all programmed cell deaths occur via apoptosis? *Proc. Natl. Acad. Sci. USA* **90**: 980-984.
Schwartz,L.M., Milligan,C.E., Bielke,W. and Robinson,S.J. (1994) Cloning cell death genes. *J. Neruobiol.* **25**: 1005-1016.
Schwartz,L.M. and Milligan,C.M. (1996) Cold thoughts of death: The role of ICE proteases in neuronal apoptosis. *Trends Neurosci.* **19**: 555-562.

Seshagiri,S. and Miller,L.K. (1997) *Caenorhabitis elegans* CED-4 stimulates CED-3 processing and CED-3 induced apoptosis. *Curr Biol.* 7: 455-460.

Shigenaga,A., Funahashi,Y., Kimura,K., *et al.* (1997) Targeted expression of *ced-3* and *Ice* induces programmed cell death in *Drosophila. Cell Death Diff.* 4: 371-1377.

Song,Z., McCall,K. and Steller,H. (1997) DCP-1, a *Drosophila* cell death protease essential for development. *Science* **275**: 536-540.

Sonnenfeld,M.J. and Jacobs,J.R. (1995a) Macrophages and glia participate in the removal of apoptotic neurons from the *Drosophila* embryonic nervous system. *J. Comp. Neurol.* **359**: 644-652.

Sonnenfeld,M.J. and Jacobs,J.R. (1995b) Apoptosis of midline glia during *Drosophila* embryogenesis: a correlation with axon contact. *Development* **121**: 569-578.

Spector,M.S., Desnoyers,S., Heoppner,D.J. and Hengartner,M.O. (1997) Interaction between the *C. elegans* cell-death regulators CED-9 and CED-4. *Nature* **275**: 1122-1226.

Spradling,A.C. and Rubin,G.M. (1982) Transposition of cloned P elements into *Drosophila* germ line chromosomes. *Science* **218**: 341-347.

Spradling,A.C. (1993) Developmental Genetics of Oogenesis. In: The Development of *Drosophila melanogaster*. Cold Spring Harbor Laboratory Press, Cold Spring Harbor. pp. 1-70.

Stemerdink,C. and Jacobs,J.R. (1997) Argos and Spitz group genes function to regulate midline glial cell number in *Drosophila* embryos. *Development* **124**: 3787-3796

Steller,H. (1995) Mechanisms and genes of cellular suicide. *Science* **267**: 1445-1449.

Sugimoto,A., Friesen,P.D. and Rothman,J. (1994) Baculovirus p35 prevents developmentally programmed cell death and rescues a *ced-9* mutant in the nematode *Caenorhabditis elegans. EMBO J.* **13**: 2023-2038.

Sulston,J.E. and Horvitz,H.R. (1977) Post-embryonic cell lineages of the nematode *Caenorhabditis elegans. Dev. Biol.* **82**: 110-156.0

Sulston,J.E., Schierenberg,E., White,J.G. and Thomsom,N., (1983) The embryonic cell lineage of the nematode *Caenorhabditis elegans. Dev. Biol.* **100**: 64-119.

Talbot,W.S., Swyryd,E.A. and Hogness,D.S. (1993) *Drosophila* tissues with different metamorphic responses to ecdysone express different ecdysone receptor isoforms. *Cell* **73**: 1323–1337.

Tepass,U., Fessler,L.I., Aziz,A. and Hartenstein,V. (1994) Embryonic origin of hemocytes and their relationship to cell death in *Drosophila. Development* **120**: 1829-1837.

Thompson,C.B. (1995) Apoptosis in the pathogensis and treatment of disease. *Science* **267**: 1456-1462.

Thornberry,N.A., Bull,H.G., Calaycay,J.R., *et al.* (1992) A novel heterodimeric cysteine protease is required for interleukin-1 beta processing in monocytes. *Nature* **356**: 768-774.

Thornberry,N.A. (1997) The caspase family of cysteine proteases. *Br. Med. Bull.* **53**: 478-490.

Thornberry,N.A., Rano,T.A., Peterson,E.P., *et al.* (1997) A combinatorial approach defines specificities of members of the caspase family and granzyme B. Functional relationships established for key mediators of apoptosis. *J. Biol. Chem.* **272**: 17907-17911

Truman,J., Thorn,R. and Robinow,S. (1992) Programmed neuronal death in insect development. *J. Neurobiol.* **23**: 1295-1311.

Truman,J.W., Taylor,B.J. and Awad,T.A. (1993) Formation of the adult nervous system.. In: The Development of *Drosophila melanogaster*. pp. 1245-1275. Cold Spring Harbor Laboratory Press, Cold Spring Harbor.

Uren,A.G., Pakusch,M., Hawkins,C.J., *et al.* (1996) Cloning and expression of apoptosis inhibitory protein homologs that function to inhibit apoptosis and/or bind tumor necrosis factor receptor-associated factors. *Proc. Natl. Acad. Sci. USA* **93**: 4974-4978.

van Leeuwen,F. and Nusse,R. (1995) Oncogene activation and oncogene cooperation in MMTV-induced mouse mammary cancer. *Semin. Cancer Biol.* **6**: 127-133.

Vaux,D.L. and Strasser,A. (1996) The molecular biology of apoptosis. *Proc. Natl. Acad. Sci. USA* **93**: 2239-2244.

Vaux,D.L., Weissman,I.L. and Kim,S.K. (1992) Prevention of programmed cell death in *Caenorhabditis elegans* by human bcl–2. *Science* **258**: 1955–1957.

Veis,D.J., Sorenson,C.M., Shutter,J.R. and Korsmeyer,S.J. (1993) Bcl-2-deficient mice demonstrate fulminant lymphoid apoptosis, polycystic kidneys, and hypopigmented hair. *Cell* **75**: 229-240.

Villa,P., Kaufmann,S.H. and Earnshaw,W.C. (1997) Caspases and caspase inhibitors. *Trends Biochem. Sci.* **22**: 388-393.

Vucic,D., Kaiser,W.J., Harvey,A.J. and Miller,L.K. (1997a) Inhibition of Reaper-induced apoptosis by interaction with inhibitor of apoptosis proteins (IAPs). *Proc. Natl. Acad. Sci. USA* **94**: 10183-10188.

Vucic,D., Seshagiri,S. and Miller,L.K. (1997b) Characterization of Reaper- and FADD-induced apoptosis in a lepitopteran cell line. *Mol. Cell Biol.* **17**: 667-676.

Vucic,D., Kaiser,W.J. and Miller,L.K. (1998) Inhibitor of apoptosis proteins physically interact with and block apoptosis induced by *Drosophila* proteins HID and GRIM. *Mol. Cell Biol.* In Press.

Wang,L., Miura,M., Bergeron,L., *et al.* (1994) Ich-1, an Ice/ced-3-related gene, encodes both positive and negative regulators of programmed cell death. *Cell* **78**: 739-750.

Warrick,J.M., Paulson,H.L., Gray-Board,G.L., *et al.* (1998) Expanded polyglutamine protein forms nuclear inclusions and causes neural degeneration in *Drosophila. Cell* **93**: 939-949.

White,E. (1996) Life, death, and the pursuit of apoptosis. *Gen. Dev.* **10**: 1-15.

White,K., Grether,M.E., Abrams,J.M., *et al.* (1994) Genetic control of programmed cell death in *Drosophila. Science* **264**: 677-683.

White,K., Tahaoglu,E. and Steller,H. (1996) Cell killing by the *Drosophila* gene *reaper*. *Science* **271**: 805-807.

Wieschaus,E., Nusslein-Volhard,C. and Kluding,H. (1984) Mutations affecting the pattern of the larval cuticle in *Drosophila Melanogaster*. III. Zygotic loci on the X-chromosome and fourth chromosome. *Wilhelm Roux's Arch. Dev. Biol.* **193**: 296-307.

Wilson,C., Pearson,R.K., Bellen,H.J., *et al.* (1989) P-element-mediated enhancer detection: an efficient method for isolating and characterizing developmentally regulated genes in *Drosophila*. *Gen. Dev.* **3**: 1301-1313.

Wing,J.P., Zhou,L., Schwartz,L.M. and Nambu,J.R. (1998) Distinct cell killing properties of the *Drosophila reaper, head involution defective*, and *grim* genes. *Cell Death and Diff.* In Press.

Woronicz,J.D., Calnan,B., Ngo,V. and Winoto,A. (1994) Requirement for the orphan steroid receptor Nur77 in apoptosis of T-cell hybridomas. *Nature* **367**: 277-281.

Wu,D., Wallen,H.D., Inohara,N. and Nunez,G. (1997) Interaction and regulation of the *Caenorhabitis elegans* death protease CED-3 by CED-4 and CED-9. *J. Biol. Chem.* **272**: 21449-21454.

Wyllie,A.H., Kerr,J.F.R. and Currie,A.R. (1980) Cell death: the significance of apoptosis. *Int. Rev. Cytol.* **68**: 251-306.

Wyllie,A.H., Morris,R.G., Smith,A.L. and Dunlop,D. (1984) Chromatin cleavage in apoptosis: association with condensed chromatin morphology and dependence on macromolecular synthesis. *J. Pathol.* **142**: 67-77.

Xie,J. Johnson,R.L., Zhang,X., Bare,J.W., *et al.* (1997) Mutations of the PATCHED gene in several types of sporadic extracutaneous tumors. *Cancer Res.* **57**: 2369-2372.

Xu,T. and Rubin,G.M. (1993) Analysis of genetic mosaics in developing and adult *Drosophila* tissues. *Development* **117**: 1223-1237.

Xue,D. and Horvitz,H.R. (1995) Inhibition of the *Caenorhabditis elegans* cell-death protease CED-3 by a CED-3 cleavage site in baculovirus p35 protein. *Nature* **377**: 248-251.

Xue,D. and Horvitz,H.R. (1997) *Caenorhabditis elegans* CED-9 protein is a bifunctional cell death inhibitor. *Nature* **390**: 305-308.

Yang,X., Chang,H.Y. and Baltimore,D. (1998) Autoproteolytic activation of pro-caspases by oligomerization. *Molec. Cell.* **1**: 319-325.

Yao,T.P., Segraves,W.A., Oro,A.E., *et al.* (1992) *Drosophila* ultraspiracle modulates ecdysone receptor function via heterodimer formation. *Cell* **71**: 63-72.

Yuan,J. and Horvitz,H.R. (1990) The *Caenorhabditis elegans* genes *ced–3* and *ced–4* act cell–autonomously to cause programmed cell death. *Dev. Biol.* **138**: 33–41.

Yuan,J., Shaham,S., Ledoux,S., *et al.* (1993) The *C. elegans* cell death gene *ced-3* encodes a protein similar to mammalian interleukin-1 beta-converting enzyme. *Cell* **75**: 641-652.

Yuan,J. (1997) Transducing signals of life and death. *Curr. Opin. Cell Biol.* **9**: 247-251.

Zhou,L., Hashimi,H., Schwartz,L.M. and Nambu,J.R. (1995) Programmed Cell Death in the *Drosophila* Central Nervous System Midline. *Curr. Biol.* **5**: 784-790.

Zhou, L., Schnitzler, A,. Agapite, J., *et al.* (1997) Cooperative functions of the *reaper* and *head involution defective* genes in the programmed cell death of *Drosophila* central nervous system midline cells. *Proc. Natl. Acad. Sci. USA* **94**: 5131-5136.

Zou,H., Henzel,W.J., Liu,X., *et al.* (1997) Apaf-1, a human protein homologous to *C. elegans* CED-4, participates in cytochrome c-dependent activation of caspase-3. *Cell* **90**: 405-413.

Chapter 9

Viral genes that modulate apoptosis.

J. Marie Hardwick, Gary Ketner and Rollie J. Clem.
Departments of Molecular Microbiology and Immunology [JMH, GK], Nuerology and Pharmacology and Molecular Sciences [JMH], John Hopkins School of Public Health and Medicine, Baltimore MD. Division of Biology, Kansas State University [RJC].

1.0 INTRODUCTION

Viruses trigger a programmed cellular suicide pathway upon infection of cells in a wide range of organisms from mammals to bacteria. Programmed cell death in response to virus infection is presumably a host defence mechanism in an attempt to eliminate infected cells and to block spread of the pathogen. Viruses activate programmed cell death by a variety of direct and indirect mechanisms that are not yet fully understood. DNA viruses with large coding capacities have acquired genes that avert the cell's attempt to die, leading to successful completion of the viral replication cycle and production of progeny virus. The study of these viruses has led to the identification of several families of genes that inhibit or delay cell death by acting at various points along the cell death pathway, from blocking initiation of receptor signalling to inhibition of caspases near the final stages of cell death. Viral apoptosis inhibitors may contribute to viral pathogenesis by mechanisms other than increasing viral load. For example, cellular anti-apoptotic proteins are known to participate in cell transformation and tumourigenesis. Thus, viral apoptosis inhibitors may also contribute to human cancer.

In contrast to those viruses that encode inhibitors of cell death, many smaller viruses replicate efficiently in cells that have activated their apoptotic pathway. Cell death triggered by these viruses, especially those that target unreplaceable post-mitotic cells, is detrimental to the host and contributes to pathogenesis. In these situations, the successful host resists the temptation to undergo apoptosis in response to a virus infection. The decision by the host to either enter or stave off the apoptotic pathway following virus infection is presumably regulated by host cell apoptosis modulators. Failure to rid the host of virus-infected cells can result in a persistent infection, but persistence of a virus such as Sindbis virus is more desirable compared to the alternative, excessive apoptosis of neurons and death of the animal. The molecular cell death pathway that modulates viral pathogenesis is just beginning to be elucidated.

2.0 VIRUS-INDUCED PROGRAMMED CELL DEATH

2.1 Baculovirus-induced apoptosis

The baculovirus *Autographa californica* multicapsid nucleopolyhedrovirus (AcMNPV) triggers the apoptotic pathway during infection of lepidopteran insect cells. This apoptotic response is normally blocked by the viral protein P35 (see below). Without P35, infected insect cells undergo apoptosis that is probably triggered by more than one type of apoptotic signal. AcMNPV-induced apoptosis can be completely blocked by treating cells with the DNA synthesis inhibitor aphidicolin (Clem and

Miller, 1994), suggesting that apoptosis is triggered by either viral DNA synthesis, the accompanying decrease in host RNA and protein synthesis, or the expression of a late viral gene product. In support of a role for virus-induced shut-off of host macromolecular synthesis, drugs that inhibit host RNA synthesis also potently induce apoptosis in the uninfected cells (Clem and Miller, 1994).

IE-1, a potent transcription factor encoded by AcMNPV, is capable of inducing apoptosis in transfected SF-21 insect cells independently of other viral genes (Prikhod'ko and Miller, 1996). However, apoptosis induced by IE-1 alone is less robust than with virus infection, suggesting that other viral components play a role. Viral DNA synthesis also has been implicated as a key factor in apoptosis (Prikhod'ko and Miller, 1996). To address this issue, a virus carrying a temperature-sensitive mutation affecting viral DNA synthesis (as well as a null p35 mutation) was tested and found competent to induce apoptosis at the non-permissive temperature albeit less efficiently (LaCount and Friesen, 1997). Thus, it appears that both early and late signals are involved in triggering apoptosis by baculovirus.

2.2 Adenovirus E1A, ADP and E4orf4

In its natural host, adenovirus infects quiescent cells which are not conducive to viral replication. However, the adenovirus E1A protein, which is expressed by a constitutively active promoter, stimulates cells to enter the cell cycle, thereby providing the virus with cell factors required for DNA replication (White, 1993; Debbas and White, 1993). However, like baculovirus, E1A triggers apoptosis in the absence of the virus-encoded apoptosis inhibitors (see below) (Debbas and White, 1993; Rao *et al.* 1992). There are multiple potential mechanisms by which the adenovirus E1A protein activates the cell death pathway.

2.2.1 E1A and p53

The E1A protein increases intracellular levels of the tumour suppressor p53 through stabilisation of p53 protein (Hinds and Weinberg, 1994; Lowe and Ruley, 1993) (Fig. 1). Elevated p53 protein levels induce cell cycle arrest in some circumstances but induce apoptosis in other circumstances. E1A-induced apoptosis in quiescent primary baby rat kidney (BRK) cells is dependent on p53 (Debbas and White, 1993). Although the molecular events by which p53 induces apoptosis are not understood, they are likely to involve its activity as a DNA-binding transcription factor. Whether p53 induces apoptosis through transcriptional activation or transcriptional repression is still debated, with reasonably convincing arguments for both hypotheses. Application of the SAGE technique identified 14 transcripts that are significantly increased in cells undergoing p53-induced apoptosis, several of which encode proteins linked to oxidative stress responses and are thought to have a causal role in cell death (Polyak *et al.* 1997). SAGE also identified several transcripts that are downregulated during p53-induced cell death, but these remain uncharacterised (Polyak *et al.* 1997). Others have reported that p53 directly activates transcription of the pro-death protein Bax (Miyashita and Reed,1995). p53 may also induce expression of death receptor (DR)-5, a member of the TNF family of membrane receptors, that triggers apoptosis upon binding to its ligand TRAIL (Wu *et al.* 1997). Still others have found that the transcription activation function of p53 is not required for induction of apoptosis (Canman and Kastan, 1997).

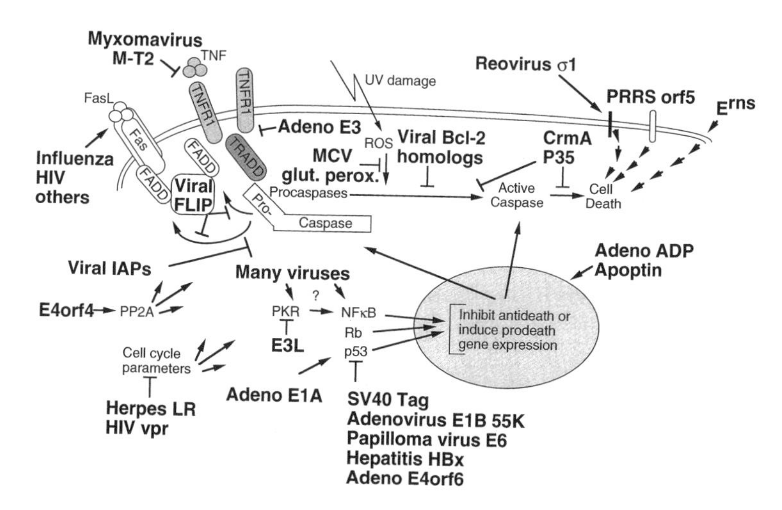

Figure 1. Virus-encoded modulators of apoptosis.

p53 also induces the expression of $p21^{WAF-1/CIP1}$ which binds to and inhibits cyclin-dependent kinases (cdk) (El-Deiry *et al.* 1993; Harper *et al.* 1993). Induction of $p21^{WAF-1/CIP1}$ by p53 is believed to contribute to p53-dependent cell cycle arrest, which allows cells to repair DNA damage before proceeding through the cell cycle. If this checkpoint is ignored in the face of DNA damage, cells enter the apoptotic pathway (Lowe *et al.* 1993). Consistent with these findings, $p21^{WAF-1/CIP1}$ and $p16^{INK}$ protect myoblasts from undergoing apoptosis during differentiation (Wang and Walsh, 1996). Furthermore, inactivation of $p21^{WAF-1/CIP1}$ leads to apoptosis in colorectal carcinoma cells (Polyak *et al.* 1996). Cleavage of $p21^{WAF-1/CIP1}$ by caspases causes a significant increase in cdk2 activity and enhancement of cell death. This finding could suggest that $p21^{WAF-1/CIP1}$ function may be separable from apoptosis; that is, unless the $p21^{WAF-1/CIP1}$ cleavage product is involved in promoting apoptosis.

2.2.2 E1A AND Rb

E1A also facilitates entry into S phase by binding to Rb (retinoblastoma tumour suppressor protein, p105) and two other Rb-related proteins p107 and p130. E1A binds to the "pocket" domain of Rb family members, displacing an E2F(1-5) transcription factor which also binds to the "pocket" domain of Rb family proteins (Hiebert *et al.* 1992; Qian *et al.* 1992; Qin *et al.* 1992). Sequestration of E2F by Rb, p107 and p130 inhibits transcription activation by one or more members of the E2F family (Ewen *et al.* 1991; Hannon *et al.* 1993; Li *et al.* 1993). Binding of E1A to Rb family members releases E2F, which heterodimerises with E2F-related proteins and acts as a transcription factor to turn on genes required for entry into S phase and regulation of the cell cycle, including dihydrofolate reductase (DHFR), thymidine kinase, thymidylate synthetase, ribonucleotide reductase, DNA pol, c-myb, c-myc, N-myc, cyclin A, cyclin E and cdc2 (Nevins, 1992; Schulze *et al.* 1995; Botz *et al.* 1996). Free E2F accumulates during G1 and rapidly disappears with the onset of S phase (Asano *et al.* 1996). However, sustained expression of E2F is potentially responsible for activation of the death pathway. Overexpression of E2F in quiescent fibroblasts or in *Drosophila* induces both entry into S phase and apoptosis (Asano *et al.* 1996; Kowalik *et al.* 1995; Shan and Lee, 1994). Thus, the attempt by quiescent fibroblasts to enter S phase triggered by E1A also triggers apoptosis perhaps in part due to E2F release. However, it is not yet known which E2F-responsive genes induce apoptosis or whether these genes are the same or distinct from those genes responsible for cell cycle progression. A more recent model suggest that E2F1 induces apoptosis by relieving active repression of pro-apoptotic genes (Hsieh *et al.* 1997).

2.2.3 E1A and p300

E1A has an alternative mechanism for driving cells through the cell cycle. In addition to the Rb family, E1A also binds to p300 and CBP, two closely related transcription factors that bind to several sequence-specific DNA binding proteins to regulate transcription (Arany *et al.* 1994). Transcriptional activation by p300 is inhibited by E1A possibly by shutting off differentiation-specific genes and deregulating the cell cycle (Webster *et al.* 1988). E1A appears to compete with a cellular factor P/CAF for the same binding site on p300 (Yang *et al.* 1996). P/CAF is a histone acetylase that may alter chromatin structure to facilitate transcription (Yang *et al.* 1996). A role for E1A-p300 interactions has been implicated in doxorubicin- and TNF-induced cell death (Sanchez-Prieto *et al.* 1995; Shisler *et al.* 1996).

The mechanisms by which E1A induces apoptosis also depends on the cell type. While p53 is required for E1A-induced apoptosis in primary BRK cells, p53 is dispensable for apoptosis in other cell types including HeLa cells (Teodoro *et al.* 1995; Chiou and White, 1997). Furthermore, analysis of E1A mutants that do not bind to Rb or p300 demonstrate that neither of these interactions is required for E1A-induced apoptosis in HeLa cells (Chiou and White, 1997). Thus, E1A can also induce apoptosis that is independent of p53, p300 and Rb implying that other mechanisms are involved in some cell types.

2.2.4 E4orf4

A mechanism by which E1A may induce p53-independent apoptosis is via the 14kD adenovirus early protein E4orf4 (Fig. 1). E1A-induced apoptosis is dependent on E4orf4 in several cell types regardless of their p53 status (Marcellus *et al.* 1996). When overexpressed in rodent cells E4orf4 is potently pro-apoptotic (Lavoie *et al.* 1998). E4orf4 binds to protein phosphatase 2A (PP2A) and impairs activation of AP1 by E1A and cAMP (Kleinberger and Shenk, 1993; Bondesson *et al.* 1996). A mutant of E4orf4 that fails to bind PP2A also fails to induce apoptosis (Shtrichman and Kleinberger, 1998). Although the mechanism by which E4orf4 induces cell death is not known, it was suggested that E4orf4 impairs or modifies the function of the PP2A phosphatase thereby altering the phosphorylation status of cellular proteins involved in apoptosis (Shtrichman and Kleinberger, 1998).

2.2.5 Adenovirus death protein (ADP)

Adenovirus is a non-enveloped virus that replicates in the nucleus and lyses the infected cell late in the replication cycle to release progeny virus. However, adenovirus mutants with a deletion in the E3 11.6K gene fail to lyse and little virus is detected in the supernatants (Tollefson *et al.* 1996a). Thus, the E3 11.6K protein, designated the adenovirus death protein (ADP), is required for cell lysis and the release of progeny virus late in infection (Fig. 1). The onset of cell lysis normally occurs at 2-3 days post infection with wild type virus, but does not begin until 5-6 days after infection with the ADP mutant (Tollefson *et al.* 1996a). Cell lysis induced by wild type virus does not show any of the classic signs of apoptotic morphology. DNA degradation occurs but no chromatin condensation or DNA laddering is observed (Tollefson *et al.* 1996a). Thus, cells infected with wild type virus appear to die by necrosis which requires ADP.

Mutants lacking functional ADP have no apparent defects in virus replication as evidenced by the presence of equivalent amounts of intracellular virus. The ADP mutants also inhibit host cell protein synthesis and elicit cytopathic effects such as cell rounding and detachment similar to wild type virus (Tollefson *et al.* 1996a). However, in contrast to wild type virus, mutant-infected cells remain viable late in infection and continue to metabolise and to synthesise viral proteins (Tollefson *et al.* 1996a,b). Nuclei become swollen with mutant virus several days after wild type virus-infected cells have lysed (Tollefson *et al.* 1996a).

ADP is a 101 amino acid protein with a central signal/anchor transmembrane domain, an N-terminal glycosylated luminal domain and a C-terminal cytoplasmic/nucleoplasmic domain (Scaria *et al.* 1992). Early in infection, ADP is detected in the endoplasmic reticulum and the Golgi but at later times it is prominently associated with the nuclear envelope (Scaria *et al.* 1992) where it may co-localise with the anti-apoptotic E1B 19K protein (see below) (Tollefson *et al.* 1996b). Though ADP leads to a necrotic-like cell death rather than apoptotic morphology, ADP appears to be the key component of a programme leading to cell lysis and could be thought of as programmed necrosis.

2.3 Reovirus σ1

The σ1 protein of reovirus is responsible for binding this non-enveloped virus to the cell surface. The interaction of strain T3Dσ1 protein with the cell surface appears to be sufficient to cause the rapid induction of apoptotic cell death in murine L929 cells (Tyler *et al.* 1995) (Fig. 1). This was determined by generating reassortant viruses with strain T1L which induces cell death far less efficiently than T3D. The rapid cell death phenotype mapped to the S1 gene segment that encoded two proteins, σ1 and σ1s. UV-inactivated T3D virus is also capable of inducing cell death, eliminating a role for the nonstructural protein σ1s. Thus, binding of σ1 protein to a yet unknown reovirus receptor may trigger the cell death pathway.

2.4 Apoptin of chicken anaemia virus (CAV)

CAV has a 2.3kb single strand circular DNA genome with 3 overlapping open reading frames encoding VP1, VP2 and VP3. The 121 amino acid VP3/apoptin protein localises to the nucleus and induces p53-independent apoptosis in the absence of other viral proteins (Noteborn *et al.* 1994; Zhuang *et al.* 1995) (Fig. 1). Interestingly, apoptin may selectively induce apoptosis in tumour cells, as long term expression of apoptin in normal human fibroblasts has no toxic effects (Danen-Van Oorschot *et al.* 1997).

2.5 Activation of the Fas pathway

Fas (CD95/APO-1), is abundantly expressed on activated mature lymphocytes and on lymphocytes infected with HTLV-1, HIV or EBV (Nagata and Goldstein, 1995). Fas and tumour necrosis factor receptor 1 (TNFR1) are members of a family of cell surface receptors and can transmit a death signal via their cytoplasmic tails. Once these receptors are brought into close proximity by binding to their trimeric extracellular ligands, FasL and TNF, respectively, they recruit several cytoplasmic proteins, including caspase-8. The death domain (DD) in the cytoplasmic tail of Fas binds to the DD of an adaptor protein, FADD. In turn, the death effector domain (DED) of FADD binds to the DED domain found in the N-terminal prodomain of caspase-8. Adaptor proteins, TRADD and FADD also are involved in recruiting caspase-8 and other factors to the TNF receptor and another related receptor TRAMP (DR3/APO-3) (Yuan, 1997). Recruiting caspase-8 to these receptors leads to caspase-8 activation perhaps by autocatalysis and subsequent activation of other downstream caspases. Fas plays an important role in apoptosis of activated T cells, and mutations in either Fas or FasL lead to autoimmune disease in a mouse model (Watanabe-Fukunaga *et al.* 1992; Cohen and Eisenberg, 1992; Ju *et al.* 1995; Brunner *et al.* 1995; Dhein *et al.* 1995).

Influenza virus infection induces the expression of Fas, and Fas-mediated cell death has been suggested to be an important mechanism of influenza virus-induced apoptosis (Wada *et al.* 1995; Takizawa *et al.* 1995). HIV-1 Tat protein upregulates FasL expression in T cells and has been reported to sensitise T cells to apoptosis triggered by T cell receptor engagement or binding of HIV-1 gp120 to CD4 (Katsikis *et al.* 1995; Westendorp, *et al.* 1995). Fas-mediated death also has been implicated in other virus infections including hepatitis B (Mochizuki *et al.* 1996). (See Fig. 1).

2.6 Activation of NFkB, PKR

While some viruses encode proteins that are sufficient to activate cell death, a specific Sindbis virus protein that induces apoptosis in all cell types has not been identified. Sindbis virus is a potent inhibitor of host cell protein synthesis, in part mediated through activation of the protein kinase PKR (p68) which phosphorylates the alpha subunit of translation initiation factor eIF-2 (Saito, 1990; Lewis and Hardwick, 1998). Activation and autophosphorylation occur when PKR binds double stranded RNA such as viral replicative intermediates. Expression of PKR using a vaccinia virus vector in which the viral PKR inhibitor, E3L, is deleted induces apoptosis in HeLa cells, while vaccinia virus containing an inactivated point mutant of PKR fails to induce cell death (Lee and Esteban, 1994). Thus, it is possible that Sindbis virus activates the cell death pathway by activating PKR and/or inhibiting host cell protein synthesis. The shut-off of host protein synthesis has in fact been postulated as a mechanism by which non-permissive polioviruses can activate the death pathway in HeLa cells (Tolskaya *et al.* 1995). Furthermore, the observation that uninfected HeLa cells undergo apoptosis following treatment with metabolic inhibitors supports the hypothesis that viruses could trigger apoptosis in this manner (Clem and Miller, 1994; Tolskaya *et al.* 1995). One possible inference from these studies is that HeLa cells require the synthesis of a protective protein to avoid activating the death pathway. In contrast, primary neurons and the BHK cell line, which both support a Sindbis virus infection, are actually protected from apoptosis by metabolic inhibitors. This implies that these cells require the expression of new genes to activate the death pathway, although other possibilities remain (Lewis and Hardwick, 1998; Martin *et al.* 1988; Estus *et al.* 1994). As a result, it seems unlikely that Sindbis virus would induce apoptosis through inhibition of cellular protein synthesis. Yet these data do not eliminate the possibility that PKR plays a role in Sindbis virus-induced cell death, since PKR may phosphorylate important targets that are distinct from those involved in translation regulation.

One potentially important target of PKR is IκB which regulates the function of the cellular transcription factor NFκB (Proud, 1995). Upon activation, IκB is phosphorylated and degraded, thereby liberating NFκB from its cytoplasmic retainer. NFκB then translocates from the cytoplasm to the nucleus where it binds specific DNA sequences and activates gene transcription (Beg and Baldwin, 1993). Overexpression of IκB or treatment of cells with oligonucleotide decoys that bind NFkB and prevent activation of downstream genes protects AT-3 cells from Sindbis virus induced-apoptosis, indicating an essential role for NFκB in the death pathway activated by Sindbis virus in AT-3 cells (Lin *et al.* 1995; 1998). Although Sindbis virus is capable of activating NFκB within 1-2 hours of infection the prodeath genes induced by NFκB have apparently already been synthesised (Lin *et al.* 1995). This is supported by the observation that NFκB activation must be inhibited prior to infection (Lin *et al.* 1998). In this manner, new transcription and translation would not be required following infection, consistent with the observation that Sindbis virus efficiently shuts off host protein synthesis.

Although NFκB activation has generally been considered a cell proliferation rather than a cell death signal, others have reported the involvement of NFκB in the induction of apoptosis (Kasibhatla *et al.* 1998) (Fig. 1). The identification of NFκB sites in the caspase-1 (ICE) protease promoter and demonstration that the p53 and TNF promoters are activated by NFκB is consistent with a role for NFκB in transcription-dependent induction of cell death (Casano *et al.* 1994; Wu and Lozano 1994; Trede *et al.* 1995). However, cell type specific activities and distinct NFκB-related transcription factors make the role of NFκB in cell death a complex problem.

2.7 HIV gp120, gp41 and Tat in apoptosis induction

A mounting body of information suggests that the loss of CD4+ T cells in HIV-infected individuals is a result of inappropriate apoptosis (reviewed by Badley *et al.* 1997). Evidence indicates that HIV can induce apoptosis of both infected as well as uninfected bystander cells, possibly by different mechanisms (Herbein *et al.* 1998). However, the mechanisms responsible for CD4 cell depletion in patients are currently unknown, but several hypotheses have been suggested. The viral glycoprotein gp120 is responsible for binding the virus to its cellular receptor, CD4 (a coreceptor for the T cell receptor). Subsequent fusion events are mediated by interactions between the viral glycoprotein and one of several cellular chemokine receptors. Other yet unidentified factors may also be important in binding and entry of HIV. HIV gp120 has been implicated in triggering apoptosis perhaps via aberrant CD4 signalling (Lu *et al.* 1994). In addition, crosslinking of the CD4 receptor by gp120 induces the expression of the prodeath receptor Fas. The chemokine receptor apparently also transmits a signal mediated by the viral glycoprotein (Weissman *et al.* 1997). HIV-induced signalling through the chemokine receptor results in chemotaxis of T cells, perhaps contributing to pathogenesis and apoptosis (Weissman *et al.* 1997). Other HIV-encoded proteins have been implicated in the induction of apoptosis. HIV Tat, a small transcription factor, has been reported to induce apoptosis of uninfected lymphocytes (Li *et al.* 1995). HIV-1 Tat protein also induces FasL expression in T cells and sensitises T cells to apoptosis triggered by the T cell receptor (Katsikis *et al.* 1995; Westendorp *et al.* 1995). Thus, modulation of cell death by HIV infections are likely to involve a number of interacting mechanisms.

2.8 Other viral glycoproteins

Porcine reproductive and respiratory syndrome (PRRS) virus is an arterivirus that causes severe disease in pigs. The 25kD glycosylated membrane protein (E) encoded by open reading frame 5 potently induces apoptosis when transiently expressed in COS-1 cells (Suarez *et al.* 1996) (Fig. 1). Interestingly, Bcl-2 did not inhibit death triggered by E. Pestiviruses (*Flaviviridae* family) cause leukopenia and immunosuppression. Glycoprotein E^{rns} of pestiviruses has RNase activity and is secreted from infected cells. Addition of E^{rns} protein, produced in insect cells, to primary lymphocytes induces apoptosis but probably without damaging the cell membrane (Bruschke *et al.* 1997) (Fig. 1).

3.0 VIRAL INHIBITORS OF THE CELL DEATH PATHWAY

3.1 Caspase inhibitors

Most apoptotic stimuli including virus infections, drug treatment, growth factor withdrawal and cell surface receptor ligation have been shown to activate a family of cellular proteases known as caspases which facilitate cell death (Henkart, 1996). Caspases are cysteine proteases that are synthesised as zymogens which require proteolytic cleavage to produce active enzyme subunits. Proteolytic activation of these proteases is carried out by other caspases evoking a protease cascade (Srinivasula *et al.* 1996). Activation of the caspase cascade is carefully regulated to prevent inappropriate induction of apoptosis. In addition to caspases themselves, a growing number of other caspase substrates have been identified including protein kinases, the retinoblastoma protein, cytoskeletal proteins, several autoantigens, Bcl-2 family members and others. Cleavage of these target proteins by caspases are likely to be important events in the apoptotic process by activating or inactivating essential functions. Cleavage of an endonuclease inhibitor ICAD/DFF triggers DNA fragmentation during apoptosis (Liu *et al.* 1997; Enari *et al.* 1998). Cleavage of Bcl-2 and Bcl-x_L by caspases converts these anti-apoptotic proteins into potent inducers of apoptosis (Cheng *et al.* 1997; Clem *et al.* 1998).

3.1.1 Poxvirus CrmA and other serpins

A role for caspases in viral infections is suggested by the finding that several caspase substrates including PARP, Bcl-2, Bcl-x_L a variety of autoantigens and caspases themselves are cleaved to their signature fragments following infection with either DNA or RNA viruses. For example, baculovirus infections activate an apoptotic cysteine protease in insect cells (Bertin *et al.* 1996), the adenovirus E1A protein activates caspases generating cleaved products of PARP (Boulakia *et al.* 1996) and Sindbis virus induces cleavage of PARP, NuMA and U1-70K proteins during infection (Nava *et al.* 1998). Furthermore, caspase inhibitors such as peptide substrates can block cell death induced by virus infections (Nava *et al.* 1998; Bjorklund *et al.* 1997). More importantly, it has become clear that the inhibition of caspases is an important mechanism by which viruses modulate cell death during infection. The pseudosubstrate caspase inhibitors encoded by some baculoviruses and poxviruses are required for efficient production of progeny virus apparently because these inhibitors stave off the cellular attempt to undergo apoptosis prior to completion of the viral life cycle. A key feature of these viral caspase inhibitors is an aspartate residue at the reactive site (P1 position), consistent with the fact that all caspases cleave their substrates following an aspartate.

Poxviruses and a murine herpesvirus encode members of the serpin superfamily (Ray *et al.* 1992; Virgin *et al.* 1997). Serpins are serine protease inhibitors (SPI) that regulate a wide range of cellular functions including immune and inflammatory responses (Potempa *et al.* 1994). Cowpox virus CrmA (SPI-2) is an unusual serpin in that it functions as a cysteine rather than serine protease inhibitor (Ray *et al.* 1992) (Fig.1). CrmA specifically inhibits caspase-1 (formerly called ICE for IL-1ß converting enzyme) which cleaves proIL-1ß to produce the active proinflammatory cytokine IL-1ß (Ray *et al.* 1992). Like other serpins, CrmA serves as a pseudosubstrate that stably complexes with caspase-1 but is ultimately cleaved and released from the protease. In contrast to the red, haemorrhagic pox on chorioallantoic membranes of embryonated eggs infected with wild type cowpox virus, mutant viruses with a defective crmA gene form white viral lesions (Pickup *et al.* 1986; Palumbo *et al.* 1989). The white pox phenotype is due to infiltration of neutrophils, indicating that CrmA inhibits the inflammatory response of the host by blocking caspase-induced activation of IL-ß. The CrmA homologue in a related virus, vaccinia, facilitates cell survival following treatment with anti-Fas antibody or with TNFα (Dobbelstein and Shenk, 1996). Thus, vaccinia virus CrmA/SPI-2 may modulate the cytotoxic T cell response during infection by blocking apoptosis following activation of Fas. Another poxvirus serpin, SPI-1, appears to inhibit poxvirus-induced apoptosis (Brooks *et al.* 1995). SPI-1 mutant virus initiates but does not complete the virus replication cycle apparently because of premature cell death. The cellular protease targeted by SPI-1 is not known.

3.1.2 Baculovirus P35

Genetic studies of baculoviruses have led to the identification of two distinct anti-apoptotic genes, *P35* and *IAP*. *Autographa californica* nuclear polyhedrosis virus (Ac*M*NPV), the widely used baculovirus expression vector, replicates in lepidopteran insect cells over a period of several days ending in necrotic cell death and release of progeny virus (Clem *et al.* 1996). A mutant of Ac*M*NPV called the annihilator (vAcAnh) was identified because of its rapid cytolytic phenotype (Clem *et al.* 1991). The annihilator mutant causes SF-21 insect cells to die within the first 24 hours. Furthermore, the dying virus-infected cells undergo a dramatic blebbing process, chromatin condensation and DNA laddering characteristic of apoptosis (Clem *et al.* 1991). The genetic determinant for the annihilator phenotype was mapped to the *p35* gene (Clem *et al.* 1991) which encodes a caspase inhibitor (Xue and Horvitz, 1995) (Fig. 1). The *P35* gene is expressed at both early and late times after infection and is required for efficient production of late viral genes and progeny virus (Clem and Duckett, 1997). Because of premature cell death, a mutation in *P35* reduces progeny virus production in SF-21 cells by approximately 100-fold and causes a serious impairment of virus infectivity in *Spodoptera frugiperda* larvae (Clem and Miller, 1993).

Like CrmA, P35 has also been used to delineate cellular apoptotic pathways. P35 inhibits apoptosis induced by a wide range of stimuli, suggesting an evolutionarily conserved molecular pathway. P35 has broad specificity for members of the caspase family and can substitute for CED-9, protecting cells that normally die during development of the worm *C.elegans* (Sugimoto *et al.* 1994). The P35 protein contains no recognisable sequence motifs that could suggest a function, and there are no known homologues except for a similar gene encoded by a closely related insect virus (Kamita *et al.* 1993). The precise mechanism by which P35 inhibits caspases is unknown but P35 probably functions as an irreversible inhibitor whose cleavage products form a stable complex with caspases (Bertin *et al.* 1996; Bump *et al.* 1995). Unlike CrmA, P35 is not a serpin as it shares no amino acid sequence homology with this family. Its reactive site is located near the N-terminus and contains several Asp residues possibly contributing to its ability to inhibit multiple caspases. Determination of the structure of P35 will be very helpful in further understanding this important inhibitor.

3.1.3 Caspase pathways in viral infections

The study of cowpox virus CrmA has significantly contributed to the concept that caspases are part of a common cell death pathway. CrmA inhibits a subset of caspases and has served to help delineate the caspase cascade where upstream proteases activate downstream proteases to facilitate cell death. At least some upstream caspases have long N-terminal prodomains that contain targeting domains for appropriate localisation of the caspase precursor following a death stimulus. The downstream caspases generally have short prodomains, although this may not be true in all cases. CrmA (or the corresponding peptide inhibitor YVAD) is a specific inhibitor of caspase-1 but also efficiently inhibits caspases 4, 5, and 8, a subset of long prodomain caspases (Zhou *et al.* 1997; Kamada *et al.* 1997). Furthermore, different upstream caspases appear to be activated by different death stimuli (Srinivasula *et al.* 1996; Cryns *et al.* 1996). For example, ionising radiation-induced apoptosis of U937 cells is inhibited by the baculovirus caspase inhibitor, P35, but not by CrmA (Datta *et al.* 1997). This was taken as evidence that CrmA-resistant proteases are involved in radiation-induced death, and that this pathway is distinct from TNF-induced apoptosis in U937 cells which is inhibited by both P35 and CrmA (Datta *et al.* 1997). The implication from these and other studies is that CrmA is specific for a subset of intracellular caspases. CrmA inhibits cell death induced by FasL, TNF-α, nerve growth factor withdrawal and extracellular matrix disruption, but does not inhibit cell death induced by DNA damaging agents or staurosporine. This is consistent with the finding that caspase-8 mediates cell death induced by Fas and the TNF receptor (Muzio *et al.* 1996). However, both the CrmA-dependent and CrmA-independent pathways lead to activation of caspase-3, a more downstream (short prodomain) protease in the cascade (Datta *et al.* 1997; Muzio *et al.* 1997). Because CrmA does not inhibit caspase-3, it is presumed to protect cells by inhibiting upstream proteases. We found that the Sindbis virus-induced death pathway is sensitive to both P35 and CrmA (Nava *et al.* 1998). Thus, Sindbis virus-induced apoptosis, which is inhibited by both CrmA and P35, may share upstream components of the death pathway mediated by FasL, TNF and nerve growth factor withdrawal that do not involve CrmA-resistant proteases (Nava *et al.* 1998).

3.2 The IAP (inhibitor of apoptosis) proteins

A genetic complementation assay to screen for anti-apoptotic baculovirus genes (based on their ability to rescue cells infected with P35-defective AcMNPV) produced a new family of apoptosis inhibitors, the IAP proteins. Three baculovirus IAP proteins, Cp-IAP, Op-IAP and the non-functional Ac-IAP, have been identified in three divergent viruses (Clem *et al.* 1996). In addition, several human and *Drosophila* homologues of the baculovirus IAP proteins have been identified (Clem and Duckett, 1997). However, the detailed molecular mechanism by which any of the IAP proteins blocks apoptosis is unknown (Fig. 1).

IAP proteins share common amino acid sequence motifs including 2-3 copies of a 60-70 amino acid motif called baculovirus iap repeat (BIR) located in the N-terminus and separated from a C-terminal RING finger by a spacer region. While the BIR motif is found in a limited number of proteins, the RING finger is found in over 50 other proteins, including ICP0, a transcription factor encoded by herpes simplex virus, the RAG-1 enzyme, the oncogenes *v-cbl*, *pml* and *myc* and the tumour suppressor BRCA1. Although the RING finger of these proteins has been implicated in protein-protein interactions, a shared biochemical function has not been identified.

A more distantly related IAP-like protein, NAIP, was identified as one of two genes mutated in spinal muscular atrophy (SMA) (Roy *et al.* 1995; Lefebvre *et al.* 1995). SMA is a motor neuron disease that results from the loss of spinal cord neurons presumably due to excessive programmed cell death. A deletion of the BIR domains occur in about half of patients with type 1 SMA and deletion correlates with severity of disease (Roy *et al.* 1995).

Although members of the extended IAP protein family have been found to interact via their BIR motifs with a variety of different proteins known to modulate apoptosis, there is no obvious common theme other than the overall observation that IAP proteins appear to prevent activation of caspases. Two mammalian homologues, IAP-1 (cIAP1, MIHB, HIAP-2) and IAP-2 (cIAP2, MIHC, HIAP-1), bind directly to TRAF2 which in turn associates with the cytoplasmic tail of the TNF receptor (Rothe *et al.* 1995). However, a third mammalian homologue (XIAP/ILP1/MIHA) fails to bind any of the known TRAFs (Clem and Duckett, 1997). These three cellular homologues were reported recently to directly bind and inhibit caspases 3 and 7 (Deveraux *et al.* 1997; Roy *et al* 1997). Baculovirus IAP proteins do not bind TRAF proteins (Uren *et al.* 1996) and apparently do not function as direct caspase inhibitors as they are unable to inhibit cell death induced by activated mammalian caspases 1 and 3, or the *Spodoptera* caspase-1 during a baculovirus infection (Seshagiri and Miller, 1997). Instead, baculovirus IAPs bind to the *Drosophila* pro-death proteins reaper and doom (Vucic *et al.* 1997; Harvey *et al.* 1997). Taken together, all of the findings from baculovirus, *Drosophila* and mammalian IAP proteins, it appears that IAP proteins will function at, or upstream of, the caspase activation step.

3.3 Bcl-2 family proteins

The activation of caspases leading to apoptosis is inhibited by members of the Bcl-2 protein family. Inhibition of caspase activation may be mediated through one or more of the assigned functions of Bcl-2 family members such as pore-formation, homo-and heterodimerisation between family members or other protein-protein interactions (Reed, 1997). In addition, Bcl-2-family proteins were recently shown to directly interact with caspases (Cheng *et al.* 1997; Clem *et al.* 1988; Xue and Horvitz, 1997). This interaction between caspases and Bcl-2 proteins may be stabilised by bridging proteins, such as p28Bap31 (Ng *et al.* 1997), MRIT (Han *et al.* 1997), *C. elegans* CED-4 and Apaf-1, a mammalian homologue of CED-4 (Pan *et al.* 1998), that physically link Bcl-2-related proteins and caspases (Irmler *et al.* 1997). These findings raise the possibility that Bcl-2-related proteins could directly inhibit caspases (Clem *et al.* 1998; Xue and Horvitz, 1997). Alternatively, Bcl-2 family proteins may prevent caspase activation through bridging protein. Several Bcl-2 homologues have been identified in viral genomes and these proteins differ from their cellular homologues in interesting ways (See Fig.1).

3.3.1 Epstein-Barr virus BHRF1

Homologues of Bcl-2 are found in a number of viruses including several oncogenic gamma herpesviruses and African swine fever virus (Afonso *et al.* 1996; Hardwick *et al.* 1997). However, the role of viral Bcl-2 homologues in the virus life cycle or in pathogenesis is not known for any of these viruses. Deletion of the Bcl-2 homologue (BHRF1) in the human herpesvirus Epstein-Barr has no detectable effect on its ability to infect and immortalise B lymphocytes in culture (Marchini *et al.* 1991; Lee and Yates, 1992). Immortalisation of cultured B lymphocytes has long been thought to, at least partially, reflect mechanisms involved in tumourigenesis. The need for a viral *bcl-2* homologue is also in question given the observation that the multi-membrane spanning EBV protein LMP-1 activates expression of the cellular *bcl-2* and the *bcl-2*-related gene *mcl-1* resulting in protection of B cells from apoptosis (Rowe *et al.* 1994; Wang *et al.* 1996). However, the connection between LMP 1 and cellular *bcl-2* expression in tumours is blurred. In addition, LMP 1 exhibits other activities that inhibit cell death and promote cell growth. LMP 1 blocks cell death by inducing A20 expression through a NFkB-dependent pathway (Laherty *et al.* 1992; Fries *et al.* 1996), and may also promote cell survival by modulating the TNFR family signalling pathway (Izumi and Kieff, 1997; Miller *et al.* 1997). If BHRF1 has a role in tumourigenesis, it would seem important that it be expressed during latency. Despite the cloning of latency-style BHRF1 transcripts by several groups (Pfitzner *et al.* 1987; Austin *et al.* 1988; Bodescot and Perricaudet 1986; Pearson *et al.*1987), the encoded protein has not been detected in tumours except in those that express lytic cycle genes.

Because BHRF1 is abundantly expressed during the lytic replication cycle, it has been postulated to have a role in blocking premature cell death triggered in the lytic phase, allowing the virus to complete its replication cycle and produce abundant progeny before the cell dies. Furthermore, BHRF1 significantly delays terminal differentiation of cultured stratified epithelial cells (Dawson *et al.* 1995). This observation draws a potential link between BHRF1 and the epithelial tumours associated with EBV infection such as nasopharyngeal carcinoma (Liebowitz, 1994; Raab-Traub, 1992), consistent with the conservation of BHRF1 in natural isolates of EBV (Heller *et al.* 1981). The identification of viral Bcl-2 homologues encoded by several animal herpesviruses will facilitate important research that is difficult to accomplish with the human viruses.

3.3.2 Kaposi's sarcoma-associated virus, HHV8

DNA sequences for a new human herpesvirus (HHV8) were identified in Kaposi's sarcoma (KS) tissues of AIDS patients through the application of representational difference analysis (Chang *et al.* 1994). This virus is more closely related to the gamma 2 herpesvirus HVS than to EBV based on DNA sequence analysis (Chang *et al.* 1994; Moore *et al.* 1996). Epidemiologic and serologic evidence together with the isolation of HHV8 DNA from KS lesions implicate HHV8 in the etiology of KS (Chang *et al.* 1994; Gao *et al.*

1996) and the prevalence of. HHV8 in the healthy population is still disputed.

Based on the similarity between HHV8 and HVS, we sequenced a segment of the HHV8 genome analogous to the position of ORF16 in HVS and identified a *bcl-2* homologue. The Bcl-2 homologue encoded by HHV8, designated KSbcl-2, also exhibits potent anti-apoptotic activity (Cheng *et al.* 1997; Sarid *et al.* 1997). Interestingly, KSbcl-2 fails to bind both Bax and Bak, suggesting the possibility that KSbcl-2, unlike its cellular homologues, may escape the negative regulatory effects of Bax and Bak (Cheng *et al.* 1997).

3.3.3 Other viral Bcl-2 homologues

We and others have identified Bcl-2 homologues in several gamma herpesviruses. Though some of these genes were not recognised at the time, the viral genome sequences were published because of their low level of homology to other cellular and viral homologues. Functional analyses has verified that Bcl-2 homologues are encoded by the oncogenic T cell-tropic herpesvirus saimiri (Smith, 1995; Nava *et al.* 1997), the mouse gamma herpesvirus 68 (Virgin *et al.* 1997) and the bovine herpesvirus 4 designated BORFB2 (Afonso *et al.* 1996; Lomonte *et al.* 1995). The role of these homologues in virus infection is unknown.

Viral homologues of *bcl-2* are not limited to herpesviruses. An unclassified pox-like virus, African swine fever virus (ASFV), also encodes a functional *bcl-2* homologue designated 5-HL (Afonso *et al.* 1996; Neilan *et al.* 1993). 5-HL protects FL5.12 cells from apoptosis induced by IL-3 withdrawal, albeit less potently than Bcl-2. Although 5-HL is abundantly expressed in ASFV-infected cells, its role in lytic versus latent virus infection is not known.

The viral Bcl-2 homologues lack the loop domain found in Bcl-2 and Bcl-x_L which contains the caspase cleavage site, suggesting that the viral homologues may not be converted to prodeath proteins by caspases similar to their cellular counterparts (Cheng *et al.* 1997; Clem *et al.* 1998). In addition, viral Bcl-2 homologues generally have a poorly conserved BH3 homology domain which is required for the prodeath activity of cleaved Bcl-2 (Cheng *et al.* 1997; Clem *et al.* 1998). Thus, viral Bcl-2 homologues may escape cellular regulatory mechanisms to which the cell homologues are subject.

3.3.4 E1B 19k protein of adenovirus

Adenovirus mutants that induce early cytopathic effects (*cyt*) and cellular DNA degradation (*deg*) following infection were described in the 1960's-1980's (Takemori *et al.* 1968,1984; Pilder *et al.* 1984; White *et al.* 1984). These morphological changes were later recognised as apoptosis and the viral mutations responsible for the apoptotic phenotype were mapped to E1B (reviewed by White, 1993,1994). Thus, E1B 19K protects infected cells by inhibiting apoptosis induced by adenovirus infection (Rao *et al.* 1992; White *et al.* 1991). Viruses with a mutation in E1B 19K produce 10-fold fewer progeny compared to wild type viruses apparently because of premature cell death (Pilder *et al.* 1984; Subramanian and Chinnadurai, 1986; White *et al.* 1986). Although E1B 19K contains only remnants of the conserved motifs that define the Bcl-2 family, it has been postulated to function by a mechanism similar to Bcl-2. The anti-apoptotic function of E1B 19K during adenovirus infection can be replaced by *bcl-2*. HeLa cells or CHO (Chinese hamster ovary) cells stably expressing the *bcl-2* gene rescue the *cyt* and *deg* phenotypes, failing to undergo apoptosis when infected with viruses lacking a functional E1B gene (Chiou *et al.* 1994; Tarodi *et al.* 1993). In addition, a recombinant adenovirus encoding *bcl-2* but lacking E1B 19K fails to induce apoptosis (Subramanian *et al.* 1995). Like *bcl-2*, E1B 19K is a potent inhibitor of apoptosis induced by a variety of other death stimuli including TNF-α (tumour necrosis factor), Fas antigen, DNA damaging agents, ultraviolet radiation, nerve growth factor deprivation and the tumour suppressor protein p53 (Debbas and White 1993; Tarodi *et al.* 1993; Subramanian *et al.* 1993; White *et al.* 1992; Gooding *et al.* 1991; Hashimoto *et al.* 1991 Martinou *et al.* 1995; Sabbatini *et al.* 1995; Lin *et al.* 1995).

Further evidence that E1B 19K is a functional homologue of Bcl-2 comes from the observation that E1B 19K protein binds to several cellular proteins with which Bcl-2 also interacts. Nip1, Nip2 and Nip3 were identified in a B cell library by yeast two-hybrid analyses using E1B 19K as bait (Boyd *et al.* 1994). The Nip proteins also bind to BHRF1 and can be co-immunoprecipitated from transfected cells with Bcl-2 (Boyd *et al.* 1994). Although the Nip proteins contain motifs with amino acid similarity to other known proteins, their role in the cell death pathway and E1B 19K/Bcl-2 function is not known. Bak, a death promoting Bcl-2 family member that ablates the protective action of Bcl-2 (Chittenden *et al.* 1995), was also retrieved as an E1B 19K-interacting protein in a yeast two-hybrid screen (Farrow *et al.* 1995). Furthermore, Bax binds to and inhibits the protective function of E1B 19K in a manner similar to Bcl-2 (Han *et al.* 1996). Another death promoting protein, Bik, which lacks definitive BH1 and BH2 domains, but contains a recognisable BH3 "death" domain, binds to and suppresses the protective activities of Bcl-2, Bcl-x_L, BHRF1 and E1B 19k (Boyd *et al.* 1995).

Upon infection, the adenovirus transforming gene E1A induces cell proliferation to provide a suitable environment for virus replication. However, the ability of E1A to induce cell proliferation cannot be separated from its ability to induce apoptosis (Rao *et al.* 1992; White *et al.* 1991). Expression of E1A alone is insufficient to transform cells and the developing cell foci die. Thus, the transforming phenotype is mediated by both E1A and E1B. E1B is required for the transforming activity of E1A by blocking the apoptotic pathway activated by E1A while permitting the E1A-activated proliferation signal. Likewise, Bcl-2 also blocks E1A-induced cell death, facilitating transformation of cultured cells. Although adenoviruses can transform cultured cells and induce tumours in newborn hamsters (Trentin *et al.* 1962), there is no compelling evidence for an etiologic role for adenoviruses in human tumours (Green *et al.* 1980; Mackey *et al.* 1976). Nevertheless, the co-operation between adenovirus anti-apoptotic and proliferative signals in cell transformation raises the possibility that other viral Bcl-2 homologues could play a role in virus-induced tumourigenesis.

3.4 Myxomavirus M-T2

The poxvirus molluscum contagiosum (MCV) replicates in the human epidermis and causes benign neoplasms in children, sexually active adults and AIDS patients. Unlike other poxviruses sequenced to date, MCV encodes a homologue of the selenoprotein glutathione peroxidase (Senkevich *et al.* 1996). The amino acid selenocysteine, containing the trace element selenium instead of sulfur, is incorporated at a UGA stop codon during translation of both viral and cellular glutathione peroxidase. Interestingly, viral glutathione peroxidase is a potent inhibitor of apoptosis (Shisler *et al.* 1997). Its overexpression in HeLa and HaCaT (immortalised human keratinocyte) cells blocks apoptotic cell death induced by ultraviolet (UV) light and by treatment with hydrogen peroxide. Thus, in contrast to poxvirus CrmA (see above), MCV glutathione peroxidase blocks apoptosis induced by UV irradiation and peroxide but not by TNF or Fas ligation (de Boer *et al.*, 1993) (Fig.1). Perhaps an enzyme that reduces peroxides is important for this virus to persist in skin cells that might otherwise be susceptible to apoptosis induced by UV exposure (Shisler *et al.* 1997). In addition, pathogenesis of this benign tumour may also be dependent on viral glutathione peroxidase to permit a net increase in cell number. However, MCV encodes at least one other apoptosis inhibitor (a FLIP protein, see below) that could potentially function in concert with viral glutathione peroxidase.

3.5 Inhibitors of the Fas and TNF receptor pathway

3.5.1 Myxoma virus

Viruses have devised several mechanisms for inhibiting the Fas and TNF receptor pathways to cell death. Myxoma virus is a poxvirus that causes an immunosuppressive disease (myxomatosis) in European rabbits and encodes several genes that modulate the immune system. Two of these, M11L, a novel transmembrane protein, and M-T2 block apoptosis induced by myxoma virus infection (Schreiber *et al.* 1997; Macen *et al.* 1996). The M-T2 protein shares homology with the TNF receptor and appears to have a dual function. A secreted form of M-T2 binds to and inhibits the action of TNF-α, while an intracellular form of the same protein interferes with TNF signalling (Schreiber *et al.* 1997) (Fig.1). The homology between the TNF receptor and M-T2 is limited to four copies of the extracellular cysteine-rich domain (CRD). The N-terminal three CRDs of M-T2 are required for extracellular binding and inhibition of TNF-α, but only two copies are required to block cell death intracellularly. The intracellular mechanism of apoptosis inhibition is presumably distinct from the extracellular mechanism since a protein containing only two CRDs is neither secreted nor binds TNF-α. It was proposed that M-T2 functions intracellularly by binding directly to the TNF receptor (Schreiber *et al.* 1997).

3.5.2 Viral FLIP proteins

Other viruses including several gamma 2 herpesviruses and molluscum contagiosum virus (MCV) encode proteins that share homology with the DED motif found in FADD and the N-terminal prodomain of caspase-8 (see above). At least some of these small viral DED-containing proteins, also called FLIPs, bind to the DED motifs of either the cellular adaptor protein FADD or the prodomain of caspase-8 (Hu *et al.* 1997; Bertin *et al.* 1997; Thome *et al.* 1997) (Fig. 1). The viral DED proteins block recruitment of cell factors to the receptor and function as potent inhibitors of cell death induced by TNF and Fas ligation as well as transfected TRADD and FADD (Hu *et al.* 1997; Thome *et al.* 1997). They also block cell death triggered through the related receptors TRAMP (DR3, APO-3) and TRAIL-R (Thome *et al.* 1997). However, they provide no protection from death triggered by staurosporine treatment or growth factor withdrawal which apparently utilise distinct pathways, thus further teasing apart these pathways (Thome *et al.* 1997). A cellular homologue of the viral DED-containing proteins was also recently identified by several groups (Wallach, 1997). However, the cellular protein comes in at least two forms derived by splicing variations, a short protein much like the viral version, and a long form that shares co-linear homology with caspases except that the protease active site is mutated. The function of the cellular versions (called by a variety of names: FLIP, FLAME, Casper, CASH or I-FLICE) were reported by some to be anti-apoptotic but others found the long form to be pro-apoptotic. Thus, it is not yet clear how the cellular homologues of the viral DED-containing proteins modulate the death pathway (Wallach, 1997).

3.5.3 Adenovirus E3 proteins

Adenovirus encodes two additional proteins that modulate the TNFR and Fas pathways. The E3-10.4K/14.5K complex was reported to inhibit Fas-induced cell death by blocking cell surface expression of Fas (Shisler *et al.* 1997). The same protein complex was also reported to inhibit TNF-induced cell death by blocking translocation of cytosolic phospholipase A(2) to membranes from the cytosol (Dimitriv *et al.* 1997). The adenovirus E3-14.7K protein also inhibits TNF-induced cell death perhaps via its interaction with a novel GTP-binding protein (Li *et al.* 1997). (See Fig.1).

3.6 Viral proteins that modulate p53 function (SV40Tag, Adenovirus E1B and E4 34K, HPV E6, HBx)

To combat the death-inducing effects of E1A, adenovirus encodes three proteins: the 55kD and 19kD products of E1B and the 34kD product of E4 ORF6. All three proteins can inhibit p53-dependent apoptosis (Rao *et al.* 1992; White *et al.* 1992; Moore *et al.* 1996) (Fig. 1). Although E1B 19K is the most potent death inhibitor, each of these proteins can cooperate with E1A to induce transformation of cells in culture and enhance tumourigenicity in animals (Rao *et al.* 1992; White *et al.* 1992; Moore *et al.* 1996; Yew and Berk, 1992). Both E1B 55K and E4 34K inhibit p53-dependent cell death by binding to p53 and inhibiting p53-mediated transcriptional activation (Sarnow *et al.* 1982; Dobner *et al.* 1996; Querido *et al.* 1997). E1B 55K binds to the N-terminal acidic activating domain of p53, presumably inactivating its function directly (Sarnow *et al.* 1982, 1984; Kao *et al.* 1990). E4 34K binds to the central and C-terminal regions of p53 and the mechanism by which it inhibits transcriptional activation is not known (Dobner *et al.* 1996). E1B 55K and E4 34K form a physical complex in infected cells which mediates their effects on late gene expression (Sarnow *et al.* 1982). However, effects on p53 function are exerted independently; each will bind and inhibit p53 in the absence of the other. The expression of E1B 55K and E4 34K are regulated by E1A. The adenovirus 243R E1A product induces accumulation of p53 when expressed in the absence of E1A 289R. Expression of E1A 289R prevents this accumulation indirectly via induction of the E1B 55K and E4 34K proteins. In contrast to transcription regulation, the E1B 55K and E4 34K proteins are both required for suppression of p53 accumulation, and seem to act at the level of modulation of p53 stability (Querido *et al.* 1997).

The mechanism by which E1B 19k blocks p53-mediated cell death is less clear. In contrast to E1B 55K, there is no evidence that E1B 19K binds to p53 or affects intracellular levels of p53. p53 can upregulate mRNA and protein levels of the death-promoting protein Bax, which in some cases correlates with apoptosis (Miyashita and Reed 1995; Han *et al.* 1996; Miyashita *et al.* 1994). Because E1B 19K has no effect on the ability of p53 to regulate gene expression, E1B 19k must function downstream of p53. Thus, it is possible that E1B 19K simply counteracts the death-promoting function of p53 by interfering with the action of Bax which is induced by p53 (Han *et al.* 1996). E1B 19K apparently has alternate methods of blocking cell death as E1B 19K inhibits p53-independent as well as p53-dependent apoptosis (Debbas and White 1993; White *et al.* 1991; Subramanian *et al.* 1995; Chen *et al.* 1995).

Likewise, other viruses encode proteins that ablate p53 function. The SV40 large T antigen binds to and inactivates p53 (Mietz *et al.* 1992) while the HPV E6 protein promotes degradation of p53 (Scheffner *et al.* 1990). Hepatitis B virus is a major risk factor in development of hepatocellular carcinoma. The hepatitis B virus X (HBx) protein is implicated in viral pathogenesis and its C-terminal domain binds to the C terminus of p53, and inhibits the DNA binding and transactivation functions of p53 (Elmore *et al.* 1997). HBx-induced functional changes in p53 may significantly contribute to the ability of HBx to block apoptosis and contribute to tumourigenesis (Elmore *et al.* 1997; Kim *et al.* 1998; Chirillo *et al.* 1997) (Fig.1).

3.7 Bovine herpesvirus LR

Like baculoviruses and adenoviruses, alpha herpesviruses are large DNA viruses that would be expected to encode inhibitors of apoptosis during the lytic replication phase to prolong cell survival and allow completion of the replication cycle. This idea is consistent with the finding that herpes simplex virus (HSV)-infected cells in culture often do not exhibit classic apoptosis, presumably because the apoptotic pathway is blocked (Koyama and Miwa, 1997; Leopardi and Roizman,1996). Furthermore, infection of HEp-2 cells with HSV serves to inhibit sorbitol-induced apoptosis, suggesting that HSV encodes a gene(s) that protects HEp-2 cells (Koyama and Miwa, 1997). Although the γ34.5 gene product and the 175kD immediate early transcription factor encoded by HSV ICP4, have been postulated to have anti-apoptotic activity, more recent evidence does not support these hypotheses. A new candidate anti-apoptotic gene of HSV is US3, which encodes a serine/threonine protein kinase (Leopardi *et al.* 1997). Mutant viruses lacking US3 induce apoptosis that is abolished when US3 is restored.

During HSV latency, only one viral gene is transcribed to produce the latency-associated transcript, LAT. The primary transcription product of LAT is an 8.3kb RNA which gives rise to several stable LAT consisting of 2kb intron RNA. This 2kb LAT RNA is localised to the nucleus, is not polyadenylated, and appears to be a stable splicing product as indicated by its lariat-like structure (Wu *et al.* 1996). Several lines of evidence suggest that LAT is important for reactivation from latency (Perng *et al.* 1996; Bloom *et al.* 1996) or for establishment of latency (Sawtell and Thompson, 1992; Thompson and Sawtell, 1997). Although LAT RNAs contain open reading frames, their encoded proteins, if any, have remained elusive. However, the related bovine herpesvirus-1 (BHV-1) latency-related RNA LR (analogous to HSV LAT) appears to encode a 41kD protein derived by splicing together shorter coding sequences (Hossain *et al.* 1995). Because this 41kD protein inhibits cell cycle progression and can be co-precipitated with cyclin A, which is induced in virus-infected cells, it has been suggested that this latency protein of BHV-1 may play a role in inhibition of apoptosis during infection of neurons (Schang *et al.* 1996) (Fig.1).

3.8 E3L of vaccinia inhibits PKR

The vaccinia virus E3L gene encodes a 25kD protein that inhibits the dsRNA-activated protein kinase (PKR, p68 or DAI). PKR is a cAMP-independent serine/threonine kinase associated with ribosomes that phosphorylates the translation initiation factor eIF-2α on Ser-51 resulting in inhibition of cellular protein synthesis (Deveraux *et al.* 1997). The N-terminal domain of PKR binds dsRNA which activates the kinase domain located in the C-terminus. PKR can be activated during infection with many viruses by virtue of the double stranded RNA intermediates that occur during viral replication, and by induction of interferon which also activates PKR (Roy *et al.* 1997). Overexpression of PKR in Hela cells leads to induction of apoptosis (Uren *et al.* 1996). The vaccinia virus E3L protein inhibits PKR activation by binding to dsRNA. Mutants of vaccinia that lack E3L also induce apoptosis which has been attributed to the induction of PKR (Uren *et al.* 1996) (Fig.1).

3.9 Cytomegalovirus IE1 and IE2

Cytomegalovirus immediate early transcription regulators IE1 and IE2, encoded by alternately spliced mRNAs, inhibit apoptotic cell death induced by TNF-α and by infection with E1B 19K-deficient adenovirus (Zhu *et al.* 1995). The ability of cytomegalovirus IE1 and IE2 to inhibit cell death is consistent with the lengthy replication cycle and the life-long persistent infections caused by this virus.

3.10 HIV vpr

The 96 amino acid Vpr protein of HIV-1 is required for importing the viral preintegration complex into nuclei of non-dividing cells. Although Vpr can induce apoptosis under some circumstances (Stewart *et al.* 1997) others have reported that Vpr suppresses apoptosis (Conti *et al.* 1998; Ayyavoo *et al.* 1997). Although the mechanism is unknown, Vpr may exert its protective activity via its ability to suppress NF-κB activation by inducing IκB, suppressing expression of cytokines, boosting Bcl-2 levels or halting cell cycle progression (Conti *et al.* 1998; Ayyavoo *et al.* 1997; Bartz *et al.* 1996) (Fig.1).

4.0 ROLE OF APOPTOSIS IN VIRAL PATHOGENESIS

4.1 Baculovirus infection of insect larvae

Clouston and Kerr proposed in 1985 that apoptosis may serve as an anti-viral defence response (Clouston and Kerr, 1985). Probably the clearest evidence to date supporting this hypothesis comes from studies of baculovirus infection of insect larvae, where there is a correlation between apoptosis and increased resistance to virus infection. AcMNPV P35 null mutants induce apoptosis in the SF-21 cell line, derived from *Spodoptera frugiperda* (fall armyworm). Not surprisingly, replication of the mutant virus is also severely curtailed in SF-21 cells (Clem and Miller, 1993; Hershberger *et al.* 1992). Similarly, *S. frugiperda* larvae are extraordinarily resistant to infection by P35 null mutant viruses (1000-fold more resistant than to infection by wild type virus). In contrast, cells derived from *Trichoplusia ni*, the cabbage looper, do not die by apoptosis when infected by P35 null mutants, the replication of the mutant viruses is completely normal. *T. ni* larvae exhibit normal susceptibility to infection by P35 mutant viruses (Clem and Miller, 1993; Hershberger *et al.* 1992). Thus, there is a correlation between the ability of cells to initiate an apoptotic response and increased resistance of larvae to infection. Additional evidence that apoptosis is responsible for the increased resistance of *S. frugiperda* larvae to P35 mutant infection is that susceptibility can be restored by expression of an unrelated baculovirus anti-apoptotic gene, the Cp-iap gene (Clem *et al.* 1994). Thus, it appears that an apoptotic response has powerful antiviral effects in *S. frugiperda* larvae. Presumably this is due to the suicide of initially infected cells, allowing the primitive immune system of the insect to clear the infection.

4.2 Sindbis virus infection of the central nervous system

Sindbis virus causes age-dependent disease similar to the closely related human encephalitic viruses, but the molecular basis for this phenomenon is not understood. Shortly after isolation of the virus from mosquitoes, it was noted that infection of newborn mice with Sindbis virus causes rapid and fatal disease, although two week old mice are able to recover. (Taylor *et al.* 1955; Griffin *et al.* 1994; Reinarz *et al.* 1971). While many factors are likely to impact on age-dependent susceptibility, we have put forth the hypothesis that endogenous regulators of apoptosis are determinants in the decision by a virus-infected neuron to either die by apoptosis or resist apoptosis following infection with Sindbis virus. Thus, neuronal cell death inhibitors may modulate the outcome of a Sindbis virus infection. In this way, Sindbis virus is in stark contrast to the baculoviruses where induction of apoptosis is a protective host response.

A variety of cultured cell types are induced to undergo the classic morphologic characteristics of apoptosis upon infection with Sindbis virus (Levine *et al.* 1993; Ubol *et al.* 1994). Likewise, TUNEL and DNA ladder assays demonstrate that Sindbis virus triggers apoptosis of neurons in Sindbis virus-infected newborn mouse brain and spinal cord (Lewis *et al.* 1996). Unlike wild type adenovirus and baculovirus, Sindbis virus replicates to high titres in cells dying by apoptosis, apparently in part because of the short viral replication cycle. Because Sindbis virus targets neurons, an unreplaceable cell population, Sindbis virus-induced apoptosis stands to be a major contributor to viral pathogenesis.

4.2.1 Bcl-2 modulates Sindbis virus-induced apoptosis in some but not all cells

The mouse age-dependent susceptibility to infection can be modelled in cultured primary neurons. Freshly explanted neurons die from a Sindbis virus infection while neurons cultured in nerve growth factor for two weeks are almost completely resistant to virus-induced apoptosis (229). However, virus can be detected in the supernatant of these matured neurons verifying that they are persistently infected, albeit at reduced levels compared to a lytic infection but similar to the persistent infection observed in mice (Levine and Griffin, 1992; Tyor *et al.* 1992). In addition, RT-PCR data indicate that matured neurons express more *bcl-2* than freshly explanted ganglia leading us to test the idea that Bcl-2 could modulate the outcome of a Sindbis virus infection (Levine *et al.* 1993). AT-3 cells stably transfected with *bcl-2* survived a Sindbis virus infection and were maintained as a persistently infected cell line for over a year (Levine *et al.* 1993). In contrast, AT-3 cells receiving only the neo vector were readily killed by Sindbis virus. Thus, a cellular inhibitor of apoptosis converted a lytic

Sindbis virus infection into a persistent infection (Fig.2). Perhaps similar mechanisms occur in infected animals (Levine *et al.* 1993).

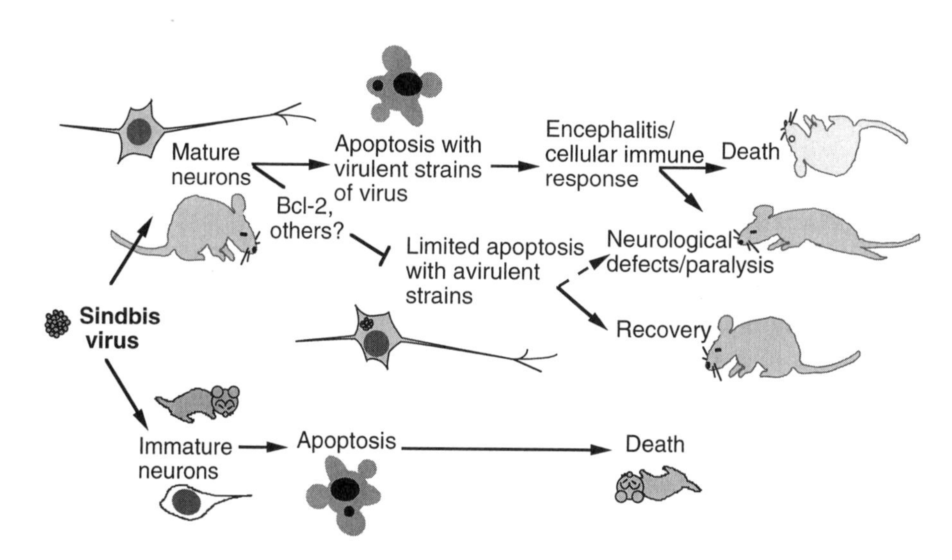

Figure 2. Apoptosis modulators alter Sinbis virus-induced disease

In contrast to AT-3 cells, expression of Bcl-2 in rat embryo fibroblasts, U937 and L929 cells had no protective effect following infection with either Sindbis virus or Semiliki Forrest virus (Grandgirard *et al.* 1998). The inability of Bcl-2 to block virus-induced cell death in these cells may be attributed to the finding that Bcl-2 is cleaved by caspases during infection of these cells. This is consistent with our finding that endogenous Bcl-2 is cleaved at Asp34 by caspases in Jurkat cells following activation of the Fas pathway, where Bcl-2 is ineffective at blocking cell death (Cheng *et al.* 1997). Similarly, we reported that endogenous Bcl-x_L is cleaved by caspases in COS cells undergoing Sindbis virus-induced apoptosis (Clem *et al.* 1998). Thus, the anti-apoptotic Bcl-2 family members may not protect in those cells where virus infection triggers cleavage of Bcl-2 family members by caspases.

4.2.2 Neurovirulent strains of Sindbis virus overcome inhibitors of apoptosis

Passage of Sindbis virus (strain AR339) in mouse brains allowed selection of a virus with increased neurovirulence, the NSV (neuroadapted) strain (Griffin and Johnson, 1977). This virus has acquired the ability to induce mortality in mice that are two weeks of age or older. The use of recombinant viruses mapped a key genetic mutation responsible for conferring neurovirulence to amino acid 55 of the E2 glycoprotein (Tucker *et al.* 1993). Thus, a single mutation of Gln55 to His55 converts an avirulent virus into one that kills older mice. To determine what effect this mutation had on the induction of apoptosis, AT-3 cells stably transfected with Bcl-2 were infected with recombinant viruses differing at E2 position 55. Strikingly, the Gln55His mutation converted Sindbis virus into one that could overcome the protective effects of Bcl-2, inducing apoptotic cell death (Ubol *et al.* 1994). Furthermore, the neurovirulent strains (His55) of Sindbis virus that kill two-week-old animals also induced abundant apoptosis of neurons in brains and spinal cords. Double labelling experiments verified that the virus-infected cells were those dying by apoptosis (Lewis *et al.* 1996). Thus, neurovirulent strains can overcome the protective effects of Bcl-2 or other endogenous cell death inhibitors. The strong correlation between mortality and induction of apoptosis is consistent with the hypothesis that Sindbis virus-induced mortality in young animals is due to induction of neuronal apoptosis by direct virus infection (Lewis *et al.* 1996) (Fig.2).

Although the neurovirulent virus of Sindbis virus replicated to higher titres in two-week-old mice compared to the avirulent strain, the mean differences in titres were 5 to 10-fold and were not statistically significant (Lewis *et al.*1996). Experimentally it is difficult to determine if the suppression in viral replication is an insignificant consequence of apoptosis inhibitors or a key mechanism for preserving cell survival. Indeed both mechanisms are likely to play a role. Taking all of the information together, our favoured hypothesis is that host cell regulators of apoptosis determine whether a given strain of Sindbis virus is capable of inducing apoptosis. At the same time, virus replication efficiency may determine the potency of the apoptotic signal. In certain situations, inhibitors of apoptosis can delay cell death without impairing Sindbis virus replication. However, suppression of viral replication by apoptosis inhibitors, may be a key event in other situations such as the establishment of a long term persistent infection. A model summarising these findings is shown in Fig.2.

4.2.3 Bcl-2 protects mice from a fatal Sindbis virus infection

Using the Sindbis virus vector system we can address the effect of apoptosis regulators on viral pathogenesis *in vivo*. Recombinant Sindbis virus expressing human *bcl-2* was inoculated into newborn mouse brains (10^3 pfu) and the effects of *bcl-2* on the outcome of a Sindbis virus infection was monitored by determining mortality, Bcl-2 protein expression, viral replication and neuronal apoptosis. Bcl-2 protected 50-90% of animals (depending on the age and strain) from fatal disease and protection correlated with expression of human Bcl-2 protein in mouse brains (Levine *et al.* 1996). Bcl-2 expression also correlated with significant reduction in the number of apoptotic neurons. In contrast, control virus vectors expressing Bcl-2 with a premature termination codon or an irrelevant gene had no effect on survival of animals or neurons (Levine *et al.* 1996). The observation that Bcl-2-deficient mice exhibit increased susceptibility to a fatal Sindbis virus infection, argues strongly that endogenous Bcl-2 may normally play a role in determining the outcome of a Sindbis virus infection (Lewis and Hardwick, 1998). The implication from this work is that AT-3 cells and CNS neurons of mice fail to activate caspases that cleave Bcl-2 upon infection with Sindbis virus.

5.0 REFERENCES

Afonso,C.L., Neilan,J.G., Kutish,G.F. *et al.* (1996) An African swine fever virus Bcl-2 homolog, 5-HL, suppresses apoptotic cell death. *J. Virol.* **70**: 4858-4863.

Arany,Z., Sellers,W.R., Livingston,D.M. *et al.* (1994) E1A-associated p300 and CREB-associated CBP belong to a conserved family of coactivators. *Cell* **77**: 799-800.

Asano,M., Nevins,J.R. and Wharton,R.P. (1996) Ectopic E2F expression induces S phase and apoptosis in Drosophila imaginal discs. *Gen. Dev.* **10**: 1422-1432.

Austin,P.J., Flemington,E., Yandava,C.N.. *et al.* (1988) Complex transcription of the Epstein-Barr virus BamHI fragment H rightward open reading frame 1 (BHRF1) in latently and lytically infected B lymphocytes. *Proc. Natl. Acad. Sci. USA* **85**: 3678-3682.

Ayyavoo,V., Mahboubi,A., Mahalingam,S. *et al.* (1997) HIV-1 Vpr suppresses immune activation and apoptosis through regulation of nuclear factor kB. *Nature Med.* **3**: 1117-1123.

Badley,A.D., Dockrell,D. and Paya,C.V. (1997) Apoptosis in AIDS. *Adv. Pharmacol.* **41**: 271-294.

Bartz,S.R., Rogel,M.E. and Emerman,M. (1996) Human immunodeficiency virus type 1 cell cycle control: Vpr is cytostatic and mediates G_2 accumulation by a mechanism which differs from DNA damage checkpoint control. *J. Virol.* **70**: 2324-2331.

Beg,A.A., and Baldwin,A.S. (1993) The IkB proteins: multifunctional regulators of Rel/NF-kB transcription factors. *Gen. Dev.* **7**: 2064-2070.

Bertin,J., Armstrong,R.C., Ottilie,S. *et al.* (1997) Death effector domain-containing herpesvirus and poxvirus proteins inhibit both Fas- and TNFR1-induced apoptosis. *Proc. Natl. Acad. Sci. USA* **94**: 1172-1176.

Bertin,J., Mendrysa,S.M., LaCount,D.J. *et al.* (1996) Apoptotic suppression of baculovirus P35 involves cleavage by and inhibition of a virus-induced CED-3/ICE-like protease. *J. Virol.* **70**: 6251-6259.

Bjorklund,H.V., Johansson,T.R. and Rinne,A. (1997) Rhabdovirus-induced apoptosis in a fish cell line is inhibited by a human endogenous acid cysteine proteinase inhibitor. *J. Virol.* **71**: 5658-5662.

Bloom,D.C., Hill,J.M., Devi-Rao,G. *et al.* (1996) A 348-base-pair region in the latency-associated transcript facilitates herpes simplex virus type 1 reactivation. *J. Virol.* **70**: 2449-2459.

Bodescot,M and Perricaudet,M. (1986) Epstein-Barr virus mRNAs produced by alternative splicing. *Nucl. Acid Res.* **14**: 7103-7114.

Bondesson,M., Ohman,K., Mannervik,M. *et al.* (1996) Adenovirus E4 open reading 4 protein autoregulates E4 transcription by inhibiting E1A transactivation of the E4 promoter. *J. Virol.* **70**: 3844-3851.

Botz,J., Zerfass-Thome,K., Spitkovsky,D. *et al.* (1996) Cell cycle regulation of the murine cyclin E gene depends on an E2F binding site in the promoter. *Mol. Cell. Biol.* **16**: 3401-3409.

Boulakia,C.A., Chen,G., Ng,F.W. *et al.* (1996) Bcl-2 and adenovirus E1B 19kDa protein prevent E1A-induced processing of CPP32 and cleavage of poly(ADP-ribose) polymerase. *Oncogene* **12**: 529-535.

Boyd,J.M., Gallo,G.J., Elangovan,B. *et al.* (1995) Bik, a novel death-inducing protein shares a distinct sequence motif with Bcl-2 family proteins and interacts with viral and cellular survival-promoting proteins. *Oncogene* **11**: 1921-1928.

Boyd,J.M., Malstrom,S., Subramanian,T. *et al.* (1994) Adenovirus E1B 19kDa and bcl-2 proteins interact with a common set of cellular proteins. *Cell* **79**: 341-351.

Brooks,M.A., Ali,A.N., Turner,P.C. *et al.* (1995) A rabbitpox virus serpin gene controls host range by inhibiting apoptosis in restrictive cells [SPI-1, not SPI-2]. *J. Virol.* **69**: 7688-7698.
Brunner,T., Mogil,R.J., LaFace,D. *et al.* (1995). Cell-autonomous Fas (CD95)/Fas-ligand interactions mediates activation-induced apoptosis in T-cell hybridomas. *Nature* **373**: 441-444.
Bruschke,C.J.M., Hulst,M.M., Moormann,R.J.M. *et al.* (1997) Glycoprotein E^{rns} of pestiviruses induces apoptosis in lymphocytes of several species. *J. Virol.* **71**: 6692-6696.
Bump,N.J., Hackett,M., Hugunin,M. *et al.* (1995) Inhibition of ICE family proteases by baculovirus antiapoptotic protein p35. *Science* **269**: 1885-1888.
Canman,C.E. and Kastan,M.B. (1997) Role of p53 in apoptosis. *Advan. Pharmacol.* **41**: 429-460.
Casano,F.J., Rolando,A.M., Mudgett,J.S. *et al.* (1994) The structure and complete nucleotide sequence of the murine gene encoding interleukin-1-beta converting enzyme (ICE). *Genomics* **20**: 474-481.
Chang,Y., Cesarman,E., Pessin,M.S. *et al.* (1994) Identification of herpesvirus-like DNA sequences in AIDS-associated Kaposi's sarcoma. *Science* **266**: 1865-1869.
Chen,G., Branton,P.E. and Shore,G.C. (1995) Induction of p53-independent apoptosis by hygromycin B - suppression by Bcl-2 and adenovirus E1B 19-KDa protein. *Exp. Cell Res.* **221**: 55-59.
Cheng,E.H-Y., Kirsch,D.G., Clem,R.J. *et al.* (1997) Conversion of Bcl-2 to a Bax-like death effector by caspases. *Science* **278**: 1966-1968.
Cheng,E.H-Y., Nicholas,J., Bellows,D.S. *et al.* (1997) A Bcl-2 homolog encoded by Kaposi's sarcoma-associated virus, human herpesvirus 8, inhibits apoptosis but does not heterodimerize with Bax or Bak. *Proc. Natl. Acad. Sci. USA* **94**: 690-694.
Chiou,S-K. and White,E. (1997) p300 binding by E1A cosegregates with p53 induction but is dispensable for apoptosis. *J. Virol.* **71**: 3515-3525.
Chiou,S-K., Tseng,C-C., Rao,L. *et al.* (1994) Functional complementation of the adenovirus E1B 19-kilodalton protein with Bcl-2 in the inhibition of apoptosis in infected cells. *J. Virol.* **68**: 6553-6566.
Chirillo,P., Pagano,S., Natoli,G. *et al.* (1997) The hepatitis B virus X gene induced p53-mediated programmed cell death. *Proc. Natl. Acad. Sci. USA* **94**: 8162-8167.
Chittenden,T., Harrington,E.A., O'Connor,R. *et al.* (1995) Induction of apoptosis by the Bcl-2 homologue Bak. *Nature* **374**: 733-736.
Clem, R.J. and Duckett,C.S. (1997) The iap genes: unique arbitrators of cell death. *Trends Cell Biol.* **7**: 337-339.
Clem,R.J. and Miller,L.K. (1993) Apoptosis reduces both the in vitro replication and the in vivo infectivity of a baculovirus. *J. Virol.* **67**: 3730-3738.
Clem,R.J. and Miller,L.K. (1994) Control of programmed cell death by the baculovirus genes p35 and iap. *Mol. Cell. Biol.* **14**: 5212-5222.
Clem,R.J., Cheng,E.H-Y., Karp,C.L. *et al.* (1998) Modulation of cell death by Bcl-x_L through caspase interaction. *Proc. Natl. Acad. Sci. USA* **95**: 554-559.
Clem,R.J., Fechheimer,M. and Miller,L.K. (1991) Prevention of apoptosis by a baculovirus gene during infection of insect cells. *Science* **254**: 1388-1390.
Clem,R.J., Hardwick,J.M. and Miller,L.K. (1996) Anti-apoptotic genes of baculoviruses. *Cell Death Diff.* **3**: 9-16.
Clem,R.J., Robson,M. and Miller,L.K. (1994) Influence of infection route on the infectivity of baculovirus mutants lacking the apoptosis-inhibiting gene p35 and the adjacent gene p94. *J. Virol.* **68**: 6759-6762.
Clouston,W.M. and Kerr,J.F.R. (1985) Apoptosis, lymphocytotoxicity and the containment of viral infections. *Med. Hypoth.* **18**: 399-404.
Cohen,P.L. and Eisenberg,R.A. (1992) The lpr and gld genes in systemic autoimmunity: life and death in the Fas lane. *Immunol. Today* **13**: 427-428.
Conti,L., Rainaldi,G., Matarrese,P. *et al.* (1998) The HIV-1 Vpr protein acts as a negative regulator of apoptosis in a human lymphoblastoid T cell line-possible implications for the pathogenesis of AIDS. *J. Exp. Med.* **187**: 403-413.
Cryns,V.L., Bergeron,L., Zhu,H. *et al.* (1996) Specific cleavage of a-fodrin during Fas and tumour necrosis factor-induced apoptosis is mediated by an interleukin-1β-converting enzyme/Ced-3 protease distinct from the poly(ADP-ribose) polymerase protease. *J. Biol. Chem.* **271**: 31277-31282.
Danen-van Oorschot,A.A.A.M., Fischer,D.F., Grimbergen,J.M. *et al.* (1997) Apoptin induces apoptosis in human transformed and malignant cells but not in normal cells. *Proc. Natl. Acad. Sci. USA* **94**: 5843-5847.
Datta,R., Kojima,H., Banach,D. *et al.* (1997) Activation of a CrmA-insensitive, p35-sensitive pathway in ionizing radiation-induced apoptosis. *J. Biol. Chem.* **272**: 1965-1969.
Dawson,C.W., Eliopoulos,A.G. Dawson,J. *et al.* (1995) BHRF1, a viral homologue of the Bcl-2 oncogene, disturbs epithelial cell differentiation. *Oncogene* **9**: 69-77.
de Boer,C.J., Loyson,S., Kluin,P.M. *et al.* (1993) Multiple breakpoints within the BCL-1 locus in B-cell lymphoma: rearrangements of the cyclin D1 gene. *Cancer Res.* **53**: 4148-4152.
Debbas,M. and White,E. (1993) Wild-type p53 mediates apoptosis by E1A, which is inhibited by E1B. *Gen. Dev.* **7**: 546-554.
Deveraux,Q.L., Takahashi,R., Salvesen,G.S. *et al.* (1997) X-linked IAP is a direct inhibitor of cell-death proteases. *Nature* **388**: 300-304.
Dhein,J., Walczak,H., Baumier,C. *et al.* (1995) Autocrine T-cell suicide mediated by APO-1/(Fas/CD95). *Nature* **373**: 438-441.

Dimitrov,T., Krajcsi,P., Hermiston,T.W. *et al.* (1997) Adenovirus E3-10.4K/14.5K protein complex inhibits tumour necrosis factor-induced translocation of cytosolic phospholipase A(2) to membranes. *J. Virol.* **71**: 2830-2837.
Dobbelstein,M.and Shenk,T. (1996) Protection against apoptosis by the vaccinia virus SPI-2 (B13R) gene product. *J. Virol.* **70**: 6479-6485.
Dobner,T., Horikoshi,N., Rubenwolf,S. *et al.* (1996) Blockage by adenovirus E4orf6 of transcriptional activation by the p53 tumor suppressor. *Science* **272**: 1470-1473.
El-Deiry,W.S., Tokino,T., Velculescu,V.E. *et al.* (1993) WAF1, a potential mediator of p53 tumour suppression. *Cell* **75**: 817-823.
Elmore,L.W., Hancock,A.R., Chang,S-F. *et al.* (1997) Hepatitis B virus X protein and p53 tumor suppressor interactions in the modulation of apoptosis. *Proc. Natl. Acad. Sci. USA* **94**: 14707-14712.
Enari,M., Sakahira,H., Yokoyama,H. *et al.* (1998) A caspase-activated DNase that degrades DNA during apoptosis, and its inhibitor ICAD. *Nature* **391**: 43-50.
Estus,S., Zaks,W.J., Freeman,R.S. *et al.* (1994) Altered gene expression in neurons during programmed cell death - identification of c-Jun as necessary for neuronal apoptosis. *J. Cell Biol.* **127**: 1717-1727.
Ewen,M., Xing,Y., Lawrence,J.B. *et al.* (1991) Molecular cloning, chromosomal mapping, and expression of the cDNA for p107, a retinoblastoma gene product-related proteins. *Cell* **66**: 1155-1164.
Farrow,S.N., White,J.H.M., Martinou,I. *et al.* (1995) Cloning of a bcl-2 homologue by interaction with adenovirus E1B 19K. *Nature* **374**: 731-733.
Fries,K.L., Miller,W.E. and Raab-Traub,N. (1996) Epstein-Barr virus latent membrane protein 1 blocks p53-mediated apoptosis through the induction of the A20 gene. *J. Virol.* **70**: 8653-8659.
Gao,S-J., Kingsley,L., Hoover,D.R. *et al.* (1996) Seroconversion to antibodies against Kaposi's sarcoma-associated herpesvirus-related latent nuclear antigens before the development of Kaposi's sarcoma. *N. Eng. J. Med.* **335**: 233-241.
Gooding,L.R., Aquino,L, Duerksen-Hughes,P.J. *et al.* (1991) The E1B 19,000-molecular-weight protein of group C adenoviruses prevents tumor necrosis factor cytolysis of human cells but not of mouse cells. *J. Virol.* **65**: 3083-3094.
Grandgirard,D., Studer,E., Monney,L. *et al.* (1998) Alphaviruses induce apoptosis in Bcl-2-overexpressing cells: evidence for a caspase-mediated, proteolytic inactivation of Bcl-2. *EMBO J.* **17**: 1268-1278.
Green,M., Wold,W.S.M., Brackmann,K.H. *et al.* (1980) Human adenovirus transforming genes: group relationships, integration, expression in transformed cells and analysis of human cancers and tonsils. p373-397. In: *Cold Spring Harbor Conference on Cell Proliferation Viruses in Naturally Occurring Tumors*. M. Essex, G. Todaro, and H. zurHausen (Eds.), Cold Spring Harbor, New York.
Griffin,D.E. and Johnson,R.T. (1977) Role of the immune response in recovery from Sindbis virus encephalitis in mice. *J. Immunol.* **118**: 1070-1075.
Griffin,D.E., Levine,B., Tyor,W.R. *et al.* (1994) Age-dependent susceptibility to fatal encephalitis: alphavirus infection of neurons. *Arch. Virol.* **9**: 31-39.
Han,D.K.M., Chaudhary,P.M., Wright,M.E. *et al.* (1997) MRIT, a novel death-effector domain-containing protein, interacts with caspases and $BclX_L$ and initiates cell death. *Proc. Natl. Acad. Sci. USA* **94**: 11333-11338.
Han,J., Sabbatini,P., Perez,D. *et al.* (1996) The E1B 19K protein blocks apoptosis by interacting with and inhibiting the p53-inducible and death-promoting Bax protein. *Gen. Dev.* **10**: 461-477.
Hannon,G.J., Demetrick,D. and Beach,D. (1993) Isolation of the Rb-related p130 through its interaction with cdk2 and cyclins. *Gen. Dev.* **7**: 2378-2391.
Hardwick,J.M. (1997) Virus-induced apoptosis. *Advan. Pharmacol.* **41**: 295-336.
Harper,J.W., Adami,G.R., Wei,N. *et al.* (1993) The p21 Cdk-interacting protein Cip1 is a potent inhibitor of G1 cyclin-dependent kinases. *Cell* **75**: 805-812.
Harvey,A.J., Bidwai,A.P. and Miller,L.K. (1997) A product of the Drosophila MOD(MDG-4) gene, induces apoptosis and binds to baculovirus inhibitor-of-apoptosis proteins. *Mol. Cell. Biol.* **17**: 2835-2843.
Hashimoto,S., Ishii,A., and Yonehara,S. (1991) The E1B oncogene of adenovirus confers cellular resistance to cytotoxicity of tumor necrosis factor and monoclonal anti-Fas antibody. *Intl. J. Immunol.* **3**: 343-351.
Heller,M., Dambaugh,T. and Kieff,E. (1981) Epstein-Barr virus DNA. IX. Variation among viral DNAs from producer and nonproducer infected cells. *J. Virol.* **38**: 632-648.
Henkart,P.A. (1996) ICE family proteases: mediators of all apoptotic cell death? *Immunity* **4**: 195-201.
Herbein,G., Vanlint,C., Lovett,J.L. *et al.* (1998) Distinct mechanisms trigger apoptosis in human immunodeficiency virus type 1-infected and in uninfected bystander T lymphocytes. *J. Virol.* **72**: 660-670.
Hershberger,P.A., Dickson,J.A. and Friesen,P.D. (1992) Site-specific mutagenesis of the 35-kilodalton protein gene encoded by Autographa californica nuclear polyhedrosis virus: cell line-specific effects on virus replication. J. Virol. **66**: 5525-5533.
Hiebert,S.W., Chellappan,S.P., Horowitz,J.M. *et al.* (1992) The interaction of RB with E2F coincides with an inhibition of the transcriptional activity of E2F. *Gen. Dev.*. **6**: 177-185.
Hinds,P.W. and Weinberg,R.A. (1994) Tumour suppressor genes. *Curr. Opin. Genet. Dev.* **4**: 135-141.
Hossain,A., Schang,L.M. and Jones,C. (1995) Identification of gene products encoded by the latency-related gene of bovine herpesvirus 1. *J. Virol.* **69**: 5345-5352.

Hsieh,J-K., Fredersdorf,S., Kouzarides,T. *et al.* (1997) E2F1-induced apoptosis requires DNA binding but not transactivation and is inhibited by the retinoblastoma protein through direct interaction. *Gen. Dev.*. **11**: 1840-1852.

Hu,S., Vincenz,C., Buller,M. *et al.* (1997) A novel family of viral death effector domain-containing molecules that inhibit both CD-95- and tumor necrosis factor receptor-1-induced apoptosis. *J. Biol. Chem.* **272**: 9621-9624.

Irmler,M., Hofmann,K., Vaux,D. *et al.* (1997) Direct physical interaction between the *Caenorhabditis elegans* death proteins CED-3 and CED-4. *FEBS Let.* **406**: 189-190.

Izumi,K.M. and Kieff,E. (1997) The Epstein-Barr virus oncogene product latent membrane protein 1 engages the tumor necrosis factor receptor-associated death domain protein to mediate B lymphocyte growth transformation and activate NF-kappa-B. *Proc. Natl. Acad. Sci. USA* **94**: 12592-12597.

Ju,S-T., Panka,D.J., Cui,H. *et al.* (1995) Fas(CD95)/FasL interactions required for programmed cell death after T-cell activation. *Nature* **373**: 444-448.

Kamada,S., Funahashi,Y. and Tsujimoto,Y. (1997) Caspase-4 and caspase-5, members of the ICE/CED-3 family of cysteine proteases, are CrmA-inhibitable proteases. *Cell Death and Different.* **4**: 473-478.

Kamita,S.G., Majima,K., and Maeda,S. (1993) Identification and characterization of the p35 gene of Bombyx mori nuclear polyhedrosis virus that prevents virus-induced apoptosis. *J. Virol.* **67**: 455-463.

Kao,C.C., Yew,P.R. and Berk,A.J. (1990) Domains required for in vitro association between the cellular p53 and the adenovirus 2 E1B 55K proteins. *Virology* **179**: 806-814.

Kasibhatla,S., Brunner,T., Genestier,L. *et al.* (1998) DNA damaging agents induce expression of Fas ligand and subsequent apoptosis in T lymphocytes via the activation of NF-kB and AP-1. *Molec. Cell* **1**: 543-552.

Katsikis,P.D., Wunderlich,E.S., Smith,C.A. *et al.* (1995) Fas antigen stimulation induces marked apoptosis of T lymphocytes in human immunodeficiency virus-infected individuals. *J. Exp. Med.* **181**: 2029-2036.

Kim,H., Lee,H. and Yun,Y. (1998) X-gene product of hepatitis B virus induces apoptosis in liver cells. *J. Biol. Chem.* **273**: 381-385.

Kleinberger,T. and Shenk,T. (1993) Adenovirus E4orf4 protein binds to protein phosphatase 2A, and the complex down regulates E1A-enhanced junB transcription. *J. Virol.* **67**: 7556-7560.

Kowalik,T.F., DeGregori,J., Schwarz,J.K. *et al.* (1995) E2F1 overexpression in quiescent fibroblasts leads to induction of cellular DNA synthesis and apoptosis. *J. Virol.* **69**: 2491-2500.

Koyama,A.H. and Miwa,Y. (1997) Suppression of apoptotic DNA fragmentation in herpes simplex virus type-1-infected cells. *J. Virol.* **71**: 2567-2571.

LaCount,D.J. and Friesen,P.D. (1997) Role of early and late replication events in induction of apoptosis by baculoviruses. *J. Virol.* **71**: 1530-1537.

Laherty,C.D., Hu,H.M., Opipari,A.W. *et al.* (1992) The Epstein-Barr virus LMP1 gene product induces A20 zinc finger protein expression by activating NF-κB. *J. Biol. Chem.* **267**: 24157-24160.

Lavoie,J.N., Nguyen,M., Marcellus,R.C. *et al.* (1998) E4orf4, a novel adenovirus death factor that induces p53-independent apoptosis by a pathway that is not inhibited by zVAD-fmk. *J. Cell. Biol.* **140**: 637-645.

Lee,M-A. and Yates,J.L. (1992) BHRF1 of Epstein-Barr virus, which is homologous to human proto-oncogene bcl-2, is not essential for transformation of B cells or for virus replication in vitro. *J. Virol.* **66**: 1899-1906.

Lee,S.B. and Esteban,M. (1994) The interferon-induced double-stranded RNA-activated protein kinase induces apoptosis. *Virology* **199**: 491-496.

Lefebvre,S, Burglen,L, Reboullet,S. *et al.* (1995) Identification and characterization of a spinal muscular atrophy-determining gene. *Cell* **80**: 155-165.

Leopardi,R and Roizman,B. (1996) The herpes simplex virus major regulatory protein ICP4 blocks apoptosis induced by the virus or by hyperthermia. *Proc. Natl. Acad. Sci. USA* **93**: 9583-9587.

Leopardi,R., Vansant,C. and Roizman,B. (1997) The herpes simplex virus 1 protein kinase U(S)3 is required for protection from apoptosis induced by the virus. *Proc. Natl. Acad. Sci. USA* **94**: 7891-7896.

Levine,B. and Griffin,D.E. (1992) Persistence of viral RNA in mouse brains after recovery from acute alphavirus encephalitis. *J. Virol.* **66**: 6429-6435.

Levine,B., Goldman,J.E., Jiang,H.H. *et al.* (1996) Bcl-2 protects mice against fatal alphavirus encephalitis. *Proc. Natl. Acad. Sci. USA* **93**: 4810-4815.

Levine,B., Huang,Q., Isaacs,J.T. *et al.* (1993) Conversion of lytic to persistent alphavirus infection by the bcl-2 cellular oncogene. *Nature* **361**: 739-742.

Lewis,J. and Hardwick,J.M. (1998) Unpublished data.

Lewis,J., Wesselingh,S.L., Griffin,D.E. *et al.* (1996) Sindbis virus-induced apoptosis in mouse brains correlates with neurovirulence. *J. Virol.* **70**: 1828-1835.

Li,C.J., Friedman,D.J., Wang,C. *et al.* (1995) Induction of apoptosis in uninfected lymphocytes by HIV-1 tat protein. *Science* **268**: 429-431.

Li,Y., Graham,C., Lacy,S. *et al.* (1993) The adenovirus E1A-associated 130kD protein in encoded by a member of the retinoblastoma gene family and physically interacts with cyclins A and E. *Gen. Dev.* **7**: 2366-2377.

Li,Y.G., Kang,J. and Horwitz,M.S. (1997) Interaction of an adenovirus 14.7-kilodalton protein inhibitor of tumor necrosis factor alpha cytolysis with a new member of the GTPase superfamily of signal transducers. *J. Virol.* **71**: 1576-1582.

Liebowitz,D. (1994) Nasopharyngeal carcinoma: the Epstein-Barr virus association. *Semin.Oncol.* **21**: 376-381.

Lin,H-J,L,, Eviner,V., Prendergast,G.C. *et al.* (1995) Activated H-ras rescues E1A-induced apoptosis and cooperates with E1A to overcome p53-dependent growth arrest. *Mol. Cell. Biol.* **15**: 4536-4544.

Lin,K-I., DiDonato,J.A., Hoffman,A. *et al.* (1998) Suppression of steady state, but not stimulus induced NF-kB activity protects from alphavirus-induced apoptosis. *J. Cell. Biol.* (In Press)

Lin,K-I., Lee,S-H., Narayanan,R. *et al.* (1995) Thiol agents and Bcl-2 identify an alphavirus-induced apoptotic pathway that requires activation of the transcription factor NF-kappa B. *J. Cell Biol.* **131**: 1149-1161.

Liu,X., Zou,H., Slaughter,C. *et al.* (1997) DFF, a heterodimeric protein that functions downstream of caspase-3 to trigger DNA fragmentation during apoptosis. *Cell* **89**: 175-184.

Lomonte,P., Bublot,M., van Santen,V. *et al.* (1995) Analysis of bovine herpesvirus 4 genomic regions located outside the conserved gammaherpesvirus gene blocks. *J. Gen. Virol.* **76**: 1835-1841.

Lowe,S.W. and Ruley,H.E. (1993) Stabilization of the p53 tumour suppressor is induced by adenovirus E1A and accompanies apoptosis. *Gen. Dev.* **7**: 535-545.

Lowe,S.W., Ruley,H.E., Jacks,T. *et al.* (1993) p53-dependent apoptosis modulates the cytotoxicity of anticancer agents. *Cell* **74**: 957-967.

Lu,Y-Y., Koga,Y., Tanaka,K. *et al.* (1994) Apoptosis induced in CD4+ cells expressing gp160 of human immunodeficiency virus type 1. *J. Virol.* **68**: 390-399.

Macen,J.L., Graham,K.A., Lee,S.F. *et al.* (1996) Expression of the myxoma virus tumor necrosis factor receptor homologue and M11L genes is required to prevent virus-induced apoptosis in infected rabbit T lymphocytes. *Virology* **218**: 232-237.

Mackey,J.K., Rigden,P.M. and Green,M. (1976) Do highly oncogenic group A human adenoviruses cause human cancer? *Proc. Natl. Acad. Sci. USA* **73**: 4657-4661.

Marcellus,R.C., Teodoro,J.G., Wu,T. *et al.* (1996) Adenovirus type 5 early region 4 is responsible for E1A-induced p53-independent apoptosis. *J. Virol.* **70**: 6207-6215.

Marchini,A., Tomkinson,B., Cohen,J.I. *et al.* (1991) BHRF1, the Epstein-Barr virus gene with homology to Bcl-2, is dispensable for B-lymphocyte transformation and virus replication. *J. Virol.* **65**: 5991-6000.

Martin,D.P., Schmidt,R.E., DiStefano,P.S. *et al.* (1988) Inhibitors of protein synthesis and RNA synthesis prevent neuronal death caused by nerve growth factor deprivation. *J. Cell Biol.***106**: 829-844.

Martinou,I., Fernandez,P-A., Missotten,M. *et al.* (1995) Viral proteins E1B19K and p35 protect sympathetic neurons from cell death induced by NGF deprivation. *J. Cell. Biol.* **128**: 201-208.

Mietz,J.A., Unger,T., Huibregtse,J.M. *et al.* (1992) The transcriptional transctivation function of wild-type p53 is inhibited by SV40 large T-antigen and by HPV-16 E6 oncoprotein. *EMBO J.* **11**: 5013-5020.

Miller,W.E., Gosialos,G., Kieff,E. *et al.* (1997) Epstein-Barr virus LMP1 induction of the epidermal growth factor receptor is mediated through a TRAF signaling pathway distinct from NF-κB activation. *J. Virol.* **71**: 586-594.

Miyashita,T. and Reed,J.C. (1995) Tumour suppressor p53 is a direct transcriptional activator of the human bax gene. *Cell* **80**: 293-299.

Miyashita,T., Krajewski,S., Krajewska,M. *et al.* (1994) Tumor suppressor p53 is a regulator of bcl-2 and bax gene expression in vitro and in vivo. *Oncogene* **9**: 1799-1805.

Mochizuki,K., Hayashi,N., Hiramatsu,N. *et al.* (1996) Fas antigen expression in liver tissues of patients with chronic hepatitis B. *J. Hepatol.* **24**: 1-7.

Moore,M., Horikoshi,N., and Shenk,T. (1996) Oncogenic potential of the adenovirus E4orf6 protein. *Proc. Natl. Acad. Sci. USA* **93**: 11295-11301.

Moore,P.S., Gao,S-J., Dominguez,G. *et al.* (1996) Primary characterization of a herpesvirus agent associated with Kaposi's sarcoma. *J. Virol.* **70**: 549-558.

Muzio,M., Chinnaiyan,A.M., Kischkel,F.C. *et al.* (1996) FLICE, a novel FADD-homologous ICE/CED-3-like protease, is recruited to the CD95 (Fas/APO-1) death-inducing signaling complex. *Cell* **85**: 817-827.

Muzio,M., Salvesen,G.S. and Dixit,V.M. (1997) FLICE induced apoptosis in a cell-free system. *J. Biol. Chem.* **272**: 2952-2956.

Nagata,S. and Golstein,P. (1995) The Fas death factor. *Science* **267**: 1449-1456.

Nava,V.E, Cheng,E.H-Y., Veliuona,M., *et al.* (1997) Herpesvirus saimiri encodes a functional homolog of the human bcl-2 oncogene. *J. Virol.* **71**: 4118-4122.

Nava,V.E., Rosen,A., Veliuona,M.A. *et al.* (1998) Sindbis virus induces apoptosis through a caspase-dependent, CrmA-sensitive pathway *J. Virol.* **72**: 452-459.

Neilan,J.G., Lu,Z., Afonso,C.L. *et al.* (1993) An African swine fever virus gene with similarity to the proto-oncogene bcl-2 and the Epstein-Barr virus gene BHRF1. *J. Virol.* **67**: 4391-4394.

Nevins,J.R. (1992) E2F: a link between the Rb tumour suppressor protein and viral oncoproteins. *Science* **358**: 424-429.

Ng,F.W.H., Nguyen,M., Kwan,T. *et al.* (1997) p28 Bap31, a Bcl-2/Bcl-X_L and procaspase-8 associated protein in the endoplasmic reticulum. *J. Cell Biol.* **139**: 327-338.

Noteborn,M.H.M.., Todd,D., Verschueren,C.A.J. *et al.* (1994) A single chicken anemia virus protein induces apoptosis. *J. Virol.* **68**: 346-351.
Palumbo,G.J., Pickup,D.J., Fredrickson,T.N. *et al.* (1989) Inhibition of an inflammatory response is mediated by a 38-kDa protein of cowpox virus. *Virology* **172**: 262-273.
Pan,G., O'Rourke,K. and Dixit,V.M. (1998) Caspase-9, Bcl-x_L and Apaf-1 form a ternary complex. *J. Biol. Chem.* **273**: 5841-5845.
Pearson,G.R., Luka,J., Petti,L. *et al.* (1987) Identification of an Epstein-Barr virus early gene encoding a second component of the restricted early antigen complex. *Virology* **160**: 151-161.
Perng,G-C., Ghiasi,H., Slanina,S.M. *et al.* (1996) The spontaneous reactivation function of the herpes simplex virus type 1 LAT gene resides completely within the first 1.5 kilobases of the 8.3-kilobase primary transcript. *J. Virol.* **70**: 976-984.
Pfitzner,A.J., Tsai,E., Strominger,J.L. *et al.* (1987) Isolation and characterization of cDNA clones corresponding to transcripts from the BamHI H and F regions of the Epstein-Barr virus genome. *J. Virol.* **61**: 2902-2909.
Pickup,D.J., Ink,B.S., Hu,W. *et al.* (1986) Hemorrhage in lesions caused by cowpox virus in induced by a viral protein that is related to plasma protein inhibitors of serine proteases. *Proc. Natl. Acad. Sci. USA* **83**: 7698-7702.
Pilder,S., Logan,J. and Shenk,T. (1984) Deletion of the gene encoding the adenovirus 5 early region 1B 21,000-molecular-weight polypeptide leads to degradation of viral and host cell DNA. *J. Virol.* **52**: 664-671.
Polyak,K., Waldman,T., He,T-C. *et al.* (1996) Genetic determinants of p53-induced apoptosis and growth arrest. *Gen. Dev.* **10**: 1945-1952.
Polyak,K., Xia,Y., Zweier,J.L. *et al.* (1997) A model for p53-induced apoptosis. *Nature* **389**: 300-305.
Potempa,J., Korzus,E. and Travis,J. (1994) The serpin superfamily of proteinase inhibitors: structure, function, and regulation. *J. Biol. Chem.* **269**: 15957-15960.
Prikhod'ko,E.A. and Miller,L.K. (1996) Induction of apoptosis by baculovirus transactivator IE-1. *J.Virol.* **70**: 7116-7124.
Proud,C.G. (1995) PKR: a new name and new roles. *TIBS* **20**: 241-246.
Qian,Y., Luckey,C., Horton,L. *et al.* (1992) Biological function of the retinoblastoma protein requires distinct domains for hyperphosphorylation and transcription factor binding. *Mol. Cell. Biol.* **12**: 5363-5372.
Qin,X.Q., Chittenden,T., Livingston,D.M. *et al.* (1992) Identification of a growth suppression domain within the retinoblastoma gene product. *Gen. Dev..* **6**: 953-964.
Querido,E., Marcellus,R.C., Lai,A. *et al.* (1997) Regulation of p53 levels by the E1B-55kDa protein and E4orf6 in adenovirus-infected cells. *J. Virol.* **71**: 3788-3798.
Raab-Traub,N. (1992) Epstein-Barr virus and nasopharyngeal carcinoma. *Sem. Cancer. Biol.* **3**: 297-307.
Rao,L., Debbas,M., Sabbatini,P. *et al.* (1992) The adenovirus E1A proteins induce apoptosis, which is inhibited by the E1B 19kDa and Bcl-2 proteins. *Proc. Natl. Acad. Sci.* **89**: 7742-7746.
Ray,C.A., Black,R.A., Kronheim,S.R. *et al.* (1992) Viral inhibition of inflammation: cowpox virus encodes an inhibitor of the interleukin-1-beta converting enzyme. *Cell* **69**: 597-604.
Reed,J.C. (1997) Double identity for proteins of the Bcl-2 family. *Nature* **387**: 773-776.
Reinarz,A.B.G., Broome,M.G. and Sagik,B.P. (1971) Age-dependent resistance of mice to Sindbis virus infection: viral replication as a function of host age. *Infect. Immun.* **3**: 268-273.
Rosen,A., Casciola-Rosen,L. and Ahearn,J. (1995) Novel packages of viral and self-antigens are generated during apoptosis. *J. Exp. Med.* **181**: 1557-1561.
Rothe,M., Pan,M-G. Henzel,W.J. *et al.* (1995) The TNFR2-TRAF signaling complex contains two novel proteins related to baculoviral inhibitor of apoptosis proteins. *Cell* **83**: 1243-1252.
Rowe,M., Peng-Pilon,M., Huen,D.S. *et al.* (1994) Upregulation of bcl-2 by the Epstein-Barr virus latent membrane protein LMP1: a B-cell-specific response that is delayed relative to NF-κB activation and to induction of cell surface markers. *J. Virol.* **68**: 5602-5612.
Roy,N., Deveraux,Q.L., Takahashi,R. *et al.* (1997) The c-IAP-1 and c-IAP-2 proteins are direct inhibitors of specific caspases. *EMBO J.* **16**: 6914-6925.
Roy,N., Mahadevan,M.S., McLean,M. *et al.* (1995) The gene for neuronal apoptosis inhibitory protein is partially deleted in individuals with spinal muscular atrophy. *Cell* **80**: 167-178.
Sabbatini,P., Chiou,S-K., Rao,L. *et al.* (1995) Modulation of p53-mediated transcriptional repression and apoptosis by the adenovirus E1B 19K protein. *Mol. Cell. Biol.* **15**: 1060-1070.
Saito,S. (1990) Enchancement of the interferon-induced double-stranded RNA-dependent protein kinase activity by Sindbis virus infection and heat-shock stress. *Microbiol. Immunol.* **34**: 859-870.
Sanchez-Prieto,R., Lleonart,M. and Ramon y Cajal,S. (1995) Lack of correlation between p53 protein level and sensitivity of DNA-damaging agents in keratinocytes carrying adenovirus E1A mutants. *Oncogene* **11**: 675-682.
Sarid,R., Sato,T., Bohenzky,R.A. *et al.* (1997) Kaposi's sarcoma-associated herpesvirus encodes a functional bcl-2 homologue. *Nature Med.* **3**: 293-298.
Sarnow,P., Hearing,P., Anderson,C.W. *et al.* (1984) Adenovirus early region 1b 58,000-dalton tumor antigen is physically associated with an early region 4 25,000-dalton protein in productively infected cells. *J. Virol.* **49**: 692-700.
Sarnow,P., Ho,Y.S., Williams,J. *et al.* (1982) Adenovirus E1B-58kd tumor antigen and SV40 large tumor antigen are physically associated with the same 54kd cellular protein. *Cell* **28**: 387-395.

Sawtell,N.M. and Thompson,R.L. (1992) Herpes simplex virus type 1 latency-associated transcription unit promotes anatomical site-dependent establishment and reactivation from latency. *J. Virol.* **66**: 2157-2169.
Scaria,A., Tollefson,A.E., Saha,S.K. *et al.* (1992) The E3-11.6K protein of adenovirus is an asn-glycosylated integral membrane protein that localizes to the nuclear membrane. *Virology* **191**: 743-753.
Schang,L.M., Hossain,A. and Jones,C. (1996) The latency-related gene of bovine herpesvirus 1 encodes a product which inhibits cell cycle progression. *J. Virol.* **70**: 3807-3814.
Scheffner,M., Werness,B.A., Huibregtse,J.M. *et al.* (1990) The E6 oncoprotein encoded by human papillomavirus types 16 and 18 promotes the degradation of p53. *Cell* **63**: 1129-1136.
Schreiber,M., Sedger,L. and McFadden,G. (1997) Distinct domains of M-T2, the myxoma virus tumor necrosis factor (TNF) receptor homolog, mediate extracellular TNF binding and intracellular apoptosis inhibition. *J. Virol.* **71**: 2171-2181.
Schulze,A., Zerfa,B.K., Spitkovsky,D. *et al.* (1995) Cell cycle regulation of cyclin A gene transcription is mediated by a variant E2F binding site. *Proc. Natl. Acad. Sci. USA* **92**: 11264-11268.
Senkevich,T.G., Bugert,J.J., Sisler,J.R. *et al.* (1996) Genome sequence of a human tumorigenic poxvirus: prediction of specific host response-evasion genes. *Science* **273**: 813-816.
Seshagiri,S. and Miller,L.K. (1997) Baculovirus inhibitors of apoptosis (IAPs) block activation of Sf-caspase-1. *Proc. Natl. Acad. Sci. USA* **94**: 13606-13611.
Shan,B. and Lee,W-H. (1994) Deregulated expression of E2F-1 induces S-phase entry and leads to apoptosis. *Mol. Cell. Biol.* **14**: 8166-8173.
Shisler,J., Duerksen-Hughes,P., Hermiston,T.M. *et al.* (1996) Induction of susceptibility to tumour necrosis factor by E1A is dependent on binding to either p300 or p105-Rb and induction of DNA synthesis. *J. Virol.* **70**: 68-77.
Shisler,J., Yang,C., Walter,B. *et al.* (1997) The adenovirus E3-10.4K/14.5K complex mediates loss of cell surface Fas (CD95) and resistance to Fas-induced apoptosis. *J. Virol.* **71**: 8299-8306.
Shisler,J..L., Senkevich,T.G., Berry,M.J. *et al.* (1997) Ultraviolet-induced cell death blocked by a selenoprotein from a human dermatotropic poxvirus. *Science* (In Press)
Shtrichman,R. and Kleinberger,T. (1998) Adenovirus type 5 E4 open reading frame 4 protein induces apoptosis in transformed cells. *J. Virol.* (In Press).
Smith,C.A. (1995) A novel viral homologue of Bcl-2 and Ced-9. *Trends Cell Biol.* **5**: 344.
Srinivasula,S.M., Ahmad,M., Fernandes-Alnemri,T. *et al.* (1996) Molecular ordering of the Fas-apoptotic pathway: the Fas/APO-1 protease Mch5 is a CrmA-inhibitable protease that activates multiple Ced-3/ICE-like cysteine proteases. *Proc. Natl. Acad. Sci. USA* **93**: 14486-14491.
Stewart,S.A., Poon,B., Jowett,J.B.M. *et al.* (1997) Human immunodeficiency virus type 1 Vpr induces apotosis following cell cycle arrest. *J. Virol.* **71**: 5579-5592.
Suarez,P., Diaz-Guerra,M., Prieto,C. *et al.* (1996) Open reading frame 5 of porcine reproductive and respiratory syndrome virus as a cause of virus-induced apoptosis . *J. Virol.* **70**: 2876-2882.
Subramanian,T. and Chinnadurai,G. (1986) Separation of the functions controlled by the adenovirus 2 lp^+ locus. *Virology* **150**: 381-389.
Subramanian,T., Tarodi,B. and Chinnadurai,G. (1995) p53-independent apoptotic and necrotic cell deaths induced by adenovirus infection: suppression by E1B 19K and Bcl-2 proteins. *Cell Growth Different.* **6**: 131-137.
Subramanian,T., Tarodi,B., Govindarajan,R. *et al.* (1993) Mutational analysis of the transforming and apoptosis suppression activities of the adenovirus E1B 175R protein. *Gene* **124**: 173-181.
Sugimoto,A., Friesen,P.D. and Rothman,J.H. (1994) Baculovirus p35 prevents developmentally programmed cell death and rescues a ced-9 mutant in the nematode Caenorhabditis elegans. *EMBO J.* **13**: 2023-2028.
Takemori,N., Cladaras,C., Bhat,B. *et al.* (1984) Cyt gene of adenovirus 2 and 5 is an oncogene for transforming function in early region E1B and encodes the E1B 19,000-molecular-weight polypeptide. *J. Virol.* **52**: 793-805.
Takemori,N., Riggs,J.L. and Aldrick,C.D. (1968) Genetic studies with tumorigenic adenoviruses. I. Isolation of cytocidal (cyt) mutants of adenovirus type 12. *Virology* **36**: 575-586.
Takizawa,T., Fukuda,R., Miyawaki,T. *et al.* (1995) Activation of the apoptotic Fas antigen-encoding gene upon influenza virus infection involving spontaneously produced beta-interferon. *Virology* **209**: 288-296.
Tarodi,B., Subramanian,T. and Chinnadurai,G. (1993) Functional similarity between adenovirus E1B 19K and Bcl-2 oncogene: mutant complementation and suppression of cell death induced by DNA damaging agents. *Intl. J. Oncol.* **3**: 467-472.
Taylor,R.M., Hurlbut,H.S., Work,T.H. *et al.* (1955) Sindbis virus: a newly recognized arthropod-transmitted virus. *Amer. J. Trop. Med. Hyg.* **4**: 844-862.
Teodoro,J.G., Shore,G.C. and Branton,P.E. (1995) Adenovirus E1A proteins induce apoptosis by both p53-dependent and p53-independent mechanisms. *Oncogene* **11**: 467-474.
Thome,M., Schneider,P., Hofmann,K. *et al.* (1997) Viral FLICE-inhibitory proteins (FLIPs) prevent apoptosis induced by death receptors. *Nature* **386**: 517-521.
Thompson,R.L. and Sawtell,N.M. (1997) The herpes simplex virus type 1 latency-associated transcript gene regulates the establishment of latency. *J. Virol.* **71**: 5432-5440.

Tollefson,A.E., Ryerse,J.S., Scaria,A. *et al.* (1996) The E3-11.6kDa adenovirus death protein (ADP) is required for efficient cell death: characterization of cells infected with adp mutants. *Virology* **220**: 152-162.
Tollefson,A.E., Scaria,A., Hermiston,T.W. *et al.* (1996) The adenovirus death protein (E3-11.6K) is required at very late stages of infection for efficient cell lysis and release of adenovirus from infected cells. *J. Virol.* **70**: 2296-2306.
Tolskaya,E.A., Romanova,L.I., Kolesnikova,M.S. *et al.* (1995) Apoptosis-inducing and apoptosis-preventing functions of poliovirus. *J. Virol.* **69**: 1181-1189.
Trede,N.S., Tsytsykova,A.V., Chatila,T. *et al.* (1995) Transcriptional activation of the human TNF-alpha promoter by superantigen in human monocytic cells: role of NF-kappa B. *J. Immunol.* **155**: 902-908.
Trentin,J.J., Tabe,Y. and Taylor,G. (1962) The quest for human cancer viruses. *Science* **137**: 835-849.
Trgovcich,J., Ryman,K., Extrom,P. *et al.* (1997) Sindbis virus infection of neonatal mice results in a severe stress response. *Virology* **227**: 234-238.
Tucker,P.C., Strauss,E.G., Kuhn,R.J. *et al.* (1993) Viral determinants of age-dependent virulence of Sindbis virus for mice. *J. Virol.* **67**: 4605-4610.
Tyler,K.L., Squier,M.K.T., Rodgers,S.E. *et al.* (1995) Differences in the capacity of reovirus strains to induce apoptosis are determined by the viral attachment protein 1. *J. Virol.* **69**: 6972-6979.
Tyor,W.R., Wesselingh,S.L., Levine,B. *et al.* (1992) Longterm intraparenchymal immunoglobulin secretion after acute viral encephalitis in mice. *J. Immunol.* **149**: 4016-4020.
Ubol,S., Tucker,P.C., Griffin,D.E. *et al.* (1994) Neurovirulent strains of alphavirus induce apoptosis in bcl-2-expressing cells: role of a single amino acid change in the E2 glycoprotein. *Proc. Natl. Acad. Sci.* **91**: 5202-5206.
Uren,A.G., Pakusch,M., Hawkins,C.J. *et al.* (1996) Cloning and expression of apoptosis inhibitory protein homologs that function to inhibit apoptosis and/or bind tumor necrosis factor receptor-associated factors. *Proc. Natl. Acad. Sci. USA* **93**: 4974-4978.
Virgin,H.W., Latreille,P., Wamsley,P. *et al.* (1997) Complete sequence and genomic analysis of murine gammaherpesvirus 68. *J. Virol.* **71**: 5894-5904.
Vucic,D., Kaiser,W.J., Harvey,A.J. *et al.* (1997) Inhibition of reaper-induced apoptosis by interaction with inhibitor of apoptosis proteins (IAPs). *Proc. Natl. Acad. Sci. USA* **94**: 10183-10188.
Wada,N., Matsumura,M., Ohba,Y. *et al.* (1995) Transcription stimulation of the Fas-encoded gene by nuclear factor for interleukin-6 expression upon influenza virus infection. *J. Biol. Chem.* **270**: 18007-18012.
Wallach,D. (1997) Placing death under control. *Nature* **388**: 123-126.
Wang,J. and Walsh,K. (1996) Resistance to apoptosis conferred by cdk inhibitors during myocyte differentiation. *Science* **273**: 359-361.
Wang,S, Rowe,M. and Lundgren,E. (1996) Expression of the Epstein-Barr virus transforming protein LMP1 causes a rapid and transient stimulation of the Bcl-2 homologue Mcl-1 levels in B-cell lines. *Can. Res.* **56**: 4610-4613.
Watanabe-Fukunaga,R., Brannan,C.I., Copeland,N.G. *et al.* (1992) Lymphoroliferation disorder in mice explained by defects in Fas antigen that mediates apoptosis. *Nature* **356**: 314-317.
Webster,K.A., Muscat,G.E.O. and Kedes,L. (1988) Adenovirus E1A products suppress myogenic differentiation and inhibit transcription from muscle-specific promoters. *Nature* **332**: 553-557.
Weissman,D., Rabin,R.L., Arthos,J. *et al.* (1997) Macrophage-tropic HIV and SIV envelope proteins induce a signal through the CCR5 chemokine receptor. *Nature* **389**: 981-985.
Westendorp,M.O., Frank,R., Ochsenbauer,C. *et al.* (1995) Sensitisation of T cells to CD95-mediated apoptosis by HIV-1 Tat and gp120. *Nature* **375**: 497-500.
White,E. (1993) Regulation of apoptosis by the transforming genes of the DNA tumour virus adenovirus. *Proc. Soc. Exp. Biol. Med.* **204**: 30-39.
White,E. (1994) Function of the adenovirus E1B oncogene in infected an transformed cells. Virology **5**: 341-348.
White,E., Cipriani,R., Sabbatini,P. *et al.* (1991) Adenovirus E1B 19-kilodalton protein overcomes the cytotoxicity of E1A proteins. *J. Virol.* **65**: 2968-2978.
White,E., Faha,B. and Stillman,B. (1986) Regulation of adenovirus gene expression in human WI38 cells by an E1B-encoded tumor antigen. *Mol. Cell. Biol.* **6**: 3763-3773.
White,E., Grodzicker,T. and Stillman,B.W. (1984) Mutations in the gene encoding the adenovirus early region 1B 19,000-molecular weight tumor antigen cause the degradation of chromosomal DNA. *J. Virol.* **52**: 410-419.
White,E., Sabbatini,P., Debbas,M. *et al.* (1992) The 19-kilodalton adenovirus E1B transforming protein inhibits programmed cell death and prevents cytolysis by tumor necrosis factor alpha. *Mol. Cell. Biol.* **12**: 2570-2580.
Wu,G.S., Burns,T.F., McDonald,E.R.III., *et al.* (1997) Killer/DR5 is a DNA damage-inducible p53-regulated death receptor gene. *Nature Gen.* **17**: 141-143.
Wu,H. and Lozano,G. (1994) NF-kappa B activation of p53. A potential mechanism for suppressing cell growth in response to stress. *J. Biol. Chem.* **269**: 20067-20074.
Wu,T-T., Su,Y-H., Block,T.M. *et al.* (1996) Evidence that two latency-associated transcripts of herpes simplex virus type 1 are nonlinear. *J. Virol.* **70**: 5962-5967.
Xue,D. and Horvitz,R. (1995) Inhibition of the *Caenorhabditis elegans* cell-death protease CED-3 by a CED-3 cleavage site in baculovirus p35 protein. *Nature* **377**: 248-251.

Xue,D. and Horvitz,R. (1997) *Caenorhabditis elegans* CED-9 protein is a bifunctional cell-death inhibitor. *Nature* **390**: 305-308.
Yang,X-J., Ogryzko,V.V., Nishikawa,J-I. *et al.* (1996) A p300/CBP-associated factor that competes with the adenoviral oncoprotein E1A. *Nature* **382**: 319-324.
Yew,P.R. and Berk,A.J. (1992) Inhibition of p53 transactivation required for transformation by adenovirus early 1B protein. *Nature* **357**: 82-85.
Yuan,J. (1997) Transducing signals of life and death. *Curr. Opin. Cell Biol.* **9**: 247-251.
Zhou,Q., Snipas,S., Orth,K. *et al.* (1997) Target protease specificity of the viral serpin CrmA. *J. Biol. Chem.* **272**: 7797-7800.
Zhu,H., Shen,Y. and Shenk,T. (1995) Human cytomegalovirus IE1 and IE2 proteins block apoptosis. *J. Virol.* **69**: 7960-7970.
Zhuang,S-M., Shvarts,A., van Ormondt,H. *et al.* (1995) Apoptin, a protein derived from chicken anemia virus, induces p53-independent apoptosis in human osteosarcoma cells. *Cancer Res.* **55**: 486-489.

Chapter 10

Therapeutic manipulation of apoptosis in cancer and neurological disease.

Alastair Watson and Pedro Lowenstein
Department of Medicine, University of Manchester, Clinical Sciences Building, Hope Hospital, Eccles Old Road, Salford M6 8HD. Molecular Medicine Unit, Room 1.302 Stopford Building, Department of Medicine, University of Manchester, Oxford Road, Manchester M13 9PT.

1.0 INTRODUCTION

This chapter explores the therapeutic implications of the discovery of the genetic control of cell death which encompasses most of medicine. It would require an entire book to cover all the diseases and therapeutic manoeuvres in which apoptosis plays a role or directs future research for effective treatment. Rather in this chapter we shall illustrate these implications by concentrating on cancer and neurological disease making only passing comment about other diseases.

Apoptosis is a fundamental biological process which can be traced back to primitive multicellular organisms, and possibly bacteria, suggesting it first appeared in early evolution (Ameisen, 1996, Chaloupka and Vinter, 1996). It regulates cell numbers and sculpts the organism during embryogenesis and development. In the immune system it is an important mechanism of killing invading micro-organisms or foreign cells and plays an essential role in the development of lymphocytes, and the reduction of mature T cell numbers (Osborne, 1996, Kabelitz, 1998). Special mechanisms have evolved for removing apoptotic cells without accidental triggering of an inflammatory response, so that apoptosis can occur without damage to neighbouring viable cells. The range of stimuli which have been found to induce apoptosis is so great that it appears to be the preferred mode of cell death when cellular injury is too severe to be repaired. Necrosis only occurs when damage to cellular membranes and loss of ATP is so great that apoptotic mechanisms are overwhelmed, although some authors have proposed a continuum between both death processes. With these thoughts in mind it is not surprising that apoptosis plays an important role in the pathogenesis of many diseases, particularly if its powerful regulatory mechanisms cease to function properly. Thus, resistance to the induction of apoptosis may contribute to the initiation of neoplastic disease, while in ischaemic brain injury or neurodegeneration, fulminant liver failure or HIV infection, reduction in apoptosis is the desired therapeutic goal. In this chapter we shall consider these disease scenarios and in addition consider immunosuppression for transplantation where novel therapeutic strategies have been developed based on apoptosis research.

2.0 CANCER

2.1 Cytotoxic, chemotherapeutic and other anti-cancer agents induce apoptosis.

Apoptosis is fundamental to cancer treatment as genes which regulate apoptosis including the *bcl-2* family, *p53*, *myc*, *ras*, *COX-2*, *CD95* and *CD95 ligand* are either deleted, mutant, or inappropriately expressed in cancer cells (Fearon and Vogelstein, 1990, Kinzler and Vogelstein, 1996, Muller *et al.*, 1997, Watson and DuBois, 1997, Watson *et al.*, 1996). Furthermore, the major mode of action of chemotherapeutic drugs and radiotherapy is to induce apoptosis in target cells. In this section of the chapter we shall consider apoptosis in the context of cancer prevention, conventional cancer treatment and novel anticancer therapeutic strategies including gene therapy.

2.1.1 Chemotherapy and radiotherapy

Recent work has demonstrated that virtually all cytotoxic drugs and radiotherapy induce apoptosis in tumour cells. This discovery highlights future avenues for therapeutic intervention, since these diverse cytotoxic treatments have been shown to stimulate a common cell death programme. For example, early studies by Ijiri and Potten demonstrated that 13 commonly used cytotoxic drugs and radiation, all induced apoptosis in the proliferative compartment of the small intestinal crypts (Ijiri and Potten, 1983, Ijiri and Potten, 1987). *In vitro* and *in vivo* studies have shown that the early apoptotic events (within 12-24 hours) observed after administration of cytotoxic drugs or radiotherapy is mediated by DNA damage-induced activation of p53, and consequently is completely absent in *p53*-knockout mice (Merritt *et al.*, 1994, Pritchard *et al.*, 1996, Lowe *et al.*, 1993a, Lowe *et al.*, 1993b). However, p53-indepentent apoptosis may occur at later times following cytotoxic insult (Merritt *et al* 1997).

Consistent with this observation is the fact that in clinical practice it has been frequently observed that tumours with mutant p53 are resistant to chemotherapy or radiotherapy. This can now be understood, since p53-dependent apoptosis cannot be activated by DNA damage in these tumours (Berns *et al.*, 1998, Lens *et al.*, 1997). RNA damage may also be a mechanism by which some anticancer drugs induce apoptosis. 5-fluorouracyl (5-FU) induces apoptosis in normal intestinal epithelium and is known to inhibit thymidylate synthetase thereby reducing available thymidine for DNA synthesis. In some cell types administration of exogenous thymidine can overcome this lack of thymidine and rescue the tissue from the damaging effects. Surprisingly, thymidine cannot prevent 5-FU-induced apoptosis in mouse intestinal epithelium *in vivo* although administration of uridine can, suggesting that RNA damage is a major determinant of 5-FU-induced apoptosis in this tissue. Curiously, 5-FU does not induce apoptosis in *p53*-knockout mice. Together these observations suggest that p53 can detect RNA damage in addition to DNA damage (Pritchard *et al.*, 1997).

However, whether the clinical response of tumours to radio- and chemotherapy can be explained simply by the induction of apoptosis alone remains to be determined. For example it is not clear why some tumour cells are more sensitive to DNA damaging agents than normal cells. This paradox may be partly explained by the multiple cellular responses following injury. In addition to activating apoptosis, p53 also upregulates $p21^{WAF1/CIP1}$ causing arrest at the G1 phase of the cell cycle (Hartwell and Kastan, 1994, Eldeiry *et al.*, 1993, Eldeiry *et al.*, 1994, Merritt *et al.*, 1997). Cells with intact $p21^{WAF1/CIP1}$ undergo stable cell cycle arrest whereas, cells with defective $p21^{WAF1/CIP1}$ undergo continuing rounds of DNA synthesis without mitosis, leading to polyploidy and apoptotic cell death (Waldman *et al.*, 1996). *In vitro*, the same overall response is observed in both $p21^{WAF1/CIP1}$ -/- and $p21^{WAF1/CIP1}$ +/+ cells, since neither cell cycle arrested cells nor apoptotic cells can divide and form clones. However, *in vivo*, tumour cells with defective $p21^{WAF1/CIP1}$ grown as xenografts in nude mice can often be completely cured by radiation whereas, those with functional $p21^{WAF1/CIP1}$ are never cured (Waldman *et al.*, 1997). An interpretation of these results is that following irradiation, there will be a range of responses of cells within the tumour. Some will sustain severe damage and die by either necrosis or apoptosis, some will experience minimal damage and be capable of normal cell division while others will sustain intermediate degrees of damage and not immediately die. It is the fate of these cells with moderate degrees of damage which is hypothesised to determine the different cure rates of $p21^{WAF1/CIP1}$ positive and negative cells. Cells with functional $p21^{WAF1/CIP1}$ undergo cell cycle arrest and remain within the tumour and may act as "feeder cells" to other cells capable of dividing and repopulating the tumour whereas cells lacking $p21^{WAF1/CIP1}$ undergo apoptosis. While provocative, these observations are almost certainly an oversimplification of the clinical response of a tumour to therapy. Cytotoxic drugs also act by other mechanisms, for example, via the direct induction of cell cycle arrest in tumour cells or by the induction of apoptosis in non-neoplastic cells within the tumour, such as vascular endothelial cells (Lamb and Friend, 1997). Thus, further study of how apoptosis fits into the integrated response of a tumour *in vivo* to anticancer therapy is required.

Induction of apoptosis plays an important role in the toxicity of chemotherapeutic agents and ionising radiation to normal tissues surrounding tumour cells. An attractive clinical goal would be to block apoptosis in normal cells without blocking apoptosis in tumours. For example, infertility due to apoptosis of oocytes is a distressing complication of chemotherapy. Daunorubicin induces apoptosis in oocytes by a mechanism involving the generation of ceramide at the cell membrane, the activation of *bax* and finally the activation of caspases: pharmacological or genetic blockade of these steps prevents daunorubicin-induced apoptosis *in vivo*. Surprisingly, p53 does not appear to play a role (Perez *et al.*, 1997), but oocytes from mice deficient in caspase-2 are resistant to chemotherapeutic agents. Thus, specific inhibitors of caspase-2 may be useful to treat this complication, especially in very young children (Bergeron *et al.*, 1998).

Another complication of some chemotherapeutic drugs is diarrhoea due to loss of intestinal epithelial cells by apoptosis. However, in this tissue p53 is a major determinant of apoptosis as ionising radiation and 5-FU does not induce apoptosis in intestinal crypts in p53 knockout mice (Merritt *et al.*, 1994, Pritchard *et al.*, 1997). Conversely, Bcl-2 blocks radiation-induced apoptosis in the small and large intestine (Merritt *et al.*, 1995). The role of the pro-apoptotic Bax in radiation-induced apoptosis is not clear. Although Bax has been shown to be upregulated in the intestinal epithelial cells after radiation (Kitada *et al.*, 1996), experiments with *bax*-null mice demonstrate that Bax is not required for p53-dependent apoptosis, at least in thymocytes, in response to ionising radiation (Knudson *et al.* 1995).

Currently the most plausible apoptosis-based therapeutic strategy to prevent damage to normal surrounding tissues would be to block the caspases, which are the final common step of all apoptotic signal transduction pathways and are expressed in many tissues (Krajewska *et al.*, 1997). However, methods would have to be found to target the anti-caspase therapy away from the tumour and to allow apoptosis associated with normal immune function and tissue homeostasis to proceed uninhibited.

2.1.2 Cyclo-oxygenases and colorectal cancer: prevention and treatment

There is a wealth of evidence that non-steroidal anti-inflammatory drugs (NSAIDs), whose principle action is to inhibit cyclo-oxygenase isoenzymes, COX-1 and COX-2, prevent colorectal cancer. Animal studies have shown that a variety of NSAIDs, including sulindac (Rao *et al.*, 1995), piroxicam (Reddy *et al.*, 1987) and aspirin (Reddy *et al.*, 1993) can prevent chemically-induced cancer in rodents. Furthermore, NSAIDs reduce the number and size of intestinal polyps in patients with familial adenomatous polyposis (FAP). Thus, the original observation by Waddell and colleagues in 1983 that sulindac reduced the number polyps in FAP patients, has been confirmed by a number of studies including two double-blinded, randomised placebo-controlled trials of sulindac (Waddell and Loughry, 1983; Labayle *et al.*, 1991; Giardiello *et al.*, 1993). In addition, three prospective cohort studies of the incidence of spontaneous colorectal cancer have demonstrated substantial reductions in the relative risk of developing colorectal cancer in patients taking aspirin. In a cohort of over one million people, men who used aspirin more than 16 times a month had a relative risk of developing colorectal cancer of 0.48 (0.30 - 0.76), and women a relative risk of 0.53 (0.32 - 0.87) (Thun *et al.*, 1991) compared to controls not consuming aspirin. These observations were confirmed in two later prospective studies (Giovannucci *et al.*, 1994, Giovannucci *et al.*, 1995). In the latter study, analysis of duration of use showed that the protective effect of aspirin did not become statistically apparent unless intake had been for 10 years or greater. This was attributed to the idea that adenomas take approximately 10 years to evolve into invasive carcinomas.

There is accumulating evidence that this protective effect of NSAIDs may be due to their ability to induce apoptosis. Apoptosis has been observed in biopsy specimens from patients with NSAID-induced colitis (Lee, 1993) and a flow cytometric study of isolated rectal epithelial cells from FAP patients treated with sulindac has shown increased apoptosis (Pasricha *et al.*, 1995). A number of studies have shown that sulindac sulphide and aspirin can induce apoptosis in a number of adenoma (Elder *et al.*, 1996) and carcinoma cell lines (Piazza *et al.*, 1995, Shiff *et al.*, 1995, Yoshikawa *et al.*, 1995). The mechanism by which NSAIDs induce apoptosis is unclear. One attractive hypothesis is that they act by inhibiting COX-2, thereby reducing the production of COX-2 products which can induce apoptosis (Watson and DuBois, 1997). In favour of this hypotheses rat intestinal cells over-expressing COX-2 are resistant to butyrate-induced apoptosis which can be restored by addition of sulindac sulphide (Tsujii and Dubois, 1995). Significantly, over-expression of COX-2 increases Bcl-2 expression. Sulindac may also be able to reduce expression of COX-2 protein (Boobol *et al.*, 1996). Sulindac has also been reported to both induce G_1/S block in addition to apoptosis, in the colorectal adenocarcinoma cell line, HT-29.

However, not all data are compatible with NSAIDs inducing apoptosis via inhibition of COX-2. Sulindac sulfide and piroxicam can induce apoptosis in HCT-15 cells which do not express *COX-2* (Hanif *et al.*, 1996). In addition to COX 2 products such as PGE_2, there is evidence that other arachidonic acid metabolites, such as the lipoxygenase products LTB4 and 12 R-HETE and also p450 products, can stimulate proliferation of HT-29 cells (Qiao *et al.*, 1995, Tang *et al.*, 1993, Korystov *et al.*, 1996, Bortuzzo *et al.*, 1996). Definitive evidence demonstrating that the protective effect of NSAIDs against colorectal cancer is due to the induction of apoptosis in colonic epithelium is lacking. Colorectal cancers are thought to arise from mutant stem cells located at the base of colonic crypts. Thus, for a stem cell to propagate a malignant clone after mutation, it is essential that the mutation does not trigger apoptosis. It is currently unknown whether

NSAIDs enhance the ability of colonic stem cells to undergo apoptosis following mutation.

With this idea in mind it is interesting to note that stem cells in the small intestine, which has a very low incidence of cancer, respond to exposure to the carcinogens N-nitroso-N-methylurea, N-nitroso-N-ethylurea, 1,2 dimethylhydrazine and N-nitrosodimethylamine by undergoing apoptosis. By contrast, colonic stem cells are far more resistant to carcinogen induced apoptosis. This presumably raises the chance of mutant colonic stem cells developing into malignant clones (Li *et al.*, 1992, Potten *et al.*, 1992). This suggests that apoptosis pathways may vary in stem cells located at different levels of the gastrointestinal tract. If so, this information may become of therapeutic use in the future.

An alternative hypothesis to explain the protective effects of NSAIDs against colorectal cancer is inhibition of malondialdehyde (MDA) formation, resulting from the metabolism of PGH2. MDA is mutagenic in bacteria and mammalian cells and carcinogenic in rodents (Mukai and Goldstein, 1976, Basu and Marnett, 1983, Spalding, 1988). It reacts with DNA to form the adduct M_1guanine which can induce transitions to adenosine and transversions to thymidine (Moriya *et al.*, 1994). This biochemistry raises the possibility that inhibition of submucosal COX-1 by NSAIDs could reduce the mutation frequency in colonic epithelium and prevent polyp initiation.

2.1.3 Antioxidants and chemotherapy

Chemotherapy for disseminated colorectal cancer relies on 5-Flurouracil (5-FU) but its efficacy remains disappointing with response rates of approximately 20% in most studies. Current therapeutic strategies rely on combining 5-FU on other agents which enhance or complement its action. Regimes in current clinical practice are the combination of 5-FU with leucovorin (tetrahydrofolate) which potentiates the binding of 5-FU to its target molecule, thymidylate synthase (Moertel *et al.*, 1990) or the combination of 5-FU with the immunomodulatory agent levamisole (Petrelli *et al.*, 1989).

Recently it has been observed that the antioxidants pyrrolidinedithiocarbamate (PDTC) and the water soluble vitamin E analogue, 6-hydroxy-2,5,7,8-tetramethylchroman-carboxylic acid, enhances 5-FU-induced apoptosis in cultured colorectal cancer cells regardless of their p53 status (Chinery *et al.*, 1997b). More impressively, the combination of 5-FU and near toxic concentrations of PDTC cures colorectal tumour xenografts in nude mice, whereas 5-FU alone only slows tumour growth. The biochemistry of PDTC is complex. It is a dithiocarbamate and is related to thiuram disulphide which is used in alcohol aversion therapy. It can act as an anti-oxidant by scavenging free copper or iron ions, thereby inhibiting the generation of reactive oxygen species (ROS). However, under some circumstances PDTC can paradoxically act as an oxidant (Orrenius *et al.*, 1996). In the Chinery study, PDTC is likely to be acting as an antioxidant, as induction of apoptosis closely correlates with reduction in hydrogen peroxide concentration within tumour cells and the effects of PDTC could be mimicked by vitamin E, which is chemically distinct from PDTC.

Studies into the mechanisms by which PDTC induces apoptosis in transformed colorectal cancer cells have proven informative. PDTC induces the cyclin-dependent kinase inhibitor $p21^{WAF1/CIP1}$ in cells with either wild-type or non-functional p53 (Chinery *et al.*, 1997b). The induction of apoptosis is substantially reduced in HCT116 cells with targeted disruption of $p21^{WAF1/CIP1}$ gene, suggesting $p21^{WAF1/CIP1}$ plays a role in the induction of apoptosis by PDTC. Furthermore PDTC does not alter the IC_{50} of 5-FU against $p21^{WAF1/CIP1}$ disabled cells. A series of experiments with mutant $p21^{WAF1/CIP1}$ reporter constructs and electrophoretic mobility shift assays indicate that the induction of $p21^{WAF1/CIP1}$ by PDTC is mediated by the C/EBPβ transcription factor. Experiments with cells stably transfected with human C/EBPβ, in either sense or antisense orientation, indicate that the apoptotic effects of PDTC can be mimicked C/EBPβ. These anti-oxidant events are mediated by protein kinase A-mediated phosphorylation of Ser^{299} in C/EBPβ which enables its translocation from the cytoplasm to the nucleus and subsequent transactivation of antioxidant responsive genes (Chinery *et al.*, 1997a). While these results are encouraging, further validation and clinical trials with antioxidants are clearly required.

Recently some new agents have been discovered which induce apoptosis in transformed cells and have potential as new anticancer drugs.

2.1.5 Betulinic Acid

Plants are a major source of new drugs. Evaluation of over 2,500 plant extracts against a panel of human cancer cell lines revealed an extract from *Ziziphsus Mauritiana*, a white birch tree, had selective cytotoxicity against melanoma and neuroectodermal tumours such as neuroblastoma and Ewing's tumour (Pisha *et al.*, 1995, Schmidt *et al.*, 1997). Its mechanism of action appears to be different from conventional chemotherapeutic drugs which induce apoptosis by activation of p53 or ligation of CD95 with its ligand. Instead it appears to interact with mitochondria in some way to directly activate caspase 3 and caspase 8. Although its cytotoxic action is blocked by Bcl-2, it can kill neuroblastoma cells which lack CD95 and are resistant to daunorubicin suggesting that it may be able to bypass some forms of drug resistance in cancer cells (Fulda *et al.*, 1997). The clinical utility of betulinic acid has yet to be established.

2.1.6 Taxol

The chemotherapeutic agent Taxol (Paclitaxel), is a microtubule stabilising agent which has activity against ovarian and breast cancers and melanoma. Experiments with cultured human fibroblasts in which p53 is wild-type or has been inactivated by SV40 T antigen or HPV-16 E6 have shown Taxol to have a much greater cytotoxic action against p53 defective cells. This is the result of induction of apoptosis and arrest at the G_2/M stage of the cell cycle (Wahl *et al.*, 1996). These results are in contrast to its effects *in vivo* where it is more cytotoxic to cells with wild type p53. The mechanisms underlying these differences remain to be fully established but *in vivo* Taxol promotes the release of a cytokine which induces apoptosis through a mechanism which can be blocked by antibodies to TNF-α (Lanni *et al.*, 1997).

2.1.7 Retinoids

Retinoids are derivatives of vitamin A and regulate a number of important biological functions including cell growth and differentiation, development and carcinogenesis. They act by binding to two classes of nuclear receptor; the retinoic acid receptor and the retinoic X receptor which regulates gene transcription by either binding to specific DNA sequences or interacting with other transcriptional regulators such as AP1 (Chambon, 1994). Clinical studies have shown that a number of retinoids have activity against leukaemia and tumours of the head and neck (Hong *et al.*, 1990, Huang *et al.*, 1988). Recently a number of novel retinoic acid derivatives have been found to induce apoptosis in non small cell lung cancer cell lines *in vitro* and to prevent lung cell tumour growth in a mouse xenograft model *in vivo* without apparent toxicity to the mouse (Lu *et al.*, 1997). These novel retinoids bind to the retinoic acid receptor-γ and, by mechanisms yet to be defined, activate caspases 2 and 3 resulting in cleavage of a range of intracellular proteins including the transcription factor SP1 (Piedrafita and Pfahl, 1997). An exciting feature of these compounds is that like PDTC and Taxol, functional p53 is not required for the induction of apoptosis suggesting they may be effective in the treatment of tumours with mutant p53.

2.2 Molecular manipulations of apoptosis

2.2.1 Gene therapy: delivery of pro- or anti-apoptotic genes, transcriptional activators, or drugs to modify the expression of such genes.

Gene therapy, the use of nucleic acids as drugs, is currently exploiting the knowledge gained from apoptosis research, for potential therapeutic applications (Bold *et al.*, 1997; Gibson and Kennedy, 1997; Nielsen and Maneval, 1998; Riley *et al.*, 1996). This knowledge has been implemented in three different ways: (a) to restore the function of mutated tumour suppresser genes to inhibit uncontrolled tumour growth, (b) to deliver apoptosis inducers directly into tumour cells, or (c) to restrict viral replication to tumour cells with particular genetic mutations.

Adenoviral vectors, as well as other virus-derived vectors have been used in gene therapy of cancer (Bilbao *et al.*, 1997; Dachs *et al.*, 1997). Interestingly, wild type human adenovirus E1B-55K protein inhibits p53 function, thus preventing infected cells undergoing apoptosis. Thus, an adenovirus mutant lacking the *E1B-55K* gene cannot replicate in *p53* wildtype cells, because apoptosis is induced before the virus can replicate (Heise *et al.*, 1997; Murphy and Levine, this volume). However, such a virus, can selectively replicate in cells with mutant *p53*. Thus, only tumour cells ought to be killed by the virus. The capacity of such a mutant to replicate and kill tumour cells with mutant *p53* without killing normal cells expressing wildtype *p53* is now being examined in clinical trials of various types of human tumours.

Various types of viral vectors have also been used to deliver tumour suppressors directly into tumour cells. Tumour suppressor mutations are frequent in human cancers. Their restitution using gene therapy vectors is, therefore, a strategy that has proven effective in many different models *in vitro*, and is currently under examination in human clinical trials (Gibson and Kennedy, 1997).

More relevant to this chapter though, apoptosis has revealed a wealth of novel gene products mediating the induction or the inhibition of apoptosis. The capacity to deliver individual gene products to cells in the body using gene therapy strategies, has spawned a large number of attempts either to induce or block apoptosis for the theoretical treatment of a large number of diseases. For example, pro-apoptotic genes have been expressed in various cancer models to achieve tumour cell killing, while a wealth of anti-apoptotic genes is being tested in animal models of neurodegeneration, in which apoptosis causes loss of large numbers of brain cells (Offen *et al.*, 1998; Xu *et al.*, 1997).

2.2.2. Gene delivery of cytotoxic or conditionally cytotoxic nucleic acids

Gene delivery has also been used to deliver various types of genes that induce apoptosis, and especially tumour cell death. Conditional cytotoxic genes have been delivered to cells via viral and non-viral vectors. The advantage of conditionally cytotoxic genes, is their lack of toxicity in the absence of the pro-drug. A robust conditionally cytotoxic system that induced apoptosis is the use of the HSV1-TK, which metabolises ganciclovir. This is a reliable cell-killing system, and clinical trials have been testing this system when expressed from various viral vectors, and in various tumour models. Its clinical efficacy remains to be assessed as results from Phase II/III clinical trials become available. Various other prodrug-metabolising enzymes are currently being tested, i.e. nitro-reductase and carboxipeptidase G2. Nitroreductase metabolises the prodrug 5-(aziridin-1-yl)-2,4-dinitrobenzamide (Greene *et al.* 1997), while carboxipeptidase G2 metabolises a pro-alkylating agent. How these systems will behave in clinical trials remains to be determined (Springer and Niculescu-Duvaz, 1996; Roth and Cristiano, 1997; Martin and Lemoine, 1996).

Alternatively, several groups are attempting to express directly cytotoxic proteins from viral vectors. Because of their intrinsic toxicity, and lower safety compared to prodrug activating systems, in these cases the cytotoxic proteins (eg. pseudomonas exotoxin,) are being expressed under the strict control of cell type (eg. tumour cells) specific promoters, inducible promoter elements, or linked to antibodies which deliver the cytotoxic proteins to the cytoplasm (Chen *et al.*, 1997; Chen and Marasco, 1996; Hart and Vile, 1995).

Specific transcriptional activators are likely to control apoptosis pathways in particular cell types (Chin and Fu, this volume). It is likely that in the future, such transcriptional activators could be directly expressed in gene therapy vectors to directly modulate the expression of the apoptotic cell death programme. Alternatively, the activity of specific transcriptional activators could be inhibited through the expression of dominant negative mutants (Campbell and Pollack, 1997). Several groups are also interested in developing small molecules capable of modifying apoptosis gene expression. Recently, small molecules like phenylbutyrate and hydroxyurea have been shown to directly modify globin gene expression (Olivieri *et al.*, 1997). Ultimately, it is likely that molecules will be developed which can be directly modify the expression of apoptosis regulatory genes and their products, at the transcriptional level.

2.2.3 CD95 and its ligand

The CD95/CD95L signal transduction system is the most important regulator of apoptosis in the immune system. CD95 ligand (also known as CD95L) is a 40 kDa type II transmembrane protein and is a member of the tumour necrosis family of cytokines. When bound to its receptor CD95 (Fas, APO-1), a 45 kDa glycosylated transmembrane protein belonging to the tumour necrosis receptor family, it rapidly induces apoptosis (Nagata, 1997). This system plays a role in deletion of T lymphocytes in the peripheral immune system, in terminating immune responses and in mediating T lymphocyte-mediated cytotoxicity (Vergelli *et al.*, 1997, Nagata, 1997, Ando *et al.*, 1997).

CD95L has also been proposed to play a role in contributing to the immune privilege of the eye and testis. In has been known for many years that transplantation of foreign cells into the eye or testis are not rejected but survive. This phenomenon of immunoprivilege has been of great interest to transplantation immunologists as an understanding of the mechanisms involved could lead to transplantation without immunosuppression (Lau and Stoeckert, 1997). Recent work suggests that cells from these sites can induce apoptosis in invading T cells and thereby suppress graft rejection. Both Sertoli cells of the testis and corneal cells of the eye express CD95L (Bellgrau *et al.*, 1995, Griffith *et al.*, 1995). Recently it has been shown that grafts of CD95 +ve cells, but not CD95 -ve cells, into the eye of normal mice are rejected whereas grafts of CD95 +ve cells into the eye of CD95L -ve mice are accepted (Griffith *et al.*, 1995). Similar results have been obtained with the testis (Bellgrau *et al.*, 1995). These observations suggest that immune privilege is the result of CD95L bearing host cells within the eye or testis inducing apoptosis of invading activated T cells bearing CD95 thus defeating the rejection reaction.

Sometimes cancers create a state of immune privilege to avoid immune attack. CD95 expression is lost or downregulated in hepatomas whereas CD95L expression is upregulated. This renders hepatomas resistant to attack by activated T cells because of their lack of CD95. In fact the tables are turned as the hepatomas cells can induce apoptosis in attacking T cells by crossing linking their own CD95L with CD95 on T cells (Strand *et al.*, 1996).

These results have suggested to transplantation immunologists that it might be possible to produce immunoprivilege in grafts if they were engineered to express CD95L. A particularly attractive application is Type 1 Diabetes Mellitus which in principle can be cured by graft of β cells from the islets of Langerhans which could be transfected with CD95L *ex-vivo*. Unfortunately, this has not proved to be the case: islet cells transfected with CD95L are more rapidly rejected than non-transfected control cells (Kang *et al.*, 1997). Analysis of this intensified rejection reaction showed that ligation of CD95 on donor cells by CD95L-bearing lymphocytes from the host was not responsible, nor was rejection dependent on B or T cells. A far more important factor turned out to be recruitment of neutrophils to the graft. The underlying mechanism has not been fully established but CD95L on donor cells may upregulate IL-8 expression which is known to have potent chemotactic properties. However, CD95L expression does not always result in neutrophil recruitment and factors such as cell type and level of CD95L expression may also be important (Lau and Stoeckert, 1997).

A more successful strategy to enable transplantation of β islet cells without immunosuppression has been to express CD95L on carrier cells which are co-transplanted with islet cells. Grafts in which syngeneic myoblasts transfected with CD95L or Sertoli cells which naturally express CD95L, have been co-transplanted with Islet cells can survive for prolonged periods without immunosuppression and can render diabetic mice normoglycaemic for periods of up to 80 days (Lau *et al.*, 1996). These results are promising and hold the prospect of freeing diabetic patients from the tyranny of insulin injections without the use of toxic immunosuppressive agents.

Under physiological conditions expression of CD95 and CD95L is tightly controlled to prevent inadvertent induction of apoptosis. However, deregulated expression is now known to be an important cause of disease. CD95 and CD95L expression is greatly increased in hepatocytes of patients with hepatitis-B related cirrhosis and acute liver failure compared to the normal liver (Galle *et al.*, 1995, Seino *et al.*, 1997). Cross-linking and activation of CD95 in the liver by systemic administration of anti-CD95 antibodies causes massive apoptosis and a histological picture similar to viral hepatitis. This is probably the most dramatic example of apoptosis outside embryonic development as the liver is literally destroyed in a few hours. Apoptosis induced by these means does not occur in transgenic mice which express the anti-apoptotic gene *bcl-2* in hepatocytes (Lacronique *et al.*, 1996). Furthermore, introduction of a cytotoxic T cell clone specific for hepatitis B surface antigen into transgenic mice which express HB_sAg in their liver, causes acute hepatitis and liver failure. Administration of a soluble form of CD95 binds to and inactivates CD95L on the cytotoxic T cells and prevents apoptosis and the development of hepatitis in the HB_sAg transgenic mice (Kondo *et al.*, 1997). Abnormal expression of CD95L has also been found in hepatocytes from patients with alcoholic hepatitis and in the thyroid glands of patients with Hashimotos thyroiditis suggesting its involvement in the pathogenesis of these conditions (Galle *et al.*, 1995, Giordano *et al.*, 1997).

Currently there are no practical ways of reducing the activity of the CD95/CD95L system for therapeutic purposes. However, the mechanism by which the CD95/CD95L system induces apoptosis has recently been elucidated and has suggested some potential therapeutic targets. The binding of CD95L to CD95 induces the formation of a "death-inducing signalling complex" (DISC) containing the proteins FADD or RIP and RAIDD leading to the recruitment and activation of caspase 8 or caspase 2 depending on the components of the DISC (Medema *et al.*, 1997, Strasser and O'Conner, 1998). Activation of either caspase leads ultimately to the apoptotic demise of the cell. The activation of caspase 8 by the DISC can be prevented by an endogenous protein called FLIP (Irmler *et al.*, 1997). Transfection of FLIP into cells has therapeutic potential against the induction of apoptosis by this CD95 mediated pathway. Ligation of CD95 and CD95L does not induce DISC formation is every tissue (Prof. Peter Krammer, personal communication). In these cases apoptosis is induced via a slower pathway which can be inhibited by Bcl-2 (Lacronique *et al.*, 1996). Thus gene therapy strategies may be developed which directly act on the CD95/CD95L signal transduction pathways. Such an approach would be particularly attractive for fulminant hepatic failure which currently has no treatment aside from liver transplantation.

By contrast, upregulation of CD95/CD95L is far simpler. Treatment of hepatoma cells lines with a range of cytotoxic drugs including bleomycin, cisplatin and mitomycin causes apoptosis. This is mediated by increased expression of both CD95 and CD95L, resulting in cell suicide by the crossing-linking of these molecules on the same cell or fratricide by crossing-linking between neighbouring cells (Friesen *et al.*, 1996, Muller *et al.*, 1997). Exposure to cytotoxic drugs generates reactive oxygen metabolites which interact with the CD95L promoter inducing transcription of CD95L mRNA(Hug *et al.*, 1997). The importance of this mechanism in the action of cytotoxic drugs remains to be established.

3.0 NEUROLOGICAL DISEASE

As in other parts of the body apoptosis is important in the brain and a number of observations have been made which suggest that apoptosis-based therapy may have potential in the treatment of neurodegenerative disease. Apoptosis appears to be an essential part of development where 20-80% of neurons die by apoptosis before adulthood (Oppenheim, 1991). The importance of caspase activity in the development of the central nervous system is illustrated by knockout mice deficient in caspase-3 which have reduced apoptosis of neurons within the brain, an excessive number of neurons and a reduced life span (Kuida *et al.*, 1996). This in itself could have therapeutic implications. Increased numbers of neurons could have serious adverse effects on brain function; this knowledge could have implications on the use of anti-apoptosis treatments during foetal life.

As in all other examples of apoptosis, activation of caspases within neurons causes cleavage of wide range of cellular proteins resulting in cellular demise. Caspase activation occurs at the end of the series of intracellular events leading to apoptosis such that, once activation has occurred the cell is fully commited to apoptosis and death is inevitable unless caspases are inhibited pharmacologically (Kroemer, 1997). It has been proposed that activation or inhibition of apoptosis could contribute to neurological cancer, stroke, trauma-induced neurodegenration, epilepsy, and various idiopathic neurodegenerative disorders. Thus, the manipulation of apoptotic pathways are likely to be of therapeutic value in neurological diseases. The applications of apoptosis research to tumour treatment strategies has already been reviewed above; thus, although it also applies to the treatment of brain tumours, it will not be reviewed further here.

3.1 Ischaemia

Various types of brain ischaemia have been shown to trigger apoptosis (Choi, 1996). Furthermore, induction of apoptosis is directly related to the duration of perfusion deficits (Vexler *et al.*, 1997). That apoptosis is responsible for neurodegeneration in ischaemic insults, was suggested initially in experiments using cycloheximide (reviewed in Silverstein, 1998). Programmed cell death is often protein synthesis-dependent and these experiments demonstrated the necessity of protein synthesis for ischaemia-induced neuronal apoptosis. This has now been further substantiated using specific inhibitors of caspases and gene delivery of anti-apoptotic genes. Such studies have shown that *bcl-2* delivered with a herpes simplex virus type 1 vector (Linnik *et al.*, 1995), or *NAIP* delivered via an adenovirus vector (Xu *et al.*, 1997), can confer neuroprotection in a stroke models, demonstrating that gene therapy strategies can be applied to stroke anti-apoptotic treatments. It is of interest to note that IL-1 receptor antagonist, expressed through an adenoviral vector was also useful in a neonatal rat model of excitotoxic damage, another putative mechanism underlying ischemic damage (Hagan *et al.*, 1996). In a neonatal model of stroke, a pan-caspase inhibitor, but not caspase 1 (ICE) inhibitors, were neuroprotective (Cheng *et al.*, 1998). Nevertheless, in other models, ICE inhibitors were effective. It is possible that apoptosis may vary at different stages of development, or different experimental paradigms. Given the multiple role of caspases, their effectiveness in preventing neurodegeneration could be due to an effective inhibition of apoptosis, or a reduction in caspase-catalysed generation of potentially neurotoxic neurochemical mediators (Silverstein, 1998). However, recent results on *caspase-2*-knockout mice have shown this enzyme may not play a central role in ischaemia-induced apoptosis (Bergeron *et al.*, 1998). Mapping of the specific pathways inducing neuronal apopotosis in individual experimental paradigms, should allow their specific manipulations in the future.

In stroke, particularly around the edge of infarcts or bleeds where injury is due to factors such as the transient release of oxidants, it may be possible to restore the normal cellular environment sufficiently rapidly to allow these dormant cells to regain normal function (Holtzman and Deshmukh, 1997). For example, the nuclear enzyme poly(ADP-ribose) polymerase (PARP) is activated by DNA single strand breaks caused by ischaemic cellular injury. PARP transfers NAD-ribose units to nuclear proteins as part of the DNA repair process. If DNA damage is sufficient, cellular NAD can be completely depleted by PARP causing shut down of glycolysis. This will lead to severe depletion of ATP and cell death. Inhibition of PARP can prevent this ischaemic cell death (Watson *et al.*, 1995). Furthermore, mice deficient in PARP are resistant to cerebral injury caused by disruption of the cerebral blood supply (Eliasson *et al.*, 1997). A combination of PARP inhibitors with caspases could in principle be effective in limiting the cerebral injury during stroke.

3.2 Neurodegenerative disorders

In early onset familial Alzheimer's disease caspase, activity may participate directly in the pathogenesis of the disease. These patients inherit mutations in genes encoding a neuronal proteins called Presenelin 1 and 2 which is cleaved abnormally in these patients by caspase 3. These abnormal cleavage products may lead to either the accumulation of amyloid-β42 or increase the sensitivity of neurons to apoptosis, although these hypotheses remain to be substantiated (Kim *et al.*, 1997).

Neurodegeneration is a major mechanism of disease in the central nervous system where it can be acute as in stroke or spinal cord injury or chronic, as in Parkinson's disease, Amyotrophic lateral sclerosis or Alzheimer's disease. Evidence is now accumulating that apoptosis is a major mechanism of neurodegeneration (Crowe *et al.*, 1997, Holtzman and Deshmukh, 1997). Recent work has suggested that inhibition of caspases could be therapeutically useful in both acute and chronic neurodegeneration. The issues involved are illustrated by a model system in which apoptosis in developing sympathetic neurons is induced by withdrawal of nerve growth factor. Blockade of caspases by Boc-aspartyl (OMe)-fluormethyl ketone results in significant inhibition of apoptosis despite the fact that many of the events in the signal transduction pathway which lead to apoptosis still occur. However, although the cells do not die they enter a dormant state with low metabolic activity and abnormal electrical activity (Deshmukh *et al.*, 1996).

In diseases which are the result of chronic neurodegeneration inhibition of caspases may also be of value. In transgenic mice, where a syndrome resembling amytrophic lateral sclerosis is produced by overexpression of superoxide dismutase, simultaneous expression of a dominant negative inhibitor of caspase-1 slows disease progression although it has no effect on the timing of disease onset (Friedlander *et al.*, 1997); the disease is also not ameliorated in caspase-2 knockout mice. In contrast, overexpression of Bcl-2 in these mice delays disease onset and prolongs neuronal survival but does not alter the duration of disease (Kostic *et al.*, 1997). Less encouraging results have been obtained in a mouse model of progressive neuronal neuropathy where overexpression of Bcl-2 has no clinical benefit, because axonal degeneration is not prevented although nerve cell loss is reduced. Thus, Bcl-2 is effective in protecting nigral neurons from neurotoxin insults, and ischemic insults, but does not protect motor neuron degeneration. These early results suggest that the success of anti-apoptotic therapy may depend on whether the clinical manifestations of the disease are a result of neuronal or axonal degeneration (Sagot *et al.*, 1995).

Surprisingly, in caspase-2 deficient mice, sympathetic neurons underwent apoptosis more effectively when deprived of NGF, and facial motor neurons died faster, but not in greater numbers, during developmentally regulated cell death (Bergeron *et al.*, 1998). Interestingly, neuronal apoptosis induced through ischaemia or mutant Cu/Zn superoxide dismutase, remained unaffected. This suggests, that at least for caspase-2, its activity can lead to the production of both anti-apoptotic, as well as pro-apoptotic molecules for neuronal cells. This could be partly explained by the generation of two caspase-2 transcripts, a long one, which induces cell death, and a truncated protein capable of antagonising cell death.

Epilepsy has been proposed to be due to excess neuronal activity, and neuronal dysgenesia and hyperplasia has been proposed as a possible pathophysiological mechanism, as in schizophrenia. Interestingly, caspase-3 deficient (knock-out) mice display neuronal hyperplasia (Kuida *et al.*, 1996), and it will be interesting to assess whether there is cortical hyperactivity in these animals. In contrast, epileptic seizures have shown to induce neuronal apoptosis in animal models (Charriaut-Marlangue *et al.*, 1996). Here too, various methods of down-modulating neuroexcitatory agents, Ca^{2+} entry, or oxidative stress, as well as apoptosis inhibitors, could be of potential therapeutic usefulness.

A question remains on how useful inhibition of apoptosis will be once the insult has triggered the neuronal cell death programme (Cheng *et al.*, 1998). Even though it has been possible to avoid neuronal cell death following the insult, either after growth factor withdrawal or ischemia, it has been observed that surviving neurons are nevertheless physiologically impaired (Deshmukh *et al.*, 1996; and see discussion in Cheng, 1998). Interestingly, similar results have been obtained in certain experimental models of Parkinsonism in which dopaminergic nigral neurons are kept alive, while their axonal innervation degenerates (Winkler *et al.*, 1996). How far neuronal cell survival will be coupled to, or independent of axonal survival, is an important area of research in neurobiology that has currently not been explored in much detail. Its crucial importance lies in the fact that for any of these strategies to make an impact on brain function, not only cell bodies, but functional neuronal circuits will have to be maintained.

4.0 CONCLUSIONS

The major role of apoptosis in human disease pathogenesis is now being defined and a large number of new therapeutic targets are being identified which are related to inducers of apoptosis, the complex signal transduction mechanisms of apoptosis or the enzyme systems which actually destroy the cell during apoptosis. Tasks for the future will be to define how apoptosis relates to other cellular processes in causing disease. Over emphasis on *in vitro* systems can produce misleading results on the importance of apoptosis. A specific question in anticancer therapy is the importance of induction of apoptosis in transformed cells relative to cell cycle arrest and mitotic failure. What is the relative importance of factors influencing apoptosis compared to factors influencing angiogenesis and the immune response (Holmgren *et al.*, 1995). One particular therapeutic question will be, is it actually possible to manipulate the signal transduction systems of apoptosis without unacceptable side-effects? In addition, can inhibition of caspases be achieved in patients in specific target tissues without resorting to complex and impractical gene therapy? Apoptosis research will certainly lead to the design of useful new therapies but, as on many occasions in the history of therapeutics, serendipity will also reveal new therapies which will only subsequently be shown to work by acting on apoptosis.

Acknowlegements.
PRL is a research fellow of the Lister Institute of Preventive Medicine.

5.0 REFERENCES

Ameisen,J.C. (1996) The Origin of Programmed Cell-Death. *Science* **272**: 1278-1279.
Ando,K., Hiroishi,K., Kaneko,T., *et al* (1997) Perforin, Fas/Fas ligand, and TNF-alpha pathways as specific and bystander killing mechanisms of hepatitis C virus-specific human CTL. *Journal of Immunology* **158**: 5283-5291.
Arai,H., Gordon,D., Nabel,E.G. and Nabel,G.J. (1997) Gene transfer of Fas ligand induces tumor regression in vivo. *Proc. Natl. Acad. Sci. USA* **94**: 13862-13867.
Basu,A.K. and Marnett,L.J. (1983) Unequivocal demonstration that malonidiadehyde is a mutagen. *Carcingenesis* **3**: 331-333.
Bellgrau,D., Gold,D., Selawry,H., *et al.* (1995) A Role For CD95 Ligand in Preventing Graft-Rejection. *Nature* **377**: 630-632.
BengzonJ., Kokaia,Z., ElmerE., *et al.* (1997) Apoptosis and profileration of dentate gyrus neurons after single and intermittent limbic seizures. *Proc. Natl. Acad. Sci.. U.S.A.* **94** :10432-10437.
Bergeron,L., Perez,G.I., Macdonald,G., *et al.* (1998) Defects in regulation of apoptosis in caspase-2-deficient mice. *Genes Dev.* **12**: 1304-1314.
Berns,E., Klijn,J.G.M., vanPutten, W.L.J., deWitte,H.H., *et al.* (1998) p53 protein accumulation predicts poor response to tamoxifen therapy of patients with recurrent breast cancer. *J. Clin. Oncol.* **16**: 121-127.
Bilbao,G, Gomez Navarro,J., Contreras,J.L. and Curiel,D.T. (1997) Advances in adenoviral vectors for cancer gene therapy. *Expert Opinion On Therapeutic Patents*, **7**: 1427-1446.
Bold,R.J. Termuhlen ,P.M. and McConkey,D.J. (1997) Apoptosis, cancer and cancer therapy. *Surgical Oncol.* **6**: 133-142.
Boobol,S.K., Dannenberg,A.J., Chadburn,A., *et al.* (1996) Cyclooxygenase-2 overexpression and tumor formation are blocked by sulindac in a murine model of familial adenomatous polyposis. *Cancer Res.* **56**: 2556-2560.
Bortuzzo,C., Hanif,R., Kashfi,K., *et al.* (1996) The Effect of Leukotriene-B and Selected HETES On the Proliferation of Colon-Cancer Cells. *Gastroenterol.* **110**: A494-A494.
Campbell,J.W. and Pollack,I.F. (1997) Growth factors in gliomas: Antisense and dominant negative mutant strategies. *J. Neuro-Oncol.* **35**: 275-285.
Chaloupka,J. and Vinter,V. (1996) Programmed cell death in bacteria. *Folia Microbiologica* **41**: 451-464.
Chambon, P. (1994) The retinoid signalling pathway: molecular and genetic analyses. *Semin. Cell Biol..* **5**: 115-125.
Charriaut-Marlangue,C., Aggoun-Zouaoui,D., Represa,A. and Ben-Air,Y. (1996). Apoptotic features of selective neuronal death in ischemia, epilepsy and gp 120 toxicity. *Trends in Neurosci.* **19**: 109-114.
Chen,S.Y. and Marasco,W.A. (1996) Novel genetic immunotoxins and intracellular antibodies for cancer-therapy. *Semin. Oncol.* **23**: 148-153.
Chen,S.Y., Yang,A.G., Chen,J.D., *et al.* (1997) Potent antitumour activity of a new class of tumour-specific killer cells. *Nature* **385**: 78-80.
Cheng,Y., Deshmukh,M., D'Costa,A., *et al.* (1998) Caspase inhibitor affords neuroprotection with delayed administration in a rat model of neonatal hypoxic-ischemic brain injury. *J. Clin. Invest.* **101**: 1992-1999.
Chinery,R., Brockman,J.A., Dransfield,D.T. and Coffey,R.J. (1997a) Antioxidant-induced nuclear translocation of CCAAT/Enhancer-binding protein beta. *J. Biol. Chem.* **272**: 30356-30361.
Chinery,R., Brockman,J.A., Peeler,M.O., *et al.* (1997b) Antioxidants enhance the cytotoxicity of chemotherapeutic agents in colorectal cancer: A p53-independent induction of p21WAF1/CIP1 via C/EBPb. *Nature Med.* **33**: 1233-1241.
Choi,D.W. (1996) Ischemia induced neuronal apoptosis. *Curr. Opin. Neurobiol.* **6**: 667-672.
Crowe,M.J., Bresnahan,J.C., Shuman,S.L., *et al.* (1997) Apoptosis and delayed degeneration after spinal cord injury in rats and monkeys. *Nature Med.*, **3**: 73-76.
Dachs,G.U., Dougherty,G.J., Stratford,I.J. and Chaplin,D.J. (1997) Targeting gene therapy to cancer: A review. *Oncol. Res.* **9**: 313-325.
Deshmukh,M., Vasilakos,J., Deckwerth,T. L., *et al.* (1996) Genetic and Metabolic Status of NGF-Deprived Sympathetic Neurons Saved By an Inhibitor of Ice Family Proteases. *J. Cell Biol.* **135**: 1341-1354.
El-deiry,W.S., Harper,J.W., O'Connor, P.M., *et al.* (1994) Waf1/Cip1 Is Induced in P53-Mediated G(1) Arrest and Apoptosis.*Cancer Res.* **54**: 1169-1174.
El-deiry,W.S., Tokino,T., Velculescu,V.E., *et al.* (1993) Waf1, a Potential Mediator of P53 Tumor Suppression. *Cell* **75**: 817-825.
Elder,D.J., Hague,A., Hicks,D.J. and Paraskeva,C. (1996) Differential growth inhibition by the aspirin metabolite salicylate in human colorectal tumor cell lines: enhanced apoptosis in carcinoma and in vitro-transformed adenoma relative to adenoma cell lines. *Cancer Res.* **56**: 2273-2276.
Eliasson,M.J.L., Sampei,K., Mandir,A.S., *et al.* (1997) Poly(ADP-ribose) polymerase gene disruption renders mice resistant to cerebral ischemia. *Nature Med.* **3**: 1089-1095.
Fearon,E.R. and Vogelstein,B. (1990) A Genetic Model For Colorectal Tumorigenesis. *Cell* **6**: 759-767.

Friedlander,R.M., Brown,R.H., Gagliardini,V., *et al.* (1997) Inhibition of ICE slows ALS in mice. *Nature* **388**: 31.
Friesen,C., Herr,I., Krammer,P.H. and Debatin,K.M. (1996) Involvement of the Cd95 (Apo-1/Fas) Receptor/Ligand System in Drug- Induced Apoptosis in Leukemia-Cells. *Nature Med.,* **2**: 574-577.
Fulda,S., Friesen,C., Los,M., *et al.* (1997) Betulinic acid triggers CD95 (APO-1/Fas)- and p53-independent apoptosis via activation of caspases in neuroectodermal tumors. *Cancer Res.* **57**: 4956-4964.
Galle,P.R., Hofmann,W.J., Walczak,H., (1995) Involvement of the Cd95 (Apo-1/Fas) Receptor and Ligand in Liver- Damage. *J. Exptl. Med.* **182**: 1223-1230.
Giardiello,F.M., Hamilton,S.R., Krush,A.J., *et al.* (1993) Treatment of colonic and rectal adenomas with sulindac n familial adenomatous polyposis. *New Engl. J. Med.* **328**: 1313-1316.
Gibson,N.W. and Kennedy,S.P. (1997) Delivery of tumor suppressor genes to reverse the malignant phenotype. *Adv. Drug Delivery Rev.* **26**: 119-133.
Giordano,C., Stassi,G., DeMaria,R., *et al.* (1997) Potential involvement of fas and its ligand in the pathogenesis of Hashimoto's thyroiditis. *Science* **275**: 960-963.
Giovannucci,E., Egan,K.M., Hunter,D.J., *et al.* (1995) Aspirin and the risk of colorectal cancer in women. *New Engl. J. Med,* **333**: 609-614.
Giovannucci,E., Rimm,E.B., Stamper,M.J., *et al.* (1994) Aspirin use and the risk for colorectal cancer and adenoma in male health professionals. *Ann Intern. Med.* **121**: 241-246.
Greene,N.K., Youngs,D.J., Neoptolemos J.P. *et al.*(1997). Sensitisiation of colorectal and pancreatic cancer cell lines to the prodrug 5-(aziridin-1-yl)-2,4-dinirobenzamide. *Cancer Gene Ther.* **4**: 229-238.
Griffith,T.S., Brunner,T., Fletcher,S.M., *et al.* (1995) Fas ligand-induced apoptosis as a mechanism of immune privilege. *Science* **270**: 1189-1192.
Hagan,P., Barks,J.D.E., Yabut,M., *et al.* (1996) Adenovirus-mediated over-expression of interleukin-1 receptor antagonist reduces susceptibility to excitotoxic brain injury in perinatal rats. *Neurosci.* **75**: 1033-1045.
Hanif,R., Pittas,A., Feng,Y., *et al.* (1996) Effects of nonsteroiddal anti-inflammatory drugs on proliferation and on induction of apoptosis in colon cancer cells by a prostaglandin-independent pathway. *Biochem. Pharmacol.* **52**: 237-245.
Hart,I.R. and Vile,R.G. (1995) Targeted gene-therapy. *Br. Med. Bull.* **51**: 647-655.
Hartwell,L.H. and Kastan,M.B. (1994) Cell cycle control and cancer. *Science* **266**: 1821-1828.
Heise,C., Sampson Johannes,A., Williams,A., *et al.* (1997) ONYX-015, an E1B gene-attenuated adenovirus, causes tumor-specific cytolysis and antitumoral efficacy that can be augmented by standard chemotherapeutic agents *Nature Med.* **3**: 639-645.
Holmgren,L., Oreilly,M.S. and Folkman,J. (1995) Dormancy of Micrometastases - Balanced Proliferation and Apoptosis in the Presence of Angiogenesis Suppression. *Nature Med.* **1**: 149-153.
Holtzman,D.M. and Deshmukh,M. (1997) Caspases: A treatment target for neurodegenerative disease? *Nature Med.* **3**: 954-955.
Hong,W.K., Lippman,S.M., Itri,L.M., *et al.* (1990) Prevention of 2nd Primary Tumors With Isotretinoin in Squamous-Cell Carcinoma of the Head and Neck. *New Eng. J. Med.,* **323**: 795-801.
Huang,M.E., Ye,Y.C., Chen,S.R., *et al.* (1988) Use of All-Trans Retinoic Acid in the Treatment of Acute Promyelocytic Leukemia. *Blood,* **72**: 567-572.
Hug, H.,Strand,S., Grambihler,A., *et al.* (1997) Reactive oxygen intermediates are involved in the induction of CD95 ligand mRNA expression by cytostatic drugs in hepatoma cells. *J. Biol. Chem.,* **272,** 28191-28193.
Ijiri,K. and Potten,C.S. (1983) Response of intestinal cells of differing topographical and hierachical status to ten cytotoxic drugs and five sources of radiation. *Br. J. Cancer* **47**: 175-185.
Ijiri,K. and Potten,C.S. (1987) Further studies on the response of intestinal crypt cells of different hierarchical status to eighteen different cytotoxic drugs. *Br. J. Cancer* **55**: 113-123.
Irmler,M., Thome,M., Hahne,M., *et al.* (1997) Inhibition of death receptor signals by cellular FLIP. *Nature* **388**: 190-195.
Kabelitz,D. (1998) Apoptosis, graft rejection, and transplantation tolerance. *Transplantation,* **65,** 869-875.
Kang,S.M., Schneider,D.B., Lin,Z.H., *et al.* (1997) Fas ligand expression in islets of Langerhans does not confer immune privilege and instead targets them for rapid destruction.*Nature Med.,* **3**: 738-743.
Kerr,J.F.R., Wyllie,A.H. and Currie,A.R. (1972) Apoptosis: a basic biological phenomenon with wide ranging implications in tissue kinetics. *Br. J. Cancer* **26**: 239-257.
Kim,T.W., Pettingell,W.H., Jung,Y.K., *et al.* (1997) Alternative cleavage of Alzheimer-associated presenilins during apoptosis by a caspase-3 family protease. *Science* **277**: 373-376.
Kinzler,K.W. and Vogelstein,B. (1996) Lessons from hereditary colorectal cancer. *Cell* **87**: 159-170.
Knudson,C.M., Tung,K.S.K., Tourtellotte,W.G., *et al.* (1995) Bax-deficient mice with lymphoid hyperplasia and male germ cell death. *Science* **270**: 96-99.
Kondo,T., Suda,T., Fukuyama,H., Adachi,M. and Nagata,S. (1997) Essential roles of the Fas ligand in the development of hepatitis. *Nature Med.* **3**: 409-413.
Korystov,Y.N., Dobrovinskaya,O.R., Shaposhnikova,V.V. and Eidus,L.K. (1996) Role of arachidonic acid metabolism in thymocyte apotposis after irradiation. *FEBS Lett.* **388**: 238-241.
Kostic,V., Jackson-Lewis,V., de Bilbao,F., *et al.* (1997) Prolonged life in a transgenic mouse model of familial anyotrophic lateral sclerosis. *Science* **277**: 559-562.

Krajewska,M., Wang,H.G., Krajewski,S., *et al.* (1997) Immunohistochemical analysis of *in vivo* patterns of expression of CPP32 (Caspase-3), a cell death protease. *Cancer Res.* **57**: 1605-1613.
Kroemer,G. (1997) The proto-oncogene Bcl-2 and its role in regulating apoptosis. *Nature Med.* **3**: 614-620.
Kuida,K., Zheng,T.S., Na,S.Q., *et al.* (1996) Decreased Apoptosis in the Brain and Premature Lethality in Cpp32- Deficient Mice. *Nature* **384**: 368-372.
Labayle,D., Fischer,D., Vielh,P., *et al.* (1991) Sulindac casuses regression of rectal polyps in familial adenomatous polyposis. *Gastroenterol.* **101**: 635-639.
Lacronique,V., Mignon,A., Fabre,M., *et al.* (1996) Bcl-2 Protects From Lethal Hepatic Apoptosis Induced By an Anti-Fas Antibody in Mice. *Nature Med.* **2**: 80-86.
Lamb,J.R. and Friend,S.H. (1997) Which guesstimate is the best guesstimate? Proedicitng chemotherapeutic outcomes? *Nature Med.* **3**: 962-963.
Lanni,J.S., Lowe,S.W., Licitra,E.J., *et al.* (1997) p53-independent apoptosis induced by paclitaxel through an indirect mechanism. *Proc. Natl. Acad. Sci. USA* **94**: 9679-9683.
Lau,H.T. and Stoeckert,C.J. (1997) FasL - Too much of a good thing? *Nature Med.* **3**: 727-728.
Lau,H.T., Yu,M., Fontana,A. and Stoeckert,C.J. (1996) Prevention of Isl*et al*lograft rejection with engineered myoblasts expressing FasL in mice. *Science* **273**: 109-112.
Lee,F.D. (1993) Importance of apoptosis in the histopathology of drug related lesions in the large intestine. *J. Clin. Path.* **46**: 118-122.
Lens,D., Dyer,M.J.S., GarciaMarco,J.M., *et al.* (1997) p53 abnormalities in Cll are associated with excess of prolymphocytes and poor prognosis. *Br. J. Haematol.,* **99**: 848-857.
Li,Y. Q., Fam,C.Y., O'C.onnor P.J. *et al.* (1992) Target cells for the cytotoxic effects of carcinogens in the murine small intestine. *Carcinogenesis,* **13**: 361-368.
Linnik,M.D., Zahos,P., Geschwind,M.D. and Federoff,H.J. (1995) Expression of Bcl-2 from a defective herpes-simplex virus-1 vector limits neuronal death in focal cerebral-ischemia. *Stroke* **26**: 1670-1674.
Lockshin,R.A. and Williams,C.M. (1965) Programmed cell death: 1. Cytology of degeneration in the intersegmental muscles of the pernyi silk moth. *J. Insect Physiol.* **11**: 123-133.
Lowe,S.W., Ruley,H.E., Jacks,T. and Housman,D.E. (1993a) p53-dependent apoptosis modulates the ctyotoxicity of anticancer agents. *Cell.* **74**: 957-967.
Lowe,S.W., Schmitt,E.M., Smith,S.W., *et al.* (1993b) p53 is required for radiation-induced apoptosis in mouse thymocytes. *Nature* **362**: 847-849.
Lu,X.P., Fanjul,A., Picard,N., *et al.* (1997) Novel retinoid-related molecules as apoptosis inducers and effective inhibitors of human lung cancer cells in vivo. *Nature Med.* **3**: 686-690.
Martin,L.A., Lemoine,N.R. (1996) Direct cell-killing by suicide genes *Cancer Met. Rev.* **15**: 301-316.
Medema,J.P., Scaffidi,C., Kischkel,F.C., *et al.* (1997) FLICE is activated by association with the CD95 death-inducing signaling complex (DISC). *EMBO J.* **16**: 2794-2804.
Merritt,A.J., Allen,T., Potten,C.S. and Hickman,J.A. (1997) Apoptosis in small intestinal epithelia from p53-null mice: evidence for a delayed, p53-independent G2/M-associated cell death after y-irradiation. *Oncogene* **14**: 2759-2766.
Merritt,A.J., Potten,C.S., Kemp,C.J., *et al.* (1994) The role of p53 in spontaneous and radiation-induced apoptosis in the gastrointestinal tract of normal and p53-deficient mice. *Cancer Res.* **54**: 614-617.
Merritt,A.J., Potten,C.S., Watson,A.J.M., *et al.* (1995) Differential Expression of Bcl-2 in Intestinal Epithelia - Correlation With Attenuation of Apoptosis in Colonic Crypts and the Incidence of Colonic Neoplasia. *J. Cell Sci.* **108**: 2261-2271.
Moertel,C.G., Fleming,T.R., Macdonald,J.S., *et al.* (1990) Levamisole and Fluorouracil For Adjuvant Therapy of Resected Colon- Carcinoma. *New Eng. J. Med.* **322**: 352-358.
Moriya,M., Zhang,W., Johnson,F. and Grollman,A.P. (1994) Mutagenic potency of exocyclic DNA adducts: Marked differences between *Escherichia coli* and simian kidney cells. *Proc. Natl. Acad. Sci.USA.* **91**: 11899-11903.
Mukai,F.H. and Goldstein,B.D. (1976) Mutagenicity of malondiadehyde, a decomposition product of peroxidised polyunsaturated fatty acids. *Science* **191**: 868-869.
Muller,M., Strand,S., Hug,H., *et al.* (1997) Drug-induced apoptosis in hepatoma cells is mediated by the CD95 (APO/Fas) receptor/ligand system and involves activation of wild-type p53. *J. Clin. Invest.* **99**: 403-413.
Nagata, S. (1997) Apoptosis by death factor. *Cell* **88,** 355-365.
Nielsen,L.L. and Maneval,D.C. (1998). P53 tumor suppressor gene therapy for cancer. *Cancer Gene Ther.* **5**: 52-63.
Offen,D., Beart,P.M., Cheung,N.S., *et al* (1998) Transgenic mice expressing human Bcl-2 in their neurons are resistant to 6-hydroxydopamine and 1-methyl-4-phenyl-1,2,3,6-tetrahydropyridine neurotoxicity. *Proc. Natl. Acad. Sci. USA.* **95**: 5789-5794.
Olivieri,N.F., Rees,D.C., Ginder,G.D., *et al.* (1997) Treatment of Thalassaemia major with phenylbutyrate and hydroxyurea. *Lancet* **350**: 491-492.
Oppenheim,R.W. (1991) Cell death during the development of the nervous system. *Annu. Rev. Neurosci.* **14**: 453-501.
Orrenius,S., Nobel,C.S.I., Vandendobbelsteen,D.J., *et al.* (1996) Dithiocarbamates and the Redox Regulation of Cell-Death. *Biochem. Soc. Trans.* **24**: 1032-1038.

Osborne,B.A. (1996) Apoptosis and the Maintenance of Homeostasis in the Immune-System. *Curr. Opin. Immunol.* **8**: 245-254.

Pasricha,P.J., Bedi,A., O'Conner,K., *et al.* (1995) The effects of sulindac on colorectal proliferation and apoptosis in familial adenomatous polyposis. *Gastroenterol.* **109**: 994-998.

Perez,G.I., Knudson,C.M., Leykin,L., *et al.* (1997) Apoptosis-associated signaling pathways are required for chemotherapy-mediated female germ cell destruction. *Nature Med.* **3**: 1228-1232.

Petrelli,N., Douglass,H.O., Herrera,L., *et al.* (1989) The Modulation of Fluorouracil With Leucovorin in Metastatic Colorectal-Carcinoma - a Prospective Randomized Phase-Iii Trial. *J. Clin. Oncol.* **7**: 1419-1426.

Piazza,G.A., Rahm,A.L.K., Krutzch,M., *et al.* (1995) Antineoplastic drugs sulindac sulfide and sulindac sulfone inhibit cell growth by inducing apoptosis. *Cancer Res.* **1995**: 3110-3116.

Piedrafita,F.J. and Pfahl,M. (1997) Retinoid-induced apoptosis and SP1 cleavage occur independently of transcription and require caspase activation. *Mol. Cell. Biol.* **17**: 6348-6358.

Pisha,E., Chai,H., Lee,I.S., *et al.* (1995) Discovery of Betulinic Acid As a Selective Inhibitor of Human-Melanoma That Functions By Induction of Apoptosis. *Nature Med.* **1**: 1046-1051.

Potten,C.S., Li,Y.Q., O'Conner,P.J. and Winton,D.J. (1992) A possible explanation for the differential cancer incidence in the intestine, based on distribution of the cytotoxic effects of carcinogens in the murine large bowel. *Carcinogenesis* **13**: 2305-2312.

Pritchard,D.M., Watson,A.J.M., Potten,C.S. and Hickman,J.A. (1996) Characterisation of 5-Fluorouracil-Induced Apoptosis in Murine Small and Large-Intestine - Studies in P53-Null and Bcl-2-Null Transgenic Mice. *Gastroenterology* **110**: A578-A578.

Pritchard,D.M., Watson,A.J.M., Potten,C.S., *et al.* (1997) Inhibition by uridine but not thymidine of p53-dependent intestinal apoptosis initiated by 5-fluorouracil: Evidence for the involvement of RNA perturbation. *Proc. Natl. Acad. Sci. USA* **94**: 1795-1799.

Qiao,L., Kozoni,V., Tsioulias,G.J., *et al.* (1995) Selected Eicosanoids Increase the Proliferation Rate of Human Colon- Carcinoma Cell-Lines and Mouse Colonocytes in-Vivo. *Biochim. Biophys. Acta-Lipids Lipid Met.* **125**: 215-223.

Rao,C.V., Rivenson,A., Simi,B., *et al.* (1995) Chemoprevention of colon carcinogenesis by sulindac, anonsteroidal anti-inflammatory agent. *Cancer Res.* **55**: 1464-1472.

Reddy,B.S., Maruyama,H. and Kelloff,G. (1987) Dose-related inhibition of colon carcinogenesis by dietary piroxicam, a non steroidal antiiflammatory drug, during different stages or rat colon tumor development. *Cancer Res.* **47**: 5340-5346.

Reddy,B.S., Rao,C.V., Rivenson,A. and Kelloff,G. (1993) Inhibitory effect of aspirin on azoxymethane-induced colon carcinogenesis. *Carcinogenesis* **14**: 1493-1497.

Riley,D.J., Nikitin,A.Y. and Lee,W.H. (1996). Adenovirus-mediated retinoblastoma gene therapy suppresses spontaneous pituitary melanotroph tumors in Rb+/- mice. *Nature Med.* **2** : 1316-1321.

Roth,J.A., Cristiano,R.J. (1997) Gene therapy for cancer: What have we done and where are we going? *J. Natl. Cancer Inst.* **89**: 21-39.

Sagot,Y., Duboisdauphin,M., Tan,S.A., *et al.* (1995) Bcl-2 Overexpression Prevents Motoneuron Cell Body Loss But Not Axonal Degeneration in a Mouse Model of a Neurodegenerative Disease. *J. Neurosci.* **15**: 7727-7733.

Sandig,V., Brand,K., Herwig,S., *et al.* (1997). Adenovirally transferred p161INK4/CDKN2 and P53 genes cooperate to induce apoptotic tumor cell death. *Nature Med.* **3**: 313-319.

Schmidt,M.L., Kuzmanoff,K.L., Linglndeck,L. and Pezzuto,J.M. (1997) Betulinic acid induces apoptosis in human neuroblastoma cell lines. *Eur. J. Cancer* **33**: 2007-2010.

Seino,K.I., Kayagaki,N., Takeda,K., *et al.* (1997) Contribution of Fas ligand to T cell-mediated hepatic injury in mice. *Gastroenterol.,* **113**: 1315-1322.

Shiff,S.J., Qiao,L., Tsai,L.L. and Rigas,B. (1995) Sulindac Sulfide, an Aspirin-Like Compound, Inhibits Proliferation, Causes Cell-Cycle Quiescence, and Induces Apoptosis in HT-29 Colon Adenocarcinoma Cells. *J. Clin. Invest.* **96**: 491-503.

Silverstein,F.S. (1998) Can inhibition of apoptosis rescue ischemic brain? *J. Clin. Invest.* **101**: 1809-1810.

Spalding,J.W. (1988) Toxicology and carcinogenesis studies of malondiadehyde sodium salt (3-hydroxy-2-propenal, sodium salt) in F344/N rats and B6C3F1 mice. *NTP Technical Report* **331**: 5-13.

Springer,C.J. and Niculescu Duvaz,I. (1996) Gene-directed enzyme prodrug therapy (GDEPT): Choice of prodrugs. *Adv. Drug Delivery Rev.* **22**: 351-364.

Strand,S., Hofmann,W.J., Hug,H., *et al.* (1996) Lymphocyte Apoptosis Induced By Cd95 (Apo-1/Fas) Ligand-Expressing Tumor-Cells - a Mechanism of Immune Evasion. *Nature Med.* **2**: 1361-1366.

Strasser,A. and O'Conner,L. (1998) Fas ligand - caught betwee Scylla and Charybdis. *Nature Med.* **4**: 21-22.

Tang,D.G., Chen,Y.Q. and Honn,K.V. (1993) Arachidonate lipoxygenases as essential regulators of cell survival and apoptosis. *Proc. Natl. Acad. Sci. USA* **93**: 5241-5246.

Thun,M.J., Namboodiri,M.M. and Heath,C.W.J. (1991) Aspirin use and the reduced risk of fatal colon cancer. *N. Eng. J.. Med.* **325**: 1593-1596.

Tsujii,M. and Dubois,R.N. (1995) Alterations in Cellular Adhesion and Apoptosis in Epithelial-Cells Overexpressing Prostaglandin-Endoperoxide-Synthase-2. *Cell* **83**: 493-501.

Vergelli,M., Hemmer,B., Muraro,P.A., *et al.* (1997) Human autoreactive CD4(+) T cell clones use perforin- or Fas/Fas ligand-mediated pathways for target cell lysis. *J. Immunol.* **158**: 2756-2761.

Vexler,Z.S., Roberts,T.P.L. Bollen,A.W., *et al.* (1997) Transient cerebral ischemia. *J. Clin. Invest.* **99**: 1453-1459.
Waddell,W.R. and Loughry,R.W. (1983) Sulindac for polyposis coli. *J. Surg. Oncol.* **24**: 83-87.
Wahl,A.F., Donaldson,K.L., Fairchild,C., *et al.* (1996) Loss of normal p53 function confers sensitization to Taxol by increasing G2/M arrest and apoptosis. *Nature Med.* **2**: 72-79.
Waldman, T., Lengauer, C., Kinzler, K. W. and Vogelstein, B. (1996) Uncoupling of S-Phase and Mitosis Induced By Anticancer Agents in Cells Lacking P21. *Nature,* **381**: 713-716.
Waldman,T., Zhang,Y.G., Dillehay,L., *et al.* (1997) Cell-cycle arrest versus cell death in cancer therapy. *Nature Med.* **3**: 1034-1036.
Watson,A.J.M., Askew,J.N. and Benson,R.S.P. (1995) Inhibition of Poly(Adp-Ribose) Polymerase Prevents Immediate Cell- Death Induced By Hydrogen-Peroxide But Not Apoptosis in the Intestinal-Cell Line HT-29. *Gastroenterol.* **109**: 472-482.
Watson,A.J.M. and DuBois,R.N. (1997) Lipid metabolism and APC: implications for colorectal cancer production. *Lancet* **349**: 444-445.
Watson,A.J.M., Merritt,A.J., Jones,L.S., *et al.* (1996) Evidence For Reciprocity of Bcl-2 and P53 Expression in Human Colorectal Adenomas and Carcinomas. *Br. J. Cancer* **73**: 889-895.
Winkler,C., Sauer,H., Lee,C.S. and Bjorklund, A. (1996) Short-term GDNF treatment provides long-term rescue of lesioned nigral dopaminergic-neurons in a rat model of parkinsons-disease. *J. Neurosci.* **16**: 7206-7215.
Xu,D.G., Crocker,S.J., Doucet,J.P, *et al.* (1997) Elevation of neuronal expression of NAIP reduces ischemic damage in the rat hippocampus. *Nature Med.* **3**: 997-1004.
Yoshikawa,R., Ichii,S., Hashimoto,T., *et al.* (1995) Effect of aspirin on the induction of apoptosis in adenocarcinoma cell lines. *Oncol Rep.* **2**: 361-364.

Index

Page numbers in *italic* indicate figures. Page numbers followed by "t" indicate tables.

U

V

W